Advanced Researches in Communication Technology

Advanced Researches in Communication Technology

Edited by **Timothy Kolaya**

C WILLFORD PRESS

New York

Published by Willford Press,
118-35 Queens Blvd., Suite 400,
Forest Hills, NY 11375, USA
www.willfordpress.com

Advanced Researches in Communication Technology
Edited by Timothy Kolaya

International Standard Book Number: 978-1-68285-163-0 (Hardback)

Printed in the United States of America.

Contents

Preface

The world has experienced a tremendous shift in communication technologies. This book highlights the various researches and advancements in the field of communication technology. Some of the concepts discussed herein are wireless multimedia networks, performance of enhanced wireless networks, data management on mobile and wireless computing, evaluation of advanced routing protocols for wireless communications, etc. As this field is emerging at a rapid pace, the contents of this book will help the readers understand the modern concepts and applications of the subject. It also brings forth new aspects and technologies for further research and analysis.

After months of intensive research and writing, this book is the end result of all who devoted their time and efforts in the initiation and progress of this book. It will surely be a source of reference in enhancing the required knowledge of the new developments in the area. During the course of developing this book, certain measures such as accuracy, authenticity and research focused analytical studies were given preference in order to produce a comprehensive book in the area of study.

This book would not have been possible without the efforts of the authors and the publisher. I extend my sincere thanks to them. Secondly, I express my gratitude to my family and well-wishers. And most importantly, I thank my students for constantly expressing their willingness and curiosity in enhancing their knowledge in the field, which encourages me to take up further research projects for the advancement of the area.

Editor

DESIGN OF DYNAMIC MAC PROTOCOL FOR WIRELESS MULTIMEDIA NETWORKS

S.P.V.Subba Rao[1] Dr.S. Venkata Chalam [2] Dr.D.Sreenivasa Rao[3]

[1] Sreenidhi Institute of Science and Technology, Dept of Electronics and Communication Engineering, Hyderabad, Andhra Pradesh, India
[2]CVR Engineering College, Dept of Electronics and Communication Engineering, Hyderabad, Andhra Pradesh, India
[3] JNTU CE Dept of Electronics and Communication Engineering, Hyderabad,   Andhra Pradesh, India
spvsr2007@gmail.com

ABSTRACT

A Dynamic MAC protocol is developed for WCDMA wireless multimedia networks. It uses multiple slots per frame allowing multiple users to transmit simultaneously using their own CDMA codes. The proposed MAC protocol is based on contention .If there is low contention users can access any slots and if there is high level contention the owners of the slots have priority to access slots .If the owners are not having any data the non owners can access slots. An adaptive power control algorithm is applied to reduce transmission power, interference level, and to maximize system capacity. If the observed traffic is high, power will be increased; if traffic is low power will be decreased. By simulation results, we show that our proposed MAC protocol achieves 100% throughput under low contention and 90% throughput under high contention and also reduces power consumption.

KEYWORDS

Wideband Code Division Multiple Access (W-CDMA), MAC protocol, Direct Sequence Spread Spectrum (DSSS), Medium Access control, Multimedia networks.

1. INTRODUCTION

Third generation systems (3G) such as Wide band code division Multiple Access are designed for wireless multimedia networks. It provides high quality image and video transmission and support for a wide range of services with higher rates and with increased network capacity. The current trend in wireless network is to provide multiple multimedia traffic class with quality of service for more number of users by allocating resource efficiently and reliably. To improve the radio resource utilization and to provide users with quality of service requirements a Medium Access Control [1] is required. Many MAC protocols are proposed for wireless multimedia networks to maximize the throughput and to transmit multiple multimedia traffic classes with required quality of service. The basic MAC protocols for traditional communication systems are designed for voice communication and are unstable at higher load conditions. Different MAC protocols for wireless multimedia network are proposed for congestion control, Interference control but there is trade off between the performance metrics of proposed protocols. In this paper a Dynamic MAC protocol for wireless multimedia networks is proposed and an adaptive power control algorithm is applied at the beginning of each frame to reduce interference. The proposed MAC protocol is based on contention. If there is low contention users can access any slots and if there is high level contention the owners of the slots have priority to access slots. If the owners are not having any data the non owners can access slots. To maximize system capacity and to reduce interference an adaptive power control algorithm is used. If the observed traffic is high power is increased, if traffic is low power is decreased. The paper is organized as follows: In section II existing works on MAC protocols are discussed. In section III Dynamic MAC protocol for wireless multimedia networks is developed. In section IV Adaptive power control mechanism is derived for multimedia traffic in WCDMA networks. The Dynamic MAC protocol is evaluated through simulations in section V. The paper is concluded in section VI.

2. RELATED WORKS.

1. Z.Tang and J.J Garcia have proposed a CATA Protocol [2] based on Contention and reservation protocols. Each slot is sub divided into five mini slots. The first four mini slots are control ones labeled as CMS1, CMS2, CMS3, and CMS4 and are used to secure and reserve time slots and the last slot labeled DMS used for transmission of data packets. It is more flexible in terms of bandwidth management when compared with allocation protocols. In this protocol more slots are used for secure and reservation and it is unstable for certain traffic loads and mobility rates.

2. Lixin Wang and Mounir Hamdi have proposed a Hybrid adaptive MAC protocol (HAMAC) [3] based on TDMA, reservation, and contention protocols. It allow the contention channel to transmit data, unlike many other proposals in which the contention channel is used only for reservation and control signaling. It can efficiently adapt to the traffic the variance in CBR, VBR, and ABR traffic due to the mobility of mobile devices. The protocol uses isochronous service features of time division multiple access protocol and a new preservation slot technique to reduce packet contention overhead for voice and CBR traffic. In this protocol low delay is achieved for light traffic load.

3. I. Chlamtac, and A. Farago have proposed a ADAPT protocol [4] based on channel allocation TDMA protocol and contention protocol .Each mobile terminal is assigned a slot in a frame considering as owner. In each slot their is sensing interval in which only the slot owner may contend for the channel by initiating hand shake and the other users (non owner) cannot transmit data.

4. Zhijun Wang, Umapathi Mani, and MiaoJu, Hao che have proposed a RAH-MAC [5] protocol based on combination of polling and contention MAC protocols .Data transmission rate is dynamically adjusted based on the channel condition and it uses variable transmission rate. More priority is given to voice traffic but not for other traffic.

5. Ian F. Akyildiz has proposed a WISPER protocol [6] based on TDMA and CDMA .Slots are filled according to the BER requirements. Here the protocol is simple to implement in that only one power level can be used for each slot rather than several power levels depending on the number of traffic classes when congestion occurs, voice packets are the first to be sacrificed.

6. C. Roobol et al have proposed a RLC/MAC protocol [7].In this protocol slots are filled according to load, traffic class and transmission rate .The BER of traffic classes are controlled using Power control algorithm .Different transmission formats specified for transmission.

7. A.Saravan, B.parthasarathy has proposed an Analytical Model MAC Multi protocol [8] based on OFDMA, TDMA and CDMA systems. The reservation and polling methods of MAC protocols are used to handle both low and high data traffics of the mobile users. In this protocol frame is divided into different slots and the slots are transmitted with users CDMA codes.

8. Rekha Patil and A. Damodaram [9] with objective of reducing call rejection rate and to minimise interference have proposed joint scheduling and power control algorithm. The algorithm is based on optimum number of users with optimum transmitting power level. The set of optimum power levels that could be used by the users for successful transmission are determined and are solved the problems in distributed power control algorithm.

9. Rachod Patachaianand, Kumbesan Sandrasegaran [10] has proposed a new adaptive power control algorithm by eliminating limitations in current power control algorithms for UMTS. When channel fading changes slowly the proposed algorithm reduces SIR variations and capable of tracking quick changes in fast fading channels where other power control algorithms to handle.

10. Rachod Patachaian and Kumbesan Sandrasegaran [11] have proposed a new adaptive power control algorithm. The algorithm uses Consecutive TPC Ratio (CTR) to adjust power control step sizes. They showed that there is correlation between user mobility and TPC sequences.

EXISTING WORKS ON MAC PROTOCOLS				
Sl.No	Algorithm	Principle / slots assignment	Advantages	Disadvantages
1.	CATA PROTOCOL	Contention and reservation protocols	More flexible in terms of bandwidth management when compared with allocation protocols	1. More slots are used for secure and reservation. 2. Un stable for certain traffic loads and mobility rates.
2.	HAMAC PROTOCOL	1.TDMA, reservation, and contention protocols 2. The protocol uses isochronous service features of time division multiple access protocol and a new preservation slot technique to reduce packet contention overhead for voice and CBR traffic.	1. Results in very low delay in case of light traffic load. 2.Dynamic bandwidth allocation strategy 3.Eliminates the reservation overhead of CBR traffic, which results in less contention	RAH-MAC is superior than HAMAC

3.	ADAPT PROTOCOL	TDMA protocol and contention protocol	Dynamically manages the band width	Only slot owners can transmit data in their slot others cannot use it and channel is not efficiently utilized
4.	WISPER protocol	1. Based on TDMA and CDMA 2. Slots are filled according to the BER requirements. 3. WISPER is a reservation-based protocol.	protocol is simple to implement in that only one power level can be used for each slot rather than several power levels depending on the number of traffic classes	When congestion occurs, voice packets are the first to be sacrificed.
5.	RAH-MAC	Combination of polling and contention MAC protocols	Data transmission rate is dynamically adjusted based on the channel condition and it uses variable transmission rate	More priority is given to voice traffic but not for other traffic.
6.	RLC/MAC	According to load, traffic class and rate	BER of traffic classes are controlled using Power control algorithm	Different transmission formats specified for transmission
7.	Analytical Model MAC Multi protocol.	OFDMA,TDMA and CDMA systems	Capacity of system increased	Results in delay for certain traffic
Existing Works on Power control				
8.	cross-layer based joint scheduling and power control algorithm	determines the optimum set of admissible users with suitable transmitting power level	solved the multiple access problems in the distributed power control algorithm	The power control is not adaptive.
9.	Adaptive step size power control with TPC command	The algorithm uses Consecutive TPC Ratio (CTR) to adjust power control step sizes. There is correlation between user mobility and TPC sequences.	capable for tracking the rapid changes of multipath fading by utilizing existing TPC commands	The power control is not based on the data traffic classes

| 10. | New Adaptive step size power control for UMTS | When channel fading changes slowly the proposed algorithm reduces SIR variations and capable of tracking quick changes in fast fading channels where other power control algorithms to handle . | Eliminates the drawbacks of conventional power control | The power control is not based on the data traffic classes |

3. DYNAMIC MAC PROTOCOL

3.1 DS-SS (CDMA) Based MAC Scheme

By multiplying the message signal b(t) by the spreading code c(t),each information bit is chopped into a number of small time increments commonly called as chips. Thus transmitted signal m(t), may be expressed as: $m(t) = c(t). \quad b(t)$ (1)

Which is a wideband signal. The received signal r(t) contains the transmitted signal m(t),noise n(t) and the interference i(t). The interference signal contains Intra cell Interference and Inter cell interference $r(t) = m(t) + i(t) + n(t) = c(t). \, b(t) + i(t) + n(t)$ (2)

Where, n(t) is Additive White Gaussian Noise (AWGN) in the receiver. The original message signal b(t) is recovered from the received signal r(t) by multiplying received signal r(t) with the code c(t) used at the transmitter .Therefore, the demodulated output z(t) at the receiver is given by

$$z(t) = c(t). \, r(t) = c2(t).b(t) + c(t). \, i(t) + c(t).n(t) \quad (3)$$

Since, $c2(t) = 1$ (the autocorrelation property of the PN code,)

$$z(t) = b(t) + c(t).i(t) + c(t).n(t) \quad\quad\quad (4)$$

3.2 WCDMA Scheduling

In Dynamic MAC protocol data traffic is calculated at each node and the node may be in low contention mode or in high contention mode. If the data traffic at a node is greater than threshold value DT_{th} then the node is said to be in high contention mode otherwise the node is said to be in low contention mode. In low contention mode any node can transmit data in any slot .In high contention mode the slots are reserved for certain traffic classes and the users who reserved the slots are called as owners .In high contention mode owners of current slots are allowed to contend for the channel, if owner does not have data the non owners are allowed to compete the channel according to priority of traffic classes. In both cases real time traffic is given more priority than non real time traffic.

FRAME									
R E Q	DATA	R E P	R E Q	DATA	R E P	R E Q	DATA	....	R E P

Where,   REQ- Request, REP-Response

Fig1: MAC frame

The time is divided into fixed size frames in the proposed protocol. A frame has N time slots and two special slots the Request (REQ) and the Reply (REP) slots which are separated into mini slots. The mini slots of REQ are used in the uplink for transmission request by the users and mini slots of REP are utilized in the downlink. The REP mini slots are modified to a matrix of CDMA codes and data slots as in fig1.The data slot and CDMA for a user are assigned by a scheduling algorithm and this data is send to the user as a REP signal by the Base Station (BS).

A REQ signal along with some control information's is send to BS by the user which is ready for transmission. The scheduling algorithm of BS enables the user to get a REP signal about the data slots and CDMA codes. Enabling the user to transmit the data together with the processing of the requests of the nodes and the scheduling is done with the help of the REP signal. In our dynamic work, each terminal transmits at the time slots during which it is allowed to transmit using its own code sequences.

The REP is divided into mini slots, each holding information of the corresponding data slot in the next frame. Each mini slot is further divided into grid, where grid is equal to the maximum number of nodes that can transmit data simultaneously in a data slot. Each of these grids is initialized with a code which the scheduler allocates to the node which succeeded in getting a reservation for that slot.

3.3. Analytical Model.

While describing the access system we take only one mobile cell into account in which there are M active nodes(or users) that generates messages to be transmitted to another node where the base station controls all the nodes within the cell. Two kinds of links are possible in this model.
   1. Uplink: this demonstrates data transmission from mobile station MS to BS.
   2. Downlink: this describes the data transmission from BS to MS.
For the analysis following assumptions are made.

1. Each node generates messages which is Poisson distributed with arrival rate λ.

2. The message length of each node is exponentially distributed.

3. The nodes cannot generate new message until all packets of current message are transmitted completely.

4. If a node completes its transmission in current frame, it cannot generate message in the same frame.

5. Let the maximum number of users that can be accommodated in the cell is N.

For the random access protocol, we use the M/M/n/n/K Markov model by obtaining the steady state equation as:

$$\overrightarrow{x}\,A = O \qquad (5)$$

Where A is the generator matrix, 'O' is a null matrix and $\overrightarrow{x}$ is a steady state probability vector

and it is equal   to $\overrightarrow{x} = \{x_0, x_1, x_2 \cdots x_n\}$ $\qquad (6)$

For this Markov chain, the recurrent non-null and the absorbing properties are satisfied. 'n' is the number of data slots and 'K' is the number of users. The average number of packets served by the system is calculated as:

$$PA = \frac{(KT)\sum_{i=0}^{n-1}\binom{K-1}{i}T^i}{\sum_{i=0}^{n}\binom{K}{i}T^i} \qquad (7)$$

Here, T is the offered traffic to the system with the arrival rate and T is given by;

$$T = \frac{\lambda}{\mu} \qquad (8)$$

Where, λ is Poisson distribution and the service rate and μ is the exponential distribution.
The probability of the packet success rate PSR is calculated as;

$$PSR = \sum_{k=0}^{c}\sum_{j=0}^{n}(1-xj)(1-Berr(k)) \qquad (9)$$

here " c" is the active number of CDMA codes allocated to the active users in a data slot and the steady state probabilities are given as;

$$x0 = \frac{1}{\sum_{i=0}^{n}\binom{K}{i}T^i} \qquad \text{And} \qquad (10)$$

$$xj = \binom{K}{j}T^j x0 \qquad (11)$$

and Berr(k) is the BER value, which is given by the relationship as;

$$Berr(k) = \frac{1}{2}erfc\left(\sqrt{\frac{Eb}{No + \frac{2}{3}Eb(\frac{k-1}{\beta p})}}\right) \qquad (12)$$

Where,

k = Number of active user.

βp = Processing gain of the spectrum.

Eb = Energy per bit in joules

No = The two-sided psd in Watts/Hz

Each node calculates the traffic by using the traditional way to calculate the system capacity for data traffic ,DT which is given by;

$$DT = \left[\frac{\beta p}{SIR}\right] \times \frac{1}{1+\kappa} \times P \times \frac{1}{\Phi} \times \beta a$$

(13)

Where, βp and βa = the processing gain by spreading the spectrum and sector antenna gain respectively.

SIR	= Signal to interference ratio
κ	= The interference from other nodes
P	= The power control factor
Φ	= The voice/data activity factor.

4. Dynamic MAC Protocol with Power Control.

The adaptive power control algorithm uses TPC commands and power determining factor. Adaptive factor is calculated based on TPC commands. Power determining factor is calculated based on traffic rate.

4.1 Adaptive Step Size Estimation

The power control step size is modified by multiplying a factor called Adaptive Factor (AF) with the fixed step size. The algorithm uses transmission power control commands (TPC) to calculate Adaptive factor based on predefined adaptive control factor which is defined by the network. Based on the received two most recent TPC commands the adaptive factor is updated.TPC command increases or decreases the transmitting power. If same TPC commands are detected the step size will be increased or else the step size will be decreased. Power Determining Factor (PDF) is computed based on the data traffic rate to determine whether the power is decreasing or increasing. PDF factor is updated according to the traffic rate

The transmit power can be represented as:

$$P(t+1) = P(t) + \lambda. \text{ sign } (SIR_{target} - SIRest) \text{ [dB]}$$

(14)

Where SIR_{est} and SIR_{target} are the estimated and target SIR respectively, λ is the power control step size and $P(t)$ is transmit power at time 't'. The term sign is the sign function: sign(x) = 1, when $x \geq 0$, and sign(x) = -1, when $x < 0$.

The transmit power is updated according to the following equation:

$$Pu(t+1) = Pu(t) + AFu(t). PDFu(t) \lambda. TPCu(t)$$

(15)

PDFu (t) is the Power Determining Factor, TPCu (t) is the TPC command of uth user at time t, related to sign (SIR $_{target}$-SIR$_{est}$), and AFu (t) is the Adaptive Factor of uth user at time 't'.

5. SIMULATION RESULTS

5.1 Simulation Setup

In this section, we simulate the Dynamic MAC protocol with power control for WCDMA cellular networks. The simulation tool used is Network simulator2. In the simulation, mobile nodes move in a 600 meter x 600 meter region for 50 seconds simulation time. Random waypoint (RWP) model of NS2 is used to obtain the initial movements and locations of the nodes. All nodes have the same transmission range of 250 meters. In our simulation channel capacity is set to 2 Mbps. The number of users simultaneously transmits data in a slot using CDMA codes is 2 to 5. The simulation parameters are given in table 1.

Area Size	600 X 600
Number of Cells	2
Slot Duration	2 msec
Radio Range	250meters
Frame Length	2 to 8 slots
CDMA codes	2 to 5
Simulation Time	50 sec
Traffic Source	CBR, VBR
Packet Size	512 bytes
Video Trace	JurassikH263-256k
Tx power, Rx power	0.66w,0.395w
Speed of mobile	25m/s
No. of users	32

Table I. Simulation Parameters

5.2. Performance Metric

The performance is mainly evaluated according to the following metrics:

Channel Utilization: It is the ratio of utilised bandwidth to total bandwidth for a traffic flow.

Throughput: It is the successful transmission of packets in a unit of time.

Average End-to-End Delay: The end-to-end-delay is averaged over all surviving data packets from the sources to the destinations.

Average Energy: It is the average energy utilized by all nodes in transmitting, receiving and forward operations.

5.3. Results of Dynamic MAC protocol

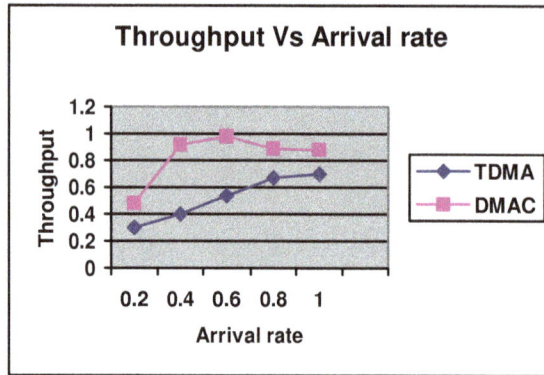

Fig 2. Throughput Vs Arrival Rate

As shown in figure 2 as the arrival rate increases throughput increases more in DMAC when compared to TDMA. We have considered 2 to 5 CDMA codes per slot. The network is in low contention mode up to 0.5 arrival rate the throughput increases and reaches a maximum value of unity. In low contention mode users can access any slots and all the users obtained slots for transmission and hence maximum throughput is achieved in this mode. The network is in high contention mode when the arrival rate is more than 0.5 and the throughput is 0.9 and is constant. In high contention if the slots are not used by the owners, the other users can access the slots.As 5 users are transmitting data in a slot in DMAC rejection rate is very small so through put is more.Whereas in TDMA, the slots are reserved for users and if the users are not having data to transmit the slots cannot be used by other users, so throughput is less in TDMA when compared to DMAC.

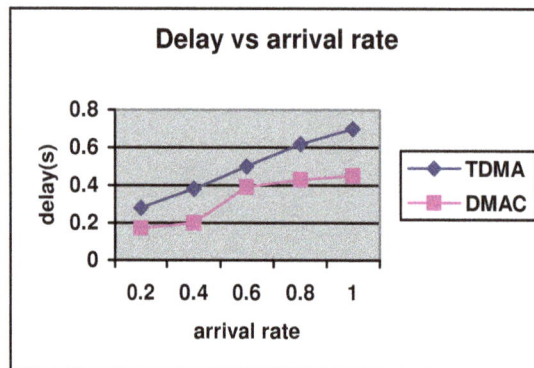

Fig3. Delay Vs Arrival Rate

Fig.3 shows the delay occurred for various rates. It can be observed that the delay increases gradually with increasing arrival. The delay is more for more number of users when the traffic is high .It shows that the delay of DMAC is significantly less than TDMA protocol. In DMAC more users can transmit data in a slot whereas in TDMA one user is allowed to transmit data in each slot .So delay is less in DMAC when compared to TDMA.

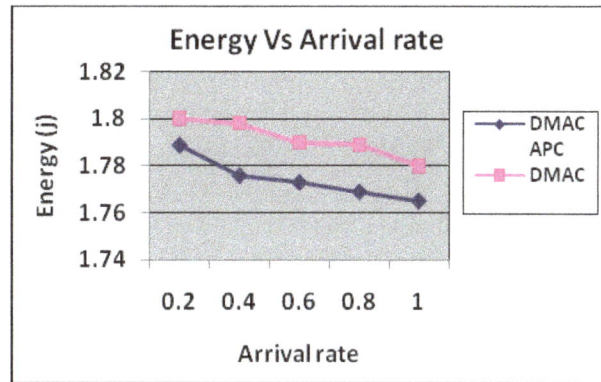

Fig4: Energy Vs arrival rate

Figure 4 shows the energy utilization with arrival rate for both DMAC-APC and DMAC. Energy utilisation is getting decreased in both DMAC-APC and DMAC as arrival rate increases. DMAC-APC uses less energy when compared with DMAC because DMAC-APC uses adaptive power control technique the transmitted power is lowered for low data traffic and power is increased for the high data traffic and hence energy is efficiently utilised. As the users transmit data without adaptive power control more energy is utilised in DMAC protocol than DMAC-APC.

6. CONCLUSION.

In this paper, we have developed a Dynamic MAC protocol for wireless multimedia networks. The proposed MAC protocol is based on contention .If there is low contention users can access any slots and if there is high level contention the owners of the slots have priority to access slot. An adaptive power control algorithm is applied on the protocol. By simulation results, we have shown that our proposed MAC protocol achieves improved throughput with reduced average delay and reduces power consumption of low and high multimedia traffic. Hence the designed protocol will well suit for multimedia data transmission on wireless multimedia network.

7. REFERENCES

[1]   N. Abramson, "Multiple access in wireless digital networks," Proceedings of IEEE, volume. 82, Set 1994, pp. 136&70.

[2]   Z.Tang and J.J Garcia –Luna _aceves. "A Protocol for Topology Dependent Transmission scheduling in Wireless Networks", Proceedings of IEEE WCNC, New Orleans, LA, 1999.

[3]   Lixin Wang and Mounir Hamdi   "HAMAC: An Adaptive Channel Access Protocol for Multimedia Wireless Networks," IEEE Proceedings of International Conference on Computer Communications and Networks, pp 408-411, 1998.

[4]   I. Chlamtac, A. Farago, A. D. Myers V. R. Syrotiuk G. Zkruba "A Performance Comparison of Hybrid and Conventional MAC Protocols for Wireless Networks,"IEEE VTC 2000 pp 201-205, 2000.

[5]   Zhijun Wang, Haoche, umapathi Mani, MiaoJu "A RAH-MAC: Rate adaptive Hybrid MAC protocol for wireless cellular Networks" Proceedings of IEEE International Conference on Computer Communications and Networks 2006.

[6]   Ian F. Akyildiz, Fellow, IEEE, Inwhee Joe ,David A. Levine, "A Slotted Code Division Multiple Access   Protocol with Bit Error Rate Scheduling for Wireless Multimedia Networks,'"IEEE/ACM Transactions on Networking, volume. 7, No. 2, pp 146-158, April 1999

[7]   C. Roobol et al., " A Proposal for an RLC/MAC Protocol for WCDMA Capable of Handling Real Time and Non Real Time Services," proceedings of IEEE VTC'98, pp. 107-111, 1998.

[8]   A.Saravan, B.parthasarathy, Analytical model Medium Access control Multiprotocol Architecture for Mobile Multimedia Networks.," proceedings of 2008 IEEE International Symposium on Ubiquitous Multimedia Computing, pp 105-111.

[9]   Rekha Patil, A. Damodaram, "A Cross-Layer Based Joint Algorithm for Power Control and Scheduling in Code Division Multiple Access Wireless Ad-Hoc Networks", proceedings of WSEAS transactions on communications, vol.8, No.1, pp.122-131, January, 2009.

[10]   Kumbesan Sandrasegaran, Rachod Patachaianand, "A New Adaptive Power Control Algorithm for UMTS", WITSP, Wollongong, Australia, pp. 1-6, 2006.

[11]   Rachod Patachaianand and Kumbesan Sandrasegaran, "An Adaptive Step Size Power Control With Transmit Power Control Command Aided Mobility Estimation" proceedings of the 4th IASTED Asian Conference Communication Systems and Networks,Thailand,2007.

[12]   S.P.V.Subbarao, S. Venkata Chalam and D.Srinivasa Rao, "A Dynamic MAC Protocol with adaptive power control for WCDMA Wireless Multimedia Networks", International Journal on Distributed and parallel systems (IJDPS) Vol.2, No.2, March 2011(PP 105-114).

[13]   S.P.V.Subbarao, S. Venkata Chalam and D.Srinivasa Rao, "A Survey on MAC protocol for Wireless Multimedia Networks "International Journal of Computer Science & Engineering Survey (IJCSES) Vol.2, No.4, November 2011 ISSN : 0976-2760 (PP 57-74)

[14]   N. Mohan, T.Ravichandran "An Efficient Multiclass Call Admission Control and Adaptive Scheduling for WCDMA Wireless Network" European Journal of Scientific Research, vol.33, No.4, pp.718-727, 2009.

[15]   Fredrik Gunnarsson "Fundamental Limitations of Power Control in WCDMA" Workshop on CDMA, Sweden, December 4, 2001.

[16]   Peter Jung , Andreas steil Paul and walter Baier "Advantages of Code Division Multiple Access and Spread Spectrum Techniques over FDMA and   TDMA in cellular Mobile Radio Applications" IEEE Transactions on Vehicular Technology, Vol 42 ,No.3, August 1993.

[17]   Harri Holma and Antti Toskala, "WCDMA for UMTS: HSPA evolution and LTE", 4th edition, wiley publication, 2007, ISBN 978-0-470-31933-8

[18]   V.Sumalatha and T.Satya Savithri "Router control mechanism for congestion avoidance in cdma based ip network", International Journal of Information Technology and Knowledge Management July-December 2010, Volume 2.

[19]   Uthman Baroudi ,Ahmed Elhakeem "Adaptive Admission/Congestion Control Policy for Hybrid TDMA /MC-Code Division Multiple Access Integrated Networks with Guaranteed QOS" proceedings of ICECS 2003.

[20]   Hai Jiang , Weihua Zhuang, and Xuemin (Sherman) Shen "Distributed Medium Access Control for Next Generation CDMAWireless Networks" IEEE Wireless Communications, Special Issue on Next Generation CDMA vs. OFDMA for 4G Wireless Applications, vol. 14, no. 3, pp.25-31, June 2007.

[21]   Junshan Zhang, Ming Hu and Ness B. Shroff "Bursty traffic over CDMA: predictive MAI temporal structure, rate control and admission control" Elsevier 2002-2003.

[22]   Liang Xu, Xuemin (Sherman) Shen and Jon W. Mark "Dynamic Fair Scheduling with QoS Constraints in Multimedia Wideband CDMA Cellular Networks" IEEE Transactions on wireless communications, Vol. 3, No. 1, January 2004.

[23]   Jennifer Price and Tara Javidi "Distributed Rate Assignments for Simultaneous Interference and Congestion Control in Code Division Multiple Access-Based Wireless Networks" UW Electrical Engineering Technical Report, 2004.

[24]   "Channel access method" from http://en.wikipedia.org/wiki/Channel_access_method

[25]   W. Ye, J. Heidemann, and D. Estrin, "An energy-efficient MAC protocol for wireless sensor networks," in Proceedings of the IEEE Computer and Communications Societies (INFOCOM '02), pp. 1567–1576, June 2002.

[26]   Nuwan Gajaweera and Dileeka Dias, "FAMA/TDMA Hybrid MAC for Wireless Sensor Networks", In proceeding of the 4th International Conference on Information and Automation for Sustainability, Colombo, Sri Lanka, 12-14 December 2008

[27]   T. van Dam, and K. Langendoen, "An adaptive energy efficient MAC protocol for wireless sensor networks," ACM SenSys 2003, pp. 171-180, November 2003.

[28]   R. Prasad, "An overview of CDMA evolution toward wideband CDMA", IEEE Communications Surveys, Vol .1 (1), 1998.

[29]   "Wideband CDMA" from http://en.wikipedia.org/wiki/W-CDMA_(UMTS)

[30]   Rajamani Ganesh, Kaveh Pahlavan and Zoran Zvonar, "Wireless multimedia network technologies", Kluwer academic publishers, 2000,

[31]   "Wideband CDMA" from http://www.mobileisgood.com/What isWCDMA.php

[32]   Juha Korhonen, "Introduction to 3G mobile communications", 2nd edition, mobile communication series, 2003

[33]   Network Simulator, http://www.isi.edu/nsnam/ns

A NOVEL MULTIBAND KOCH LOOP ANTENNA USING FRACTAL GEOMETRY FOR WIRELESS COMMUNICATION SYSTEM

Rajeev Mathur[1], Sunil Joshi[2], Krishna C Roy[3]

[1]Department of ECE, Suresh Gyan Vihar University, Jaipur, Rajasthan, India
`Rmathur_2000@yahoo.com`

[2]College of Engineering & Technology, MPUAT, Udaipur, India
`suniljoshi7@rediffmail.com`

[3]Pecific Institute of Technology, Udaipur, India
`roy.krishna@rediffmail.om`

ABSTRACT

The paper present a novel multi-band compact antenna designed on the theory of fractal geometry. The antenna is fabricated on a FR4 substrate. The performance of the proposed antenna design is analyzed and the results are compared with the simulations using IE-3D tool. The relevant antenna performance parameters of the proposed design viz. resonant bands, return loss, bandwidth and gain are reported and discussed. The VSWR of the antenna is less than 2 for six resonant bands in the vicinity of 1.15 GHz, 2.0 GHz, 3.17 GHz, 3.6 GHz, 4.17 GHz and 5.91 GHz. The performance results exhibited by the proposed antenna makes it extremely useful for the future generation of wireless broadband communication systems.

KEYWORDS

Fractal Antenna, Multiband, Return Loss, Koch Dipole, Loop Antenna.

1. INTRODUCTION

'Fractal' term was first coined by Benoit Mandelbrot in 1983 to classify the structure whose dimensions were not whole numbers. A mathematical description of dimension is based on how the "size" of an object behaves as the linear dimension increases. In one dimension consider a line segment, if the linear dimension of this line segment is doubled then obviously the length (characteristic size) of the line has doubled. In two dimensions, if the linear dimensions of a rectangle is doubled then the characteristic size, the area, increases by a factor of 4. In three dimensions, if the linear dimension of a box are doubled then its volume increases by a factor of 8. This relationship between dimension D, linear scaling L and the resulting increase in size S can be generalised and represented mathematically as [1]

$$S = L^D \qquad \qquad \dots 1$$

This is just telling us mathematically what we know from everyday experience. If we scale a two dimensional object for example then the area increases by the square of the scaling. If we scale a three dimensional object the volume increases by the cube of the scale factor. Rearranging the above expression in terms of logarithmic expression as below

$$D = \log (S)/ \log (L) \qquad\qquad .... 2$$

This relationship holds for all Euclidean shapes. But in natural world there are many shapes which do not conform to the integer based description of dimensions. There are objects which appear to be curves which cannot be described with integer number. There are shapes that lie in a plane i.e. two dimensional (D=2 in the expression), but if they are linearly scaled by a factor L, the area does not increase by L squared but by some non integer amount. These geometries are called fractals! [1]

Fractals are used to describe the branching of tree leaves and plants, the sparse filling of water vapour that forms clouds, the random erosion that carves mountain faces, that jaggedness of coastlines and bark, and many more examples in nature[1]. One of the properties of fractals geometry is that it can have an infinite length while fitting in a finite volume. The radiation characteristic of any electromagnetic radiator depends on electrical length of the structure [2]. Using the property of fractal geometry, we may increase the electrical length of an antenna, keeping the volume of antenna same. Thus a new configurations for radiators and reflectors may be developed to give better performance in terms of gain, bandwidth etc. There are an infinite number of possible geometries that are available to try as a design of fractal antenna. One of the important benefits of fractal antenna is that we get more than one resonant band.

The Simplest example of antenna using fractal geometry is given by the Von Koch, researcher. The method of creating this shape is to repeatedly replace each line segment with the following 4 line segments. The process starts with a single line segment and continues for ever. The first few iterations of this procedure are shown in Figure 1. First five iterations in the construction of the Koch curve are illustrated. Fractal dimension contains information about the self-similarity and the space-filling properties. The Fractal similarity Dimension (FD) is defined as [5]:

$$FD = \frac{\log (N)}{\log (1/\varepsilon)} = \frac{\log (5)}{\log(3)} = 1.46 \qquad .... 3$$

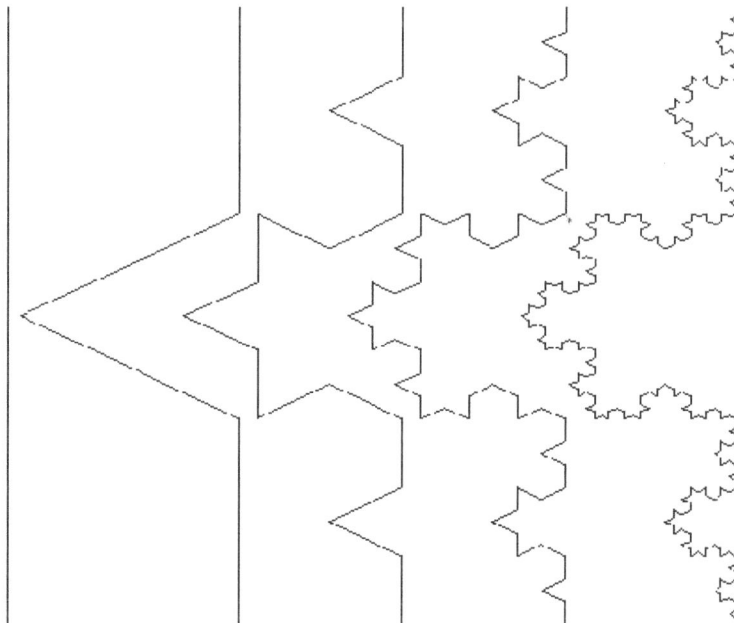

Figure 1. Koch fractal geometry.[5]

Where N is the total number of distinct copies, and ($1/\varepsilon$) is the reduction factor value which means how will the length of the new side be, with respect to the original side length. Fractal shapes thus are defined as self similar shapes which are independent of size or scaling.[5]

2. RELATED WORK

Cohen N.L. have proposed a novel Koch monopole fractal antenna for the use in defence application. he concluded that the design space for the fractal antenna afford vast new opportunities in design and application, many realised and proven beyond theory [9]. Fractal antenna can obtain radiation pattern and input impedance similar to longer antenna, yet takes less area due to the many contour of shapes. Various fractal antenna design techniques is discussed by Nemanja POPRZEN & Mico GACANOVIC. Koch Loop, Minkowski Loop, Siepinski Seive have been studies and two course of action have been concluded. Firstly, many more examples of fractals geometries could be applied to antenna and secondly, correlation could be drawn between fractal dimension and antenna performance. Figure 2, Figure 3, Figure 4 and Figure 5 below shows the various antennas so far studied [10].

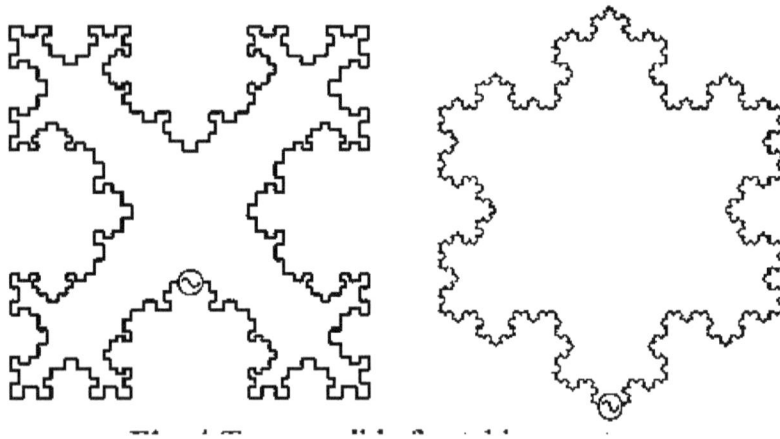

Figure 2. Fractal Loop Antennas [9].

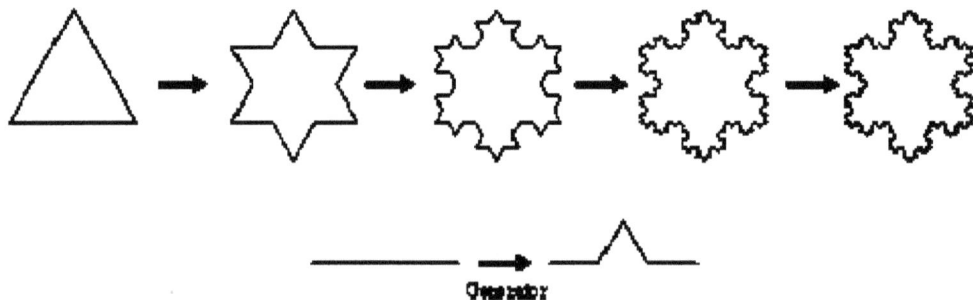

Figure 3. First Four Iteration of koch fractal antenna [10].

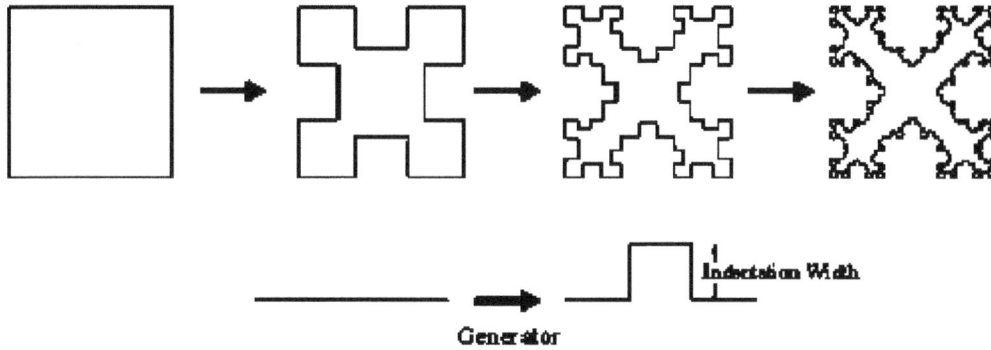

Figure 4. First four iteration of Minkowski Loop antenna [10].

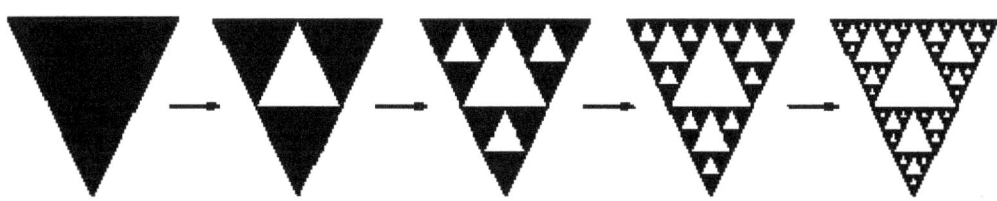

Figure 5. First four iteration of Siepinski Seive antenna [10].

Behavior of Koch monopole antenna has been analysed mathematically and experimentally by Carles Puente and Angel Cardama and it was observed that as the number of iteration in fractal antenna is increased, the Q of the antenna approaches the fundamental limits for small antenna. [11]. It has been also observed that in spite of small size fractal antenna prove to be good radiator. Ultimate application of this antenna is in mobile terminals where reduction of size is ultimate goal. It is possible to employ antenna that fits in small volume, but still have efficient performance. [11]

Many research groups are working on design of antenna based on fractal geometry which could prove to be an efficient radiator in wireless mobile communications applications.

3. PROPOSED ANTENNA DESIGN

The width of the a microstrip patch antenna is calculted by [8]:

$$W = \frac{1}{2f_r\sqrt{\mu_0\epsilon_0}}\sqrt{\frac{2}{\epsilon_r + 1}} = \frac{v_0}{2f_r}\sqrt{\frac{2}{\epsilon_r + 1}}$$

.... 4

The actual length and effective length of patch antenna is found as [8]

$$L = \frac{1}{2f_r\sqrt{\epsilon_{\text{reff}}}\sqrt{\mu_0\epsilon_0}} - 2\Delta L$$

.... 5

$$L_{\text{eff}} = L + 2\Delta L$$

.... 6

The dielectric constant, loss tangent and substrate height of designed antenna is choosen as 4.4, 0.025 and 1.588 mm respectively, for FR-4 substrate. The computed values of W and L_{eff} are 72.43 and 80.51 mm respectively.

The conventional dipole design technique is adopted to design the proposed Koch loop antenna. For a 2 GHz frequency, wavelength is 150 mm, a Dipole antenna length must be half of wavelength. The four dipoles are then arranged in the form of loop thereby increasing its physical length which comes out to 300 mm. Resonant frequency now for loop will be 1 GHz.

Figure 6. 2nd iteration of koch dipole

2nd iteration of koch dipole is as shown in Figure 6. A final antenna is designed with the dimension further reduced to 1/3rd of 2nd iteration i.e. 25mm. The dipole width is choosen 2mm. This antenna is a simple planar structure with effective permittivity of substrate to be 4.4. Height of substrate is 1.588mm with loss tangent of 0.025. Ground Plane is considered to be infinite for simulation purpose; however, practically ground plane taken is 80mm X 80mm. CPW feed is chosen for this antenna. SMA connector of @50 ohms impedance is connected at feed port 1 and 2 as shown in Figure 7.

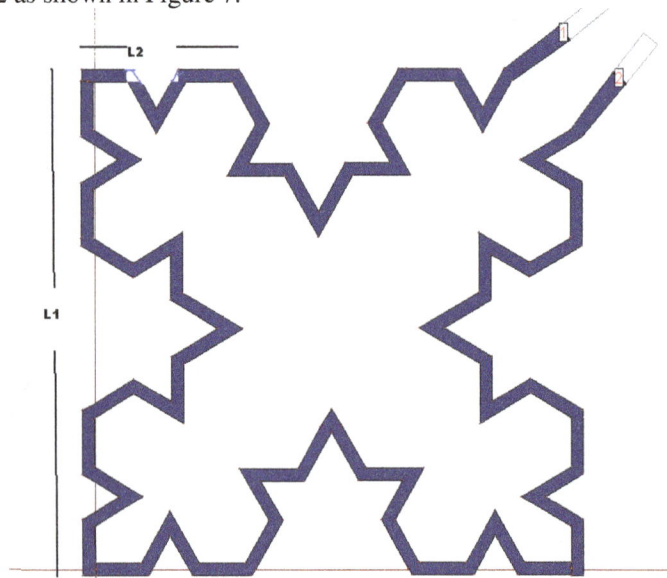

Figure 7. Koch Loop Antenna with Lengths L1 = 75mm, L2=25mm.

4. FABRICATION OF PROPOSED ANTENNA

A Prototype structure of this antenna is fabricated in the lab using photolithography technique. Mask of the antenna is prepared and than complete structure was developed as shown in the Figure 8. Commonly available substrate FR4 is used with copper cladding of 0.0004mm. The dimensions of the fabricated antenna are as given by Table 1.

Table 1: Dimensions of Koch Loop Antenna

Ltotal	Wtotal	Width of Strip	L1	L2
80mm	80mm	2mm	75mm	25mm

Figure 8. Fabricated Fractal Antenna

5. RESULT & DISCUSSION

The resonant properties of proposed antenna have been obtained by designing the antenna structure using commercially available EM tool IE3D. The return loss profile is as shown in Figure 9, showing 7 bands with return loss well below -10 dB. The central frequencies of these bands are mentioned in the Table 2. Also, as shown in Figure 10, the VSWR obtained for these bands is found to be of the order of 2. It is observed that each small iterative element acts as a separate radiating dipole element leading to multiple resonant bands in addition to the fact that the entire loop acts as a radiating element. Besides, each small element contributes towards the increase in electrical length of antenna to increase radiating field E_θ. The axial ratio of the antenna is observed to be zero revealing it to be a linearly polarised antenna.

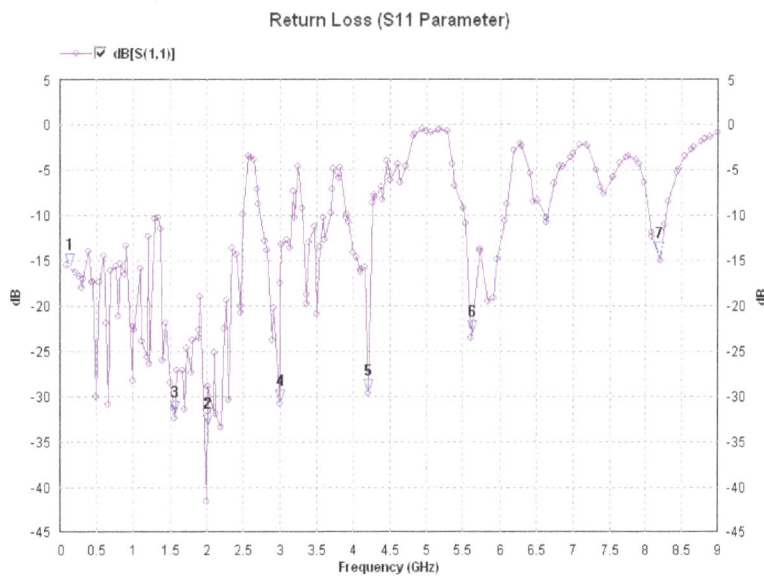

Figure 9. Return Loss obtained by simulation.

Table 2: Resonant Frequencies of Koch Loop Antenna

Points	Frequency	S11 in dB
1	0.50	-30.76
2	2.01	-42.30
3	1.56	-32.04
4	3.00	-30.84
5	4.2	-29.70
6	5.625	-23.13
7	8.19	-14.55

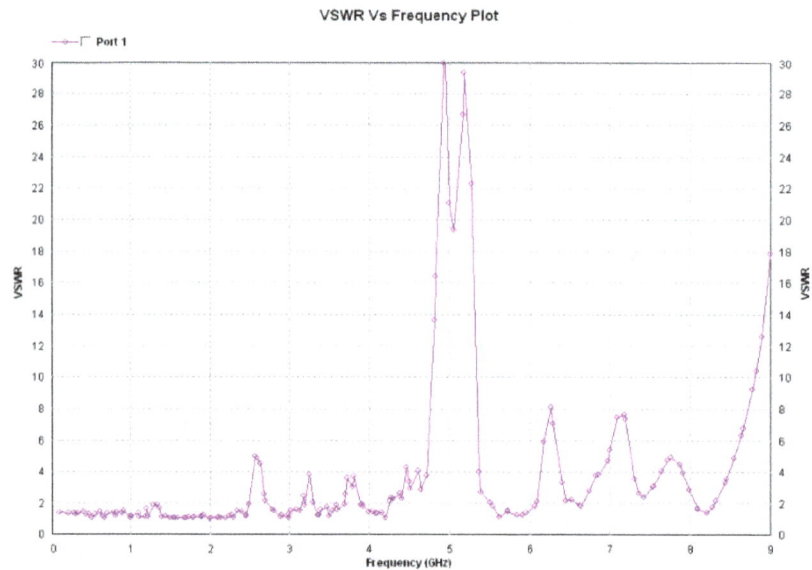

Figure 10. VSWR obtained by simulation

The measurement set up for testing the antenna performance is shown in Figure 11, which includes Vector Network Analyser (VNA) of Anritsu make , Signal Generator, Computer system and designed antenna. The VNA was first calibrated using calibration device and then coaxial feed is given to this antenna through SMA connector.

Figure 12 shows the measured return loss profile of the antenna. At design frequency of 2.00 GHz, it is obtained as -32 dB i.e. minimum. We have obtained multibands with small bandwidth. The measured VSWR, as shown in Figure 13, is also within the arrange 1-2

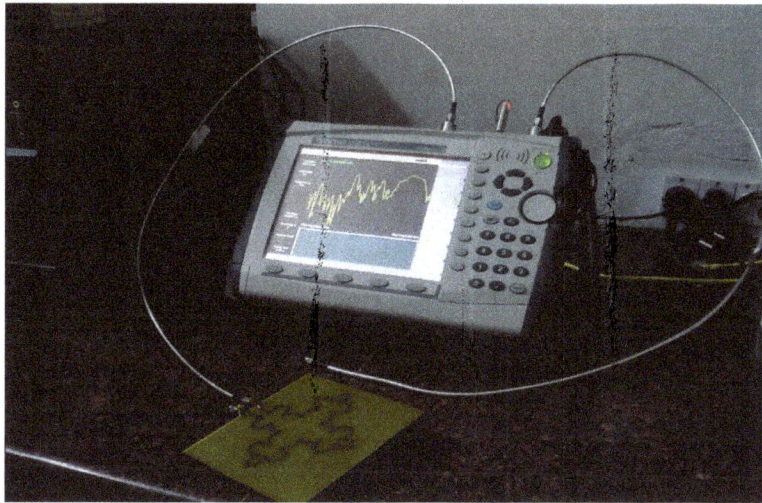

Figure 11. Laboratory setup for measurement of return loss and VSWR.

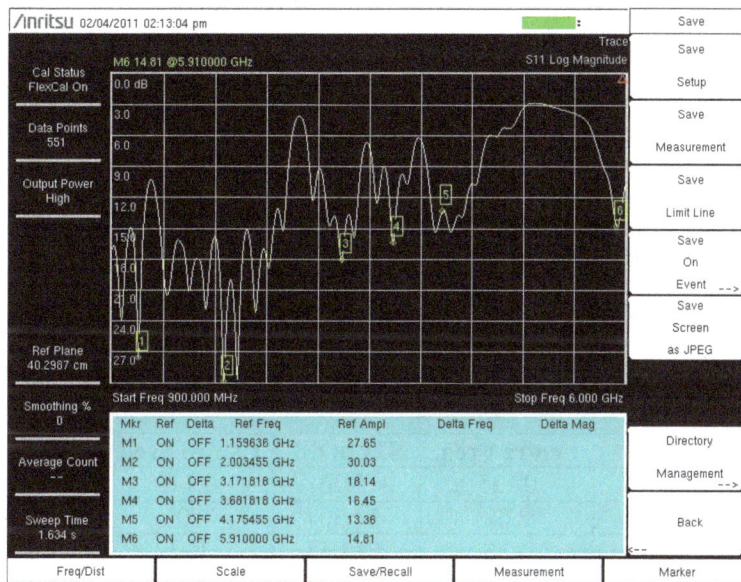

Figure 12. Measurement of return loss on VNA

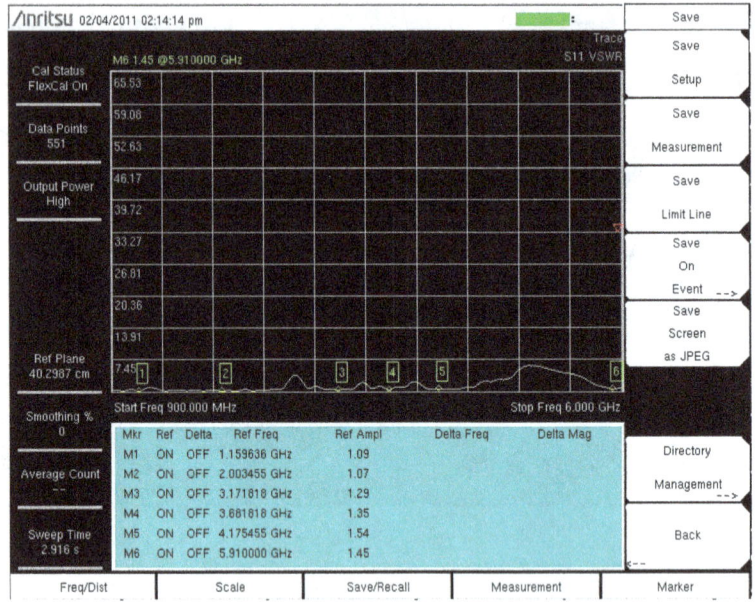

Figure 13. Measurement of VSWR on VNA

Comparisons of the simulated and experimental results were made and we found that there is a close agreement between the two as shown by Table 3. The slight variation in results may be due to environmental conditions which could not be considered in simulation. Also during fabrication process, fringing edges of the patches may have irregularities due to which fringing field gets disturbed, resulting in shift in resonant frequencies. It has been observed that as we increase the iterations number of frequency band also increases.

Table 3: Comparison between Simulated results and Measured Return Loss

Band No	Simulated results		Measured Results	
	Centre Freq.	S11 in dB	Centre Freq.	S11 in dB
I	0.135	-15.761	0.135	-15.04
II	2.01	-33.304	2.03	-30.04
III	1.56	-32.047	1.16	-27.85
IV	3.00	-30.845	3.17	-18.14
V	4.2	-29.70	4.17	-13.36
VI	5.625	-23.13	5.91	-14.01
VII	8.19	-14.554	8.19	**

Radiation pattern are simulated and investigated for all the five frequency bands as shown in Figure 14. It is deduced that as the frequency is increasing radiation pattern changes to provide higher directivity and gain. Overall gain of this antenna is good at higher frequency bands as compared to the lower frequency bands. For lower frequency bands, upto 4 GHz gain is below 4 dBi and for higher frequency ranges upto 8.2 GHz, gain is above 4 dBi. Highest gain was observed at frequency of 8.2 GHz i.e. 7.92 dBi.

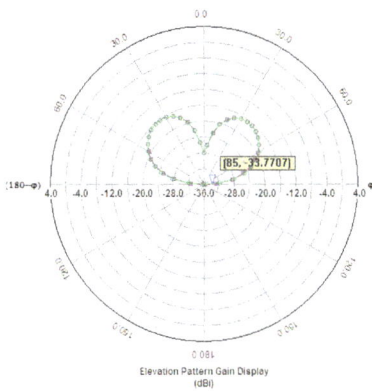

(a) Frequency = 2 GHz.

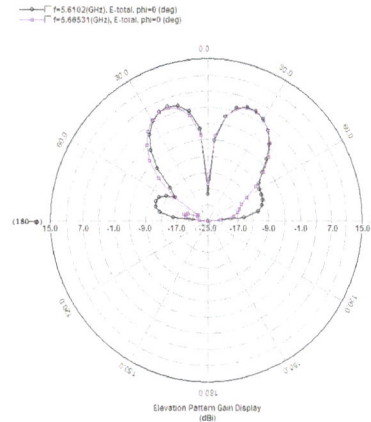

(b) Frequency = 3 GHz.

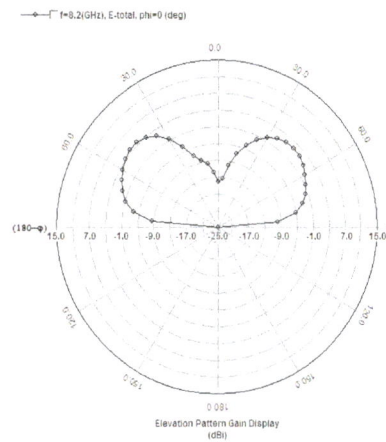

(c) Frequency = 4.2 GHz

(d) Frequency = 5.61 & 5.66 GHz

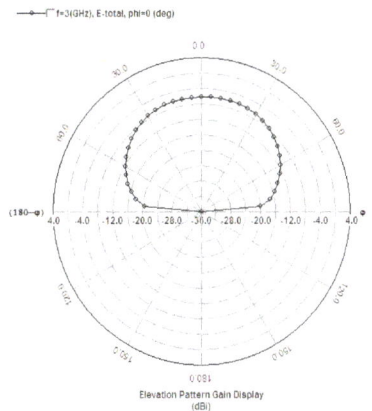

(e) Frequency = 8.2 GHz

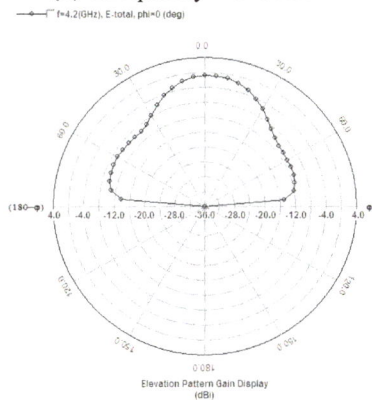

Figure 14. Radiation pattern for 5 bands

6. CONCLUSION

A novel prototype structure for Koch Loop Antenna was developed and experimentally proven to be adequate in terms of return loss. Seven resonant bands have been obtained by simulation & measurement on VNA, for this antenna. The VSWR of the designed antenna is less then 2 for all 7 resonant bands of 135MHz, 1160 MHz, 2030 MHz, 3170, 4171 MHz, 5910 MHz. and 8190MHz.

Experimentally it has been observed that fractal antenna is very good radiator as we measured return loss of -30 dB on VNA, it is obtained at the frequency for which Koch dipole is designed. Other bands observed are below and above this central frequency , it is because of the variation in the length of dipole. We may conclude that we obtain more than one resonant band due to the facts, firstly , each small element acts as a separate radiating dipole element; secondly, entire loop as a radiating element. Besides, each small element contributes towards the increase in electrical length of antenna to increase radiating field E_θ.

Designed Koch Loop Antenna has possibility of being optimized in terms of return loss and number of narrow frequency bands. It is observed that by varying the width of strip of or small variations in the geometry of the antenna does not change the frequency characteristics of the antenna. The range of the frequency bands is within the wireless communication bands of Wi-fi, WiMAX, Bluetooth and wireless LAN etc.

ACKNOWLEDGEMENT

We wish to acknowledge, Dr. S.S. Pattnaik (NITTTR, Chandigarh) for his support and Dr. O P N Calla for the motivation to do research in this area.

REFERENCES:

[1]   T. Tiehong and Z. Zheng, " A Novel Multiband Antenna: Fractal Antenna", Electronic letter, Proceedings of ICCT – 2003, pp: 1907-1910.

[2]   D. H. Werner and S. Ganguly, "An Overview of Fractal Antennas Engineering Research",IEEE Antennas and Propagation Magazine, vol. 45, no. 1, pp. 38-57, February 2003.

[3]   J. Gianvitorio and Y. Rahmat, "Fractal Antennas: A Novel Antenna Miniaturization Technique and Applications", IEEE Antennas and Propagation Magazine, vol. 44, No. 1, pp: 20-36, 2002.

[4]   K. Falconer, "Fractal Geometry: Mathematical Foundation and Applications", John Wiley, England, 1990.

[5]   S.H Zainud-Deen, K.H. Awadalla S.A. Khamis and N.d. El-shalaby, March 16-18, 2004. Radiation and Scattering from Koch Fractal Antennas. 21st National Radio Science Conference (NRSC), B8 - 1-9.

[6]   P. S. Addison, "Fractals and Chaos: An Illustrated Course", Institute of Physics Publishing Bristol and Philadelphia, 1997.

[7]   G. J. Burke and A. J. Poggio "Numerical Electromagnetic Code (NEC)-Program description", January, 1981, Lawrence Livermore Laboratory.

[8]   C. A. Balanis, "Antenna Theory: Analysis and Design", 2nd ed., Wiley, 1997.

[9]   Cohen N.L. 2005, New era in military antenna design, Defense Electronics.

[10]   Nemanja POPRZEN & Mico GACANOVIC, Fractal antenna: Design, Characteristics and Application.

[11]   Carles Puente and Angel Cardama, "The Koch Monopole: A small fractal antenna", IEEE transaction on antenna and propagation, vol 48, no. 11 Nov 2000.

3

QUALITATIVE ANALYSIS OF THE QOS PARAMETERS AT THE LAYER 1 AND 2 OF MOBILE WIMAX 802.16E

Arathi.R.Shankar [1] Adarsh Pattar[2] and V.Sambasiva Rao[3]

[1]Department of Electronics &Communication BMS College of Engineering, B'lore
arathi.rshekhar@gmail.com

[2] BMS College of Engineering, B'lore, India
techie04@gmail.com

[3] Department of Electroncics, PESIT, B'lore, India
vsrao@pes.edu

ABSTRACT

IEEE 802.16/WiMAX is the network which is designed with quality of service in mind. In this paper the authors have made an attempt to qualitatively analyse the performance factors at the physical layer of mobile WiMAX for Rayleigh and Rician channels using different modulation techniques. The comparison of quality of service parameters like BER and power spectral density between different channels in WiMAX physical layer is made. Orthogonal frequency division multiple access technique is adapted by WiMAX on its physical layer.. Scalable OFDM has been implemented for subcarriers ranging from 256 to 2048 and the performance factors have not changed. Quality of Service provisioning at the MAC layer is done by using Packet scheduling technique using RR & Fairness queue algorithms for rtPS & nrtPS service flows. Comparison of different scheduling schemes are made that ensure QoS with respect to delay, under the context of different service flows as defined in the WiMAX standard.

KEYWORDS

QoS,, OFDMA, Physical Layer, MAC, Scheduling, Round Robin and Fairness Queue

1. INTRODUCTION

Quality of service includes 6 primary components: Support, Operability, Accessibility, Retain ability, Integrity and Security. The IEEE 802.16 standard includes the QoS mechanism in the Physical (PHY) layer (layer 1) architecture and also the MAC layer (layer2).

This work focuses on the analysis of QoS in WiMAX networks. The details of the network's PHY layer QoS implementation are presented.

WiMAX IEEE Standard 802.16 also known as Air Interface for Fixed Broadband Wireless Access Systems operates in the 10 -66 Ghz frequency band and its extension 802.16a allows the usage of lower frequencies (2 -11 G Hz) many of which are unregulated, Additional standards 802.16a to 802.16e offers Quality of service, interoperability, to develop access points and support for mobile as well as fixed broadband. WiMAX can provide two flavours of wireless services, depending on the frequency range of operation, ie;LOS and NLOS operations .The standard operating between 10 – 66GHz requires LOS operations, while lower frequency bands 2-11GHz enable NLOS operations. The standard defines three different air interfaces that can be used to provide a reliable end-to-end link [3][4].

- SCa: A single-carrier modulated air interface.

- OFDM: Orthogonal-frequency division multiplexing (OFDM) with 256 carriers.

Multiple access of different SSs is time-division multiple access (TDMA)-based.

- OFDMA: A 2048-carrier OFDM scheme.

Mobile WiMAX is a rapidly growing broadband wireless access technology based on IEEE 802.16-2004 and IEEE 802.16e-2005 air-interface standards[7][8]. The WiMAX forum has provided mobile WiMAX system profiles that define the features of the IEEE standard .Mobile WiMAX . The technology is heading towards the 4G networks..

The features of mobile WiMAX

1. OFDMA: In non-line-of-sight (NLOS) environments to improve the multipath performance, the mobile WiMAX air interface uses Orthogonal Frequency Division Multiple Access (OFDMA) as the radio access method.
2. High data rates: High downlink and uplink high data rates can be achieved by the use of multiple-input multiple-output (MIMO) antenna techniques along with flexible sub channelization schemes, and different adaptive modulation and coding schemes are used.
3. Quality of Service: QoS is the fundamental criteria of the IEEE 802.16 medium access control (MAC) layer. Service flows of the MAC Layer can be mapped to the corresponding service flows of the IP layer to achieve end to end QoS.
4. Scalability: Scalable OFDMA (S-OFDMA) feature of Mobile WiMAX , enable it o operate in scalable bandwidths from 1.25 to 20 MHz to adapt for various spectrum allocations worldwide.
5. Security: The security aspects of Mobile WiMAX are well taken care off by EAP, AES, CMAC HMAC protection schemes.
6. Mobility: To support real-time applications such as Voice over Internet Protocol (VoIP) the mobile WiMAX uses optimized handover schemes with latencies less than 50 ms . [13][9].

The paper is categorised as follows – in Section-2 we present the proposed algorithm . Section-3 Simulation details and results.

The paper is concluded in section 4.

1.1 RELATED WORK:

The simulation model which has been used here is different from the related work done [17]. In the current work scheduling is performed on the mobile WiMAX compared to the fixed WiMAX used in [17].The packet loss and the delay violation rate is reduced . Scalable OFDMA, is implemented so that the carrier is divided into multiple sub carriers depending upon the application using different size of FFT's varying from 256 to 2048 .

2. PROPOSED SCHEDULING ALGORITHM

In this work, based on the per connection requests from SSs the uplink bandwidth is allocated to the BS . The bandwidth request messages should report the bandwidth requirement of each connection in SS because of multiple connections that can prevail in each SS[17]. The bandwidth allocated per connection is distributed to each SS. The SS in turn ,allocates the resource from its BS to the various connections according to their QoS specifications. Hence an additional scheduler will be required in each SS.After that the allocated bandwidth per connection is pooled together and granted to each SS.

.

1.Scheduling algorithm for UGS queues:

UGS service gets the highest prority because it generates fixed size data packets on a periodic basis and this service has a critical delay and delay jitter requirement. So, the UGS queues are given highest pririty by the SS scheduler.

2.Scheduling algorithm for rtPS queues:

The end-to-end delay of rtPS service is reduced to a significant amount because, each packet entering the rtPS queues should be assigned with a delivery deadline equal to t + tolerated delay, where t is the arrival time and tolerated delay is the Maximum Latency for such a service flow. The packet transmitted first will be the one with the least time deadline.

3.Scheduling algorithm for nrtPS and BE queues :

Deficit Fair Priority Queue (DFPQ) algorithm is used for nrtPS and BE services [17]. The algorithm is is suitable for datagram networkswith variable packet sizes.

ii. Since this algorithm requires prerequisite knowledge of packet size, it is used for the uplink traffic at SS scheduler.

iii. The algorithm is flexible and the scheduler can provide minimum bandwidth for every non real time services such as nrtPS and BE connection. Hence an acceptable throughput is maintained.

In every service round, the nrtPS queue is given higher pririty than the BE.In the algorithm, Q is assigned to each queue i. Queue i (Q[i]) represent the maximum number of bits that can be serviced in the first round. The scheduler visits every nonempty queue and analysis the number of bandwidth requests in the queue. If there are excess packets in the I th queue, after servicing the excess bits are stored in a queue state variable called Deficit Counter (DC[i]) and the scheduler serves the next non-empty queue. The bandwidth used by this flow is the sum of the value of DC[i] in the eprevious round added to Q[i]. Every flow has the Q[i] ie the Maximum Sustained traffic rate (rmax). Here connections with larger quantum get more service. The simulation model which has been used here is different from the related work done [17]. In the current work scheduling is performed on the mobile WiMAX compared to the fixed WiMAX used in [17].The packet loss and the delay violation rate is reduced . Scalable OFDMA, is implemented so that the carrier is divided into multiple sub carriers depending upon the application using different size of FFT's varying from 256 to 2048

3. SIMULATION AND ANALYSIS

3.1 PHYSICAL LAYER SIMULATION

An attempt was made to write an event driven simulation in MATLAB (7.10).Mersenne Twister - Random Number Generator (RNG) Algorithm is used. Mersenne twister generator generates a random number using a pseudorandom algorithm. It has a large linear feedback shift register. In order to generate a random number, this algorithm is used in rand function in MATLAB. Noise is Gaussian and Rayleigh fading is considered. Cyclic prefix is used. Plotting of BER Vs SNR is made to evaluate the performance. Confidence intervals used for 32times. Adaptive modulation techniques are employed here.

The adaptive modulation techniques that WiMAX uses are BPSK, QPSK, 16-QAM and 64-QAM. All modulation techniques are implemented in order to get the results on different models.

Based on these modulation techniques the following
parameters were investigated.
• Bit Error Rate (BER)
• Signal to Noise Ratio (SNR)
• Power Spectral Density (PSD)
• Probability of Error (Pe)

OFDM WITH ADAPTIVE MODULATION TECHNIQUES IN PURE AWGN FOR RAYLEIGH & RICIAN CHANNEL

The initial results were observed in the pure AWGN channel condition using adaptive modulation techniques and the performance of these techniques were compared while using the 256 multicarrier OFDM waves.

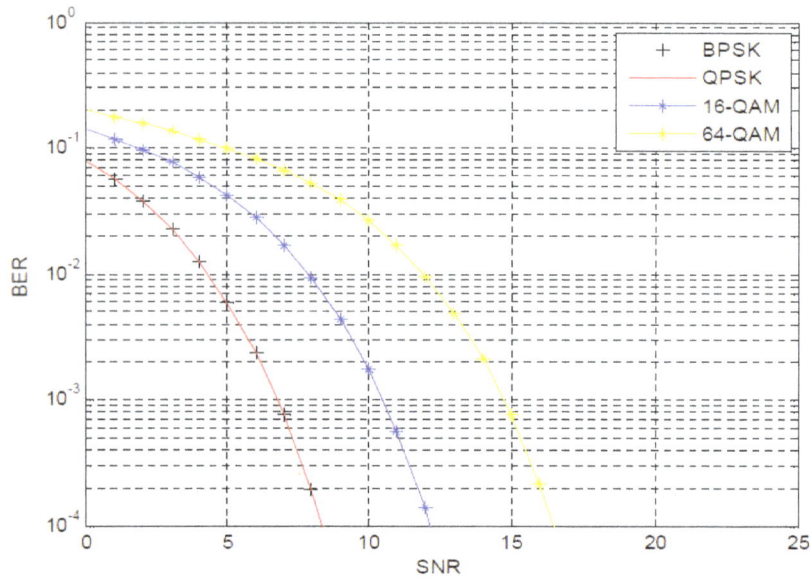

Fig 1. Adaptive Modulation Techniques in PURE AWGN for Rayleigh channel.

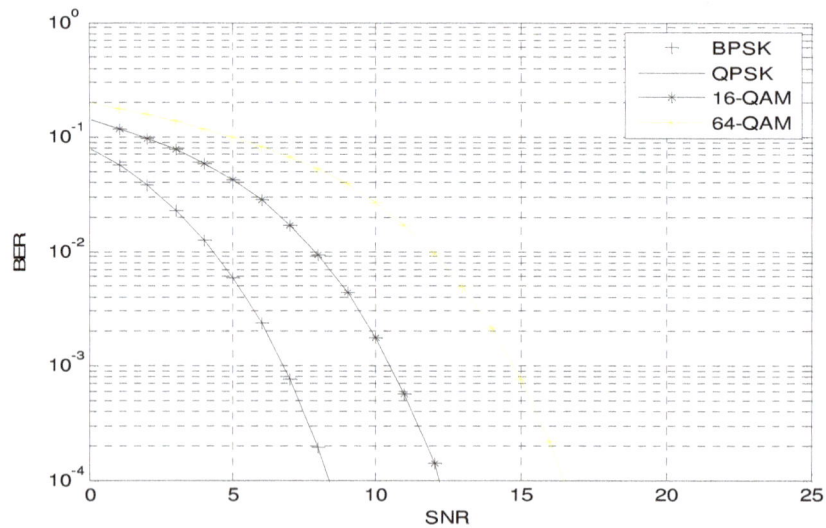

Fig 2 .OFDM with Adaptive Modulation Technique in PURE AWGN

WHEN BER = 10-3

Table 1 Adaptive Modulation In Pure AWGN , SNR and Bits/Symbol Comparison

Fading Channel	Modulation	SNR	Bits/Symbol
Rayleigh channel	BPSK	7	1
	QPSK	7	2
	16QAM	10.6	4
	64QAM	14.8	6
Rician channel	BPSK	7.4	1
	QPSK	7.4	2
	16QAM	10.8	4
	64QAM	15.0	6

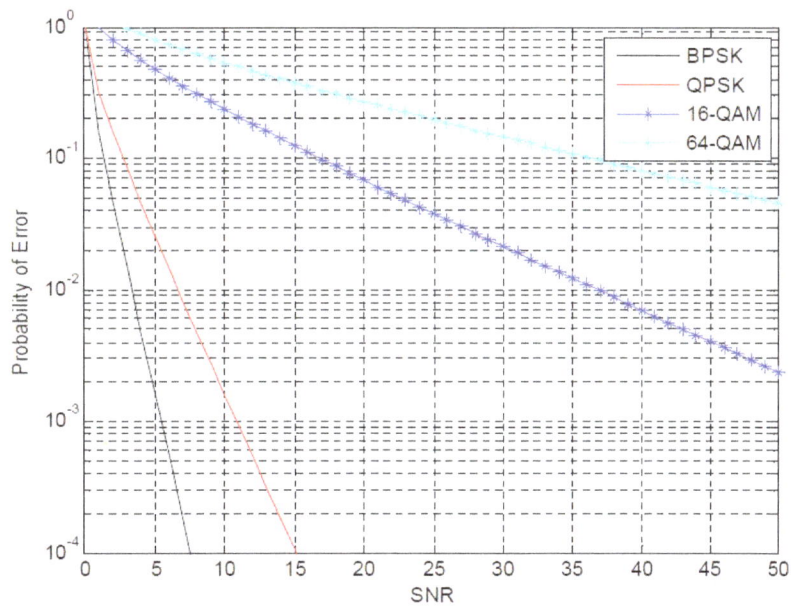

Fig .3 Probability of Error (Pe) for Adaptive Modulation
For Rician channel

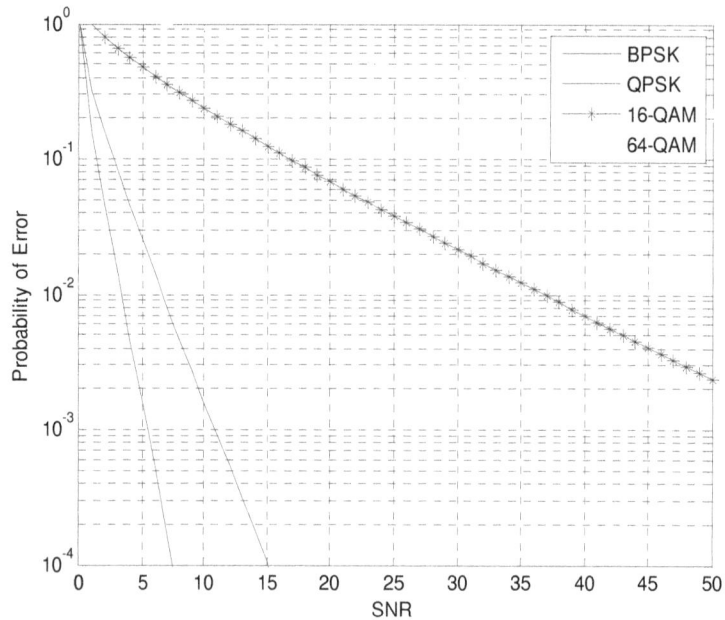

Fig 4 Probability of Error (Pe) for Adaptive Modulation
for Rayleigh channel

Due to noise and fading effects in the channel with some hardware losses at the transmitter and the receiver ends , error will be introduced in the system .

When BER = 10^{-1}

Table 2 Probability of Error (Pe) for Adaptive Modulation
Comparison SNR and Bits/Symbol

Fading Channel	Modulation	SNR	Bits/Symbol
Rayleigh channel	BPSK	2	1
	QPSK	3	2
	16QAM	16.7	4
	64QAM	36.5	6
Rician channel	BPSK	2.2	1
	QPSK	3.4	2
	16QAM	17	4
	64QAM	60	6

EFFECT OF SNR ON OFDM SYSTEM WITH RESPECT TO POWER SPECTRAL DENSITY FOR RAYLEIGH & RICIAN CHANNEL

In OFDM system the input and output signals power spectral densities differences are solely dependent on the channel conditions and the SNR levels. If the SNR level is high then the difference of the input signal with the output signal almost intermingle each others while it increases by reducing the SNR levels.

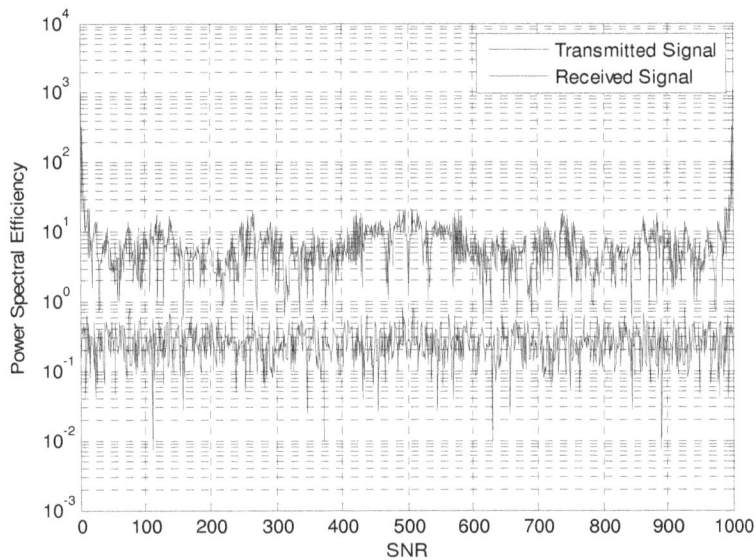

Fig. 5 Effect of SNR level 100 on OFDM system
With respect to power spectral density

SCALABLE OFDM

The scalable OFDM is implemented for the subcarriers by varying the FFT size from
128 ,256, 512 up to 2048. The simulation was performed on all 3 types of channels
i.e. AWGN,Rayleigh and Rician channels. The performance of the channel is analysed
w.r.t probability of error, BER and power spectral efficiency and the simulation results
show the same performance as the number of carriers /FFT size is scaled up from 128
to 2048 Bit error

PROBABILITY CURVE FOR QAM USING OFDM

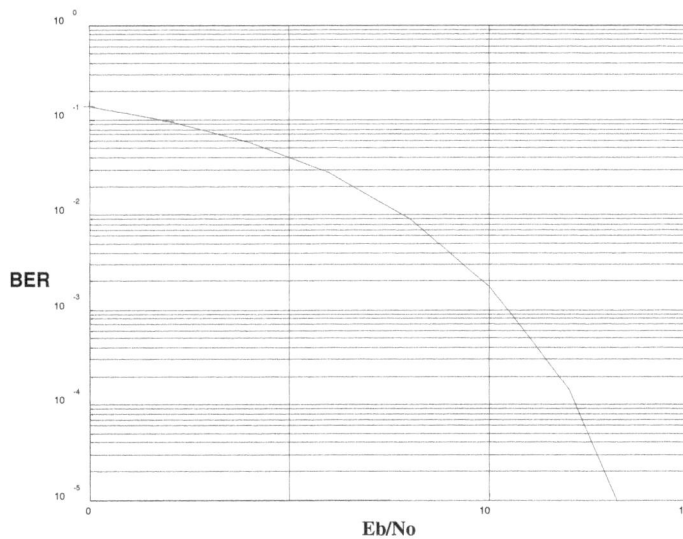

Fig .6 Scalable OFDM applied to 256 carriers through AWGN channel

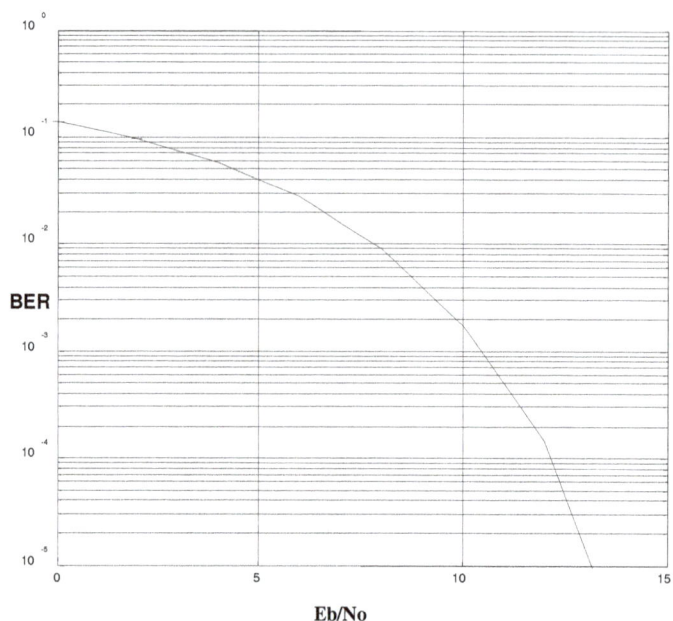

Fig .7 Scalable OFDM applied to 2048 carriers through
AWGN channel

3.2 MAC LAYER SIMULATION ENVIRONMENT AND ANALYSIS

MATLAB under version 2009a is used to effectively use the   proposed   scheduler,   at the
IEEE 802.16 MAC layer protocol [19]. Simulations are conducted using a system which is of
TDD – OFDM type and the MAC layer application parameters are indicated in table 3 The
system is of TDD-OFDM type and the MAC layer application parameters are as shown in
Table 3 and the network configuration is as shown in   Figure 8 .The operating bandwidth is
4.3 M Hz and 10 ms is considered as the frame duration. Since the standard does not specify
the values for the QoS parameters, we have assumed these values for the performance
analysis

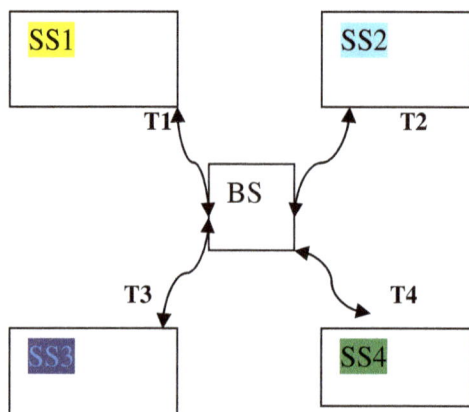

Figure 8. Proposed model architecture

T1:Traffic load1

T2:Traffic load2

T3:Traffic load3

T4:Traffic load4

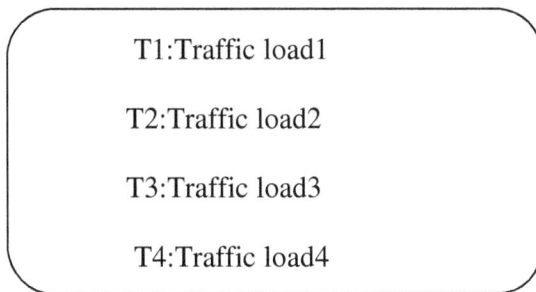

Table 3. MAC layer configuration parameters

Service	Maximum sustain rate	Minimum Reserved rate	Delay In milisecs
UGS	256	-	-
RTPS	1024	512	20
NRTPS	1024	512	
BE	-	256	-

QoS parameters such as delay, delay violation rate are considered to validate our proposed scheduling scheme. The number of packets whose delay is larger than the ratio of Maximum Latency to the total amount of packets that have been received from network is called as Delat Violation Rate[13].

SIMULATION RESULT AND DISCUSSIONS

A 802.16 network is simulated with a single Base Station and 4 Subscriber stations.All the subscriber stations are assigned with 4 types of service flows like UGS, rtPS, nrtPS and BE, catering to different types of traffic flows. Simulation is performed with and without Scheduler and the effect of SS scheduler is studied.

Here, if the simulation is performed without scheduler, then Base Station assigns the bandwidth depending on the service flow , else if the case is with scheduler, then SS scheduler designates bandwidth to individual connection [17]. Here, UGS, rtPS, nrtPS and BE are the types of service classes that are considered. Since UGS generates fixed size data packets on a periodic basis and hence delay is negligible and throughput is constant ..

Figure. 9 shows the different services and their associated delays (with and without SS scheduler).

Low prority services suffer longer delay.

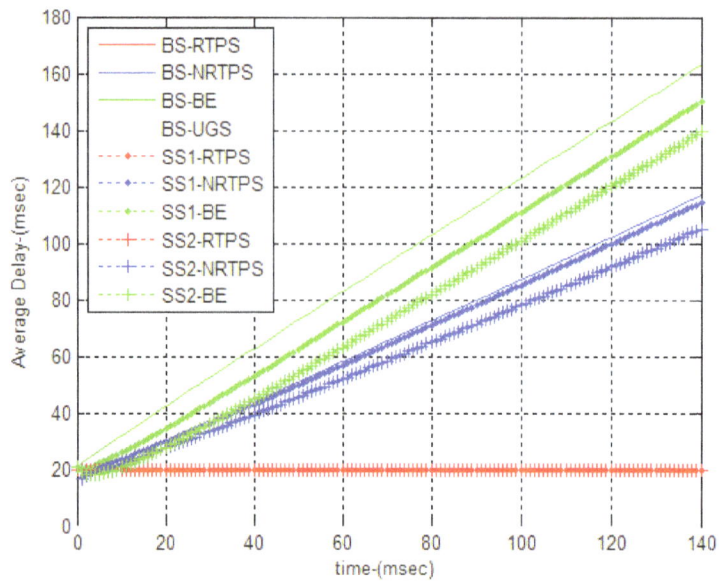

Figure .9 service delay comparison

The rtPS performance is analysed under the same number of background SS as given in Figure 10.

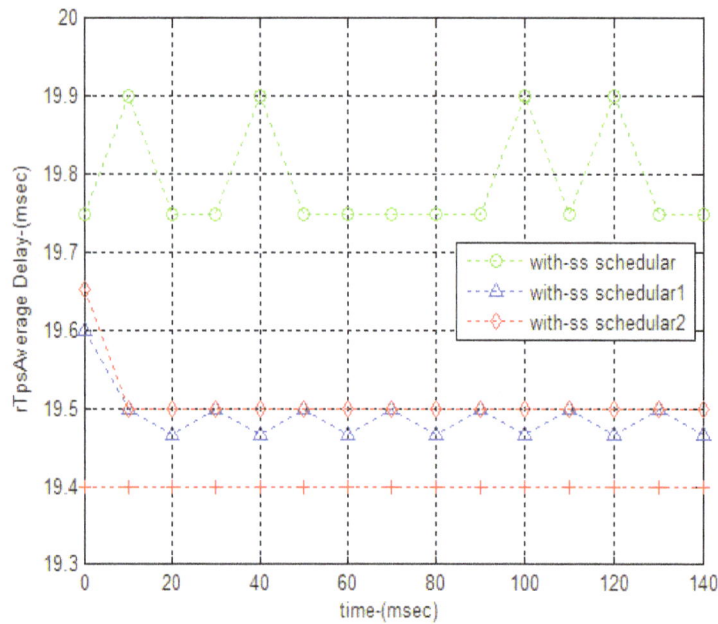

Figure.10   RTPS service delay

From Figure.11 we can see that the SS scheduler can effectively reduce the QoS violation rate of rtPS service flow

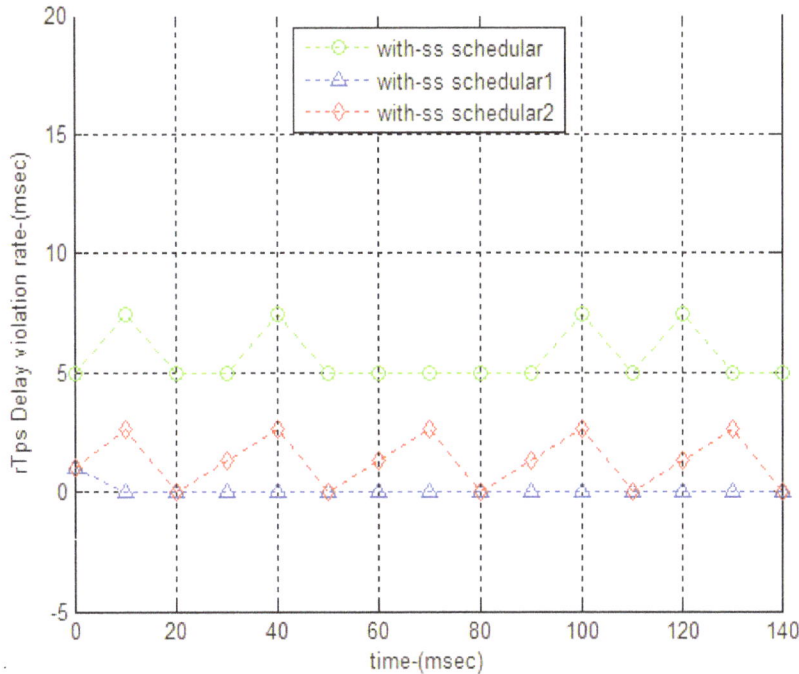

Figure.11 Packet drop comparison

4 . CONCLUSIONS

We have investigated four parameters: Bit Error Rate (BER), Signal to Noise Ratio (SNR), Power Spectral Density (PSD), Probability of Error (Pe).
We conclude that the probability of BER decrease as SNR increases more in Rician channel compared to Rayleigh channel in PURE AWGN. BPSK has the lowest BER while the 64-QAM has highest BER than others.
s
Cyclic prefix that is introduced , reduces Inter Symbol Interference and the impact of it is lower BER but the trade off is for the increase in the complexity of the system.

A hybrid packet scheduling scheme for mobile WiMAX has been implemented. A network model was developed to check the performance of the proposed scheme. Simulation results show that the proposed scheme is the best choice for QoS scheduling in Wi MAX in terms of delay, packet erro rate of the system compared to the schemes proposed in earlier methods. As a result of simulation, it can be concluded that the BS scheduler can provide each service flow with minimum required bandwidth and distribute excess bandwidth among all connections. The SS Scheduler provide differentiated QoS support for all types of service flows and reduce delay for real time applications and guarantee throughput for non- real time applications. With this the lower priority services can be eliminated of bandwidth and resource starvation.Hence tight QoS guarantee is provided for all service types.

4.1 FUTURE WORK:

To analyze the end-to end QoS variations in mobile WiMAX systems according to the QoS functions.

The end-to-end QoS may not be guaranteed even if the MAC layer QoS is definitely guaranteed. Therefore, cross-layer optimized design is necessary not only between MAC and PHY layers but also between the upper layer and MAC layer to guarantee the end-to-end QoS requirements. Schedulers can be designed taking into consideration the QoS performance factors.

REFERENCES

[1]   IEEE 802.16-2004, "IEEE standard for Local and Metropolitan Area Networks — Part 16: Air Interface for Fixed Broadband Wireless Access Systems," Oct. 2004.

[2]   Rohit A. Talwalkar*, Mohammad Ilyas," Analysis of Quality of Service (QoS) in WiMAX networks" ICON 2008 , p 1-8.

[3]   Jeffery G.Andrews, Arunabha Ghosh, Rias Muhamed, "Fundamentals of WiMAX-Understanding Broadband Wireless Networking", Pearson Education, March (2007).

[4]   John Wiley & Sons," WiMAX:"Technology for Broadband WirelessAccess", Published Online: 22 JAN 2007 Print ISBN: 9780470028087Online ISBN: 9780470319055

[5]   WiMAX Forum, "WiMAX System Evaluation Methodology V2.1," Jul. 2008, 230 pp.

[6]   Maode Ma,Mieso K Denko, Yan Zhang," Wireless Quality of Service"@2009 CRC Press, by Taylor & Francis group,LLC

[7]   Ho-Ting Wu, Kai-Wei Ke, and Chi-Fong Yang," The Design of QoS provisioning mechanisms for Mobile WiMAX Networks", 2009 International Conference on New Trends in Information and Service Science, 978-0-7695-3687-3/09 $25.00 © 2009 IEEE, DOI 10.1109/NISS.2009.49.

[8]   Omar Arafat, K. Dimyati, "Performance Parameter of Mobile WiMAX : A Study on the Physical Layer of Mobile WiMAX under Different Communication Channels & Modulation Technique," Computer Engineering and Applications, International Conference on, vol. 2, pp. 533-537, 2010 Second International Conference on Computer Engineering and Applications, 2010.

[9]   M. Ma J. Lu & C.P. Fu ,"Hierarchical scheduling framework for QoS& service in WiMAX point-to-multi-point networks",Published in IET Communications Revised on 7th August 2009

[10]   Amir Esmailpour and Nidal Nasser ,"Packet Scheduling Scheme with Quality of Service Support for Mobile WiMAX Networks",The 9th IEEE International Workshop on Wireless Local Networks (WLN 2009) Zürich, Switzerland; 20-23 October 2009

[11]   Jianhua He, Ken Guild, Kun Yang, and Hsiao-Hwa Chen ,"Modeling Contention Based Bandwidth Request Scheme for & IEEE 802.16 Networks" IEEE COMMUNICATIONS LETTERS, VOL. 11, NO. 8, AUGUST 2007

[12]   K.Vinay, N.Sreenivasulu, D.Jayaram and D.Das, "Performance evaluation of end-to-end delay by hybrid scheduling algorithm for QoS in I EEE 802.16 network", Proceedings of International Conference on Wireless and Optical Communication Networks, 5 pp., April (2006).

[13]   J.Sun, Y. Yao, H. Zhu "Quality of Service Scheduling For 802.16 Broadband Wireless AccessSystem", Advanced system technology telecom lab (Beijing) china, IEEE, (2006).

[14]   Chakchai So-In, Raj Jain, and Abdel-Karim Tamimi "Scheduling in IEEE 802.16e Mobile WiMAX Networks:Key Issues and a Survey" ieee journal on selected areas in communications, vol. 27, no. 2, february 2009

[15] Kalikivayi Suresh, Iti Saha Misra and Kalpana saha (Roy), "Bandwidth and Delay Guaranteed Call Admission Control Scheme for QOS Provisioning in IEEE 802.16e Mobile WiMAX" Proceedings of IEEE GLOBECOM, pp.1245-1250, December (2008).

[16] Chi-Hong Jiang, Tzu-Chieh Tsai, "CAC and Packet Scheduling Using Token bucket for IEEE 802.16 Networks", Consumer Communications and Networking Conf., 2006. CCNC 2006. 3rd IEEE Volume 1, pp. 183-187, Issue, 8-10 Jan. (2006).

[17] Prasun Chowdhury and Iti Saha Misra, "A Comparative Study of Different Packet Scheduling Algorithms with Varied Network ServiceLoad in IEEE 802.16 Broadband Wireless Access Int. Conf. Advanced Computing & Communications, Bangalore, Dec(2009).

[18] Ravichandiran,Dr.Pathuru Raj,Dr.Vaithiyanathan," Analysis and Modification of Scheduling Algorithm for IEEE 802.16e(Mobile WiMAX)" 2010 Second International Conference on Communication Software and Networks" 978-0-7695-3961-4/10 $26.00 © 2010 IEEEDOI 10.1109/ICCSN.2010.87

[19] Introduction to MATLAB by Rudra prathap

[20] Rakesh Jha, Hardik Patel, Dr Upena Dalal, Wankhede & Vishal A; "WiMAX System Simulation and Performance Analysis under the Influence of Jamming, Published inScientific Research journal, 2010, WET, July 2010

[21] D. S. Shu'aibu, S. K..Syed Yusof and N. Fisal; "Link Aware Unsolicited Grant Service Packet Scheduling for Mobile WiMAX" 978-1-4577-0005-7/11/$26.00 ©2011 IEEE

[22] Aldhaheri,R.W. Al-Qahtani,A.H.   Dept. of Electr. & Comput. Eng., King Abdul Aziz Univ.Jeddah, Saudi Arabia;"Performance analysis of fixed and mobile WiMAX MC-CDMA-based system" WirelessCommunication Systems(ISWCS), 2010 7th International Symposium

[23] Omar Arafat,K. Dimyati ,Department of Electrical Engineering University of Malaya;"Performance Parameter of Mobile WiMAX :A Study on the Physical Layer of Mobile WiMAX under Different Communication Channels & Modulation Technique"2010 Second International Conference on Computer Engineering and Applications

[24] ZhenTao Sun,Abdullah Gani,XiuYing Sun,Ning Liu;"Improving QoS Of WiMAX By On_Demand Bandwidth Allocation Based On PMP Mode" JOURNAL OF COMPUTERS,OCTOBER 2011

[25] Sunghyun Cho,Jae-Hyun Kim & Jonghyung Kwun;"End-to-End QoS Model for Mobile WiMAX Systems"2011 35th IEEE Annual Computer Software and Applications Conference Workshops

4

FAULT-TOLERANT MULTIPATH ROUTING SCHEME FOR ENERGY EFFICIENT WIRELESS SENSOR NETWORKS

Prasenjit Chanak,Tuhina Samanta,Indrajit Banerjee

Department of Information Technology
Bengal Engineering and Science University, Shibpur, Howrah-711103, India
prasenjit.chanak@gmail.com, t_samanta,ibanerjee,@it.becs.ac.in

ABSTRACT—*Themain challengein wireless sensor network is to improve the fault tolerance of each node and also provide an energy efficient fast data routing service. In this paper we propose an energyefficient node fault diagnosis and recovery for wireless sensor networks referred as fault tolerant multipath routing scheme for energy efficientwireless sensor network (FTMRS).The FTMRSis based on multipath data routing scheme. One shortest path is use for main data routing in FTMRS technique and other two backup paths are used as alternative path for faulty network and to handle the overloaded traffic on main channel.Shortest path data routing ensures energy efficient data routing. The performance analysis of FTMRSshows better results compared to other popular fault tolerant techniques in wireless sensor networks.*

KEYWORDS— *Wireless sensor network (WSN), fault tolerance (FT), load balance,multipath routing.*

1. INTRODUCTION

Wireless sensor network is a collection of hundreds and thousands of low cost, low power smart sensing devices. Sensor nodes are deployed in a monitoring area. They collect data from monitoring environment and transmit to base station (BS) by multi-hope or single hope communication. In WSN, fault occurrence probability is very high compare to traditional networking [1]. On the other handnetworks maintenance and nodes replacement is impossible due to remote deployment. These features motivate researchers to make automatic fault management techniques in wireless sensor networks. As a result now a day's different types fault detection and fault tolerance techniquesare proposed [2], [3]. Kim M, et al., proposed a multipath fault tolerant routing protocol based on the load balancing in 2008 [4]. In this paper, authors diagnose node failures along any individual path and increase the network persistence. The protocol constructs path between different nodes. Therefore, protocol leads to high resilience and fault tolerance and it also control message overhead. Li. S and Wu. Z proposed a node-disjoint parallel multipath routing algorithm in 2006[5]. This technique uses source delay and onehop response mechanism to construct multiple paths concurrently. In 2010 Yang Y. et al., [6] proposed a network coding base reliable disjoint and braided multipath routing. In this technique the authors construct disjoint and braided multipath to increase the network reliability. It also uses network coding mechanism to reduce packet redundancy when using multipath delivery. Y. Challal et al., proposed secure multipath fault tolerance technique known as SMRP/SEIF in 2011[7]. This technique introduces fault tolerant routing scheme with a high level of reliability through a secure multipath communication topology. Occurrences of fault in wireless sensor network are largely classified in two groups; (i) transmission fault,and (ii) node fault. The node fault [8], [9], [10] is further classified into five groups. These arepower fault, sensor circuit fault, microcontroller fault, transmitter circuit fault and receive circuit fault as discussed in [11].Energy efficiency is a prime metric in WSN performance analysis. This motivates us to propose an algorithm for fault tolerant energy efficient routing.

In this paper, we propose a fault tolerant routing which involves fault recovery process with fault detection scheme, referred to as energy efficient fault tolerantmultipath routing scheme for wireless sensor network (FTMRS). In FTMRS technique every sensor node transmitsits data to a base station through shortest path. If data or node fault occurs in the network, these are recovered very fast.The data are transmitted to base station withminimum time and energy loss. The FTMRS also controls the data traffic when data are transmitted to cluster head or base station(BS).

The rest of this paper is organized as follows. Section 2 describe proposed load balanced model.In section 3, we propose architecture for FTMRS. The proposed methodology for FTMRS is discussed in section 4. Performances and comparison result are showed in section 5. Finally, the paper is concluded in section 6.

2.   PROPOSED LOAD BALANCING MODEL

In FTMRS technique, we use standard data communication model originally proposed in [12]. In FTMRS technique cluster size are calculated with the help of*theorem 1* and *theorem 2*. *Theorems 1*establishes a relation between number of message passing through a node and nodes energy.The *theorems 2* establish a relation between numbers of nodes connection of a particular node with number of message passing in a particular time. The load of a node is directly affected by the number of node connected to it. If number of node connection is increased then load on that particular node isincreased. On the other hand if load of a node is increased, energy loss of the sensor node is increased.

Definition 1: The load$P_j$of a node is depending on number of data packet receives and transmitsby a particular node. The data load on a particular node is depends on number of sensor nodes connected with it andamount of data sensed by this particular node. The S_p denote a data packet receive by a single connection and$S_d$denote a data packet transmitby single connection.$P_j = \sum_{i=0}^{n} S_p + S_d$

Theorem 1: If initial energy of a sensor node is U, then partial derivative of the total energy of a sensor node expressed in terms of number of message passing λ_j at node j is equal to the loadP_j at sensor node j. This theorem expressed symbolically as $\frac{\partial U}{\partial \lambda_j} = P_j$

Proof: -Consider a series of loads $P_1, P_2, P_3, \dots P_j, \dots P_n$are acting on node 1, 2... j, ..., n who areproducing number of message $\lambda_1, \lambda_2 \dots \lambda_j, \dots \lambda_n$.

Now impose a small increment $\delta\lambda_j$ to the message passing at the node j keeping all other load unchanged. As a consequence, the increments in the loads are$\delta P_1, \delta P_2, \dots \delta P_j, \dots, \delta P_n$. The increment in the number of message at node$j$ and consequent increments in loadsat all the neighbour nodes. Therefore,$P_j \delta\lambda_j = \lambda_1 \delta P_1 + \lambda_2 \delta P_2 + \cdots + \lambda_j \delta P_j + \cdots + \lambda_n \delta P_n.\frac{\delta U}{\delta \lambda_j} = P_j$In the limit $\delta\lambda_j \to 0$, the above equation becomes $\frac{\partial U}{\partial \lambda_j} = P_j$

Theorem 2: Partial derivative of the energy loss in sensor nodes expressed in terms of load with respect toany loadP_j at any sensor nodes j is equal to the number of message passing $\lambda_j$through thej^{th} node. This theorem may be expressed mathematically as $\frac{\partial U}{\partial P_j} = \lambda_j$

Proof: -Consider a series of loads $P_1, P_2, P_3, \dots P_j, \dots P_n$ acting on a node j and for this section the messagepassed are $\lambda_1, \lambda_2 \dots \lambda_j, \dots \lambda_n$. Now impose a small increment δP_j to the load at the node j keeping all other factors unchanged.As a consequence, the message passing increasesby$\delta\lambda_1, \delta\lambda_2, \dots \delta\lambda_j, \dots, \delta\lambda_n$. However, due to increments in the load at node j,there is a consequent increment in message passing in all the neighbouring nodes. Therefore,$P_1 \delta_1 +$

$P_2\delta_2 + \cdots + P_j\delta_j + \cdots + P_n\delta_n = \lambda_j\delta P_j$, $\frac{\delta U}{\delta P_j} = \lambda_j$ in the limit $\delta P_j \to 0$, the above equation becomes $\frac{\partial U}{\partial P_j} = \lambda_j$.

3. ARCHITECTURE FOR ENERGY EFFICIENTFAULT TOLERANT MULTIPATH ROUTINGSCHEME (FTMRS):

In FTMRS technique, cluster sizes are calculated based on the cluster head load using theorem 1 and theorem 2. The clusters head load depends on the number of message received in cluster head and number of data transmitted from cluster head.

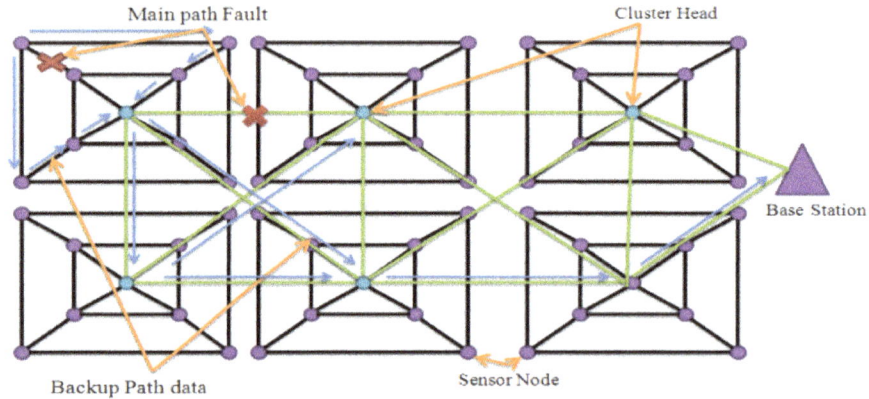

Figure1: Alternative path data routing in FTMRS

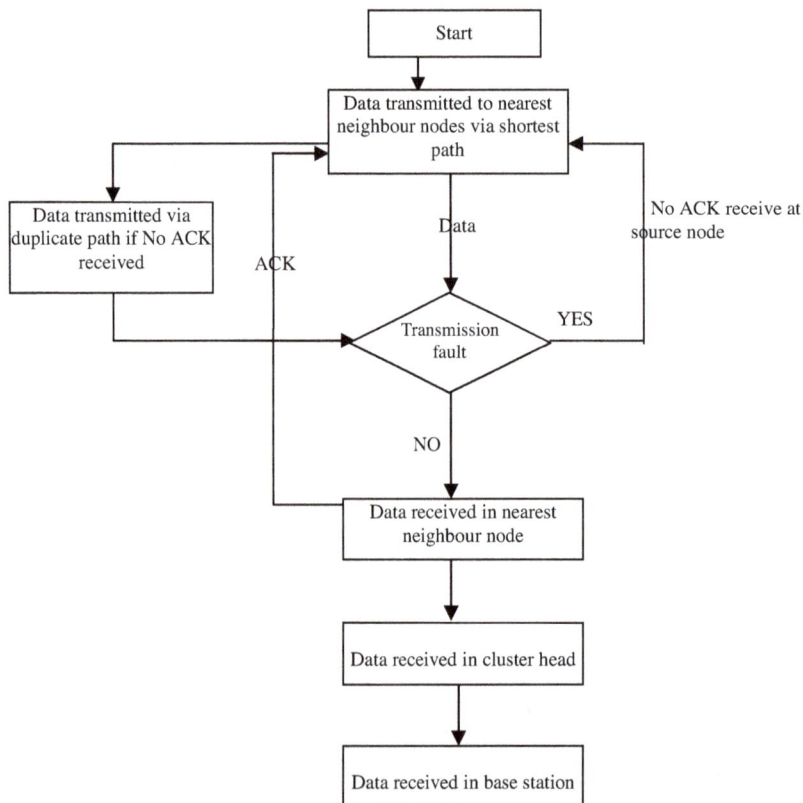

Figure 2: FTMRS architecture for fault tolerance

3.1 Fault Tolerance Data Routing Model

In FTMRS technique sensor nodes are arranged into small clusters. Every cluster contains a cluster head and cluster member node. Every cluster member node is capable of sending data over multiple paths. Cluster member nodes send their data in a shortest path to CH.Other paths are used for duplicated data transmission. Within a cluster a cluster member node sends its data to other cluster member nodesthrough three alternative paths. One of them is shortest, which is responsible for fast data transmission to cluster head. However, due to any external or internal problem shortest path fails,then next available shortest alternativepathis used to recover the faulty data transmission (Figure1).

In FTMRS technique when data are reaching to the neighbouring destination node via data routing path, they first check their received data and their own sensed data. If these two are same then neighbour nodes are not forwarding the received data to others. If a node receives different data then receiver node sends receiving data toward the cluster head with shortest path.

In FTMRS technique, clusters head and base station arealso connected to each other with the help of multiple (three) data path(Figure1). The shortest path is mainly used for fast energy efficient data routing towards base station. Other two backup paths are used for duplicate data routing, which makes the network path fault tolerant.

4. PROPOSED METHOD FOR FTMRS

In this section, we briefly describe our proposedFTMRS technique. This section is divided into two sub section one is fault tolerant data routing another is Energy efficient routing methodology.

4.1 Fault Tolerance Data Routing

In FTMRS technique, every cluster member node transmits data to cluster head.Cluster head collects all cluster member data. Cluster head after data aggregation transmits to base station.

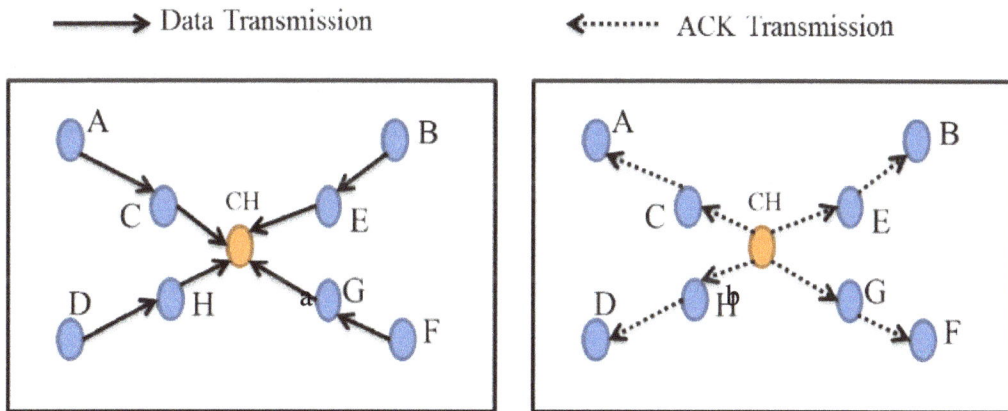

Figure3: Data transmission policy in FTMRS technique

In Figure 3.a. Cluster member nodes A, B, D, F transmits their data to cluster head's nearest to cluster member nodes C, E, H, and G respectively. If any node failure or transmission fault does not occurs, then after data is receivedby the cluster head's nearest member nodes C, E, H, and G send acknowledgement messages (Figure 3.b).In the same way, cluster head nearest member nodes C, E, H and G send their data to cluster head CH.

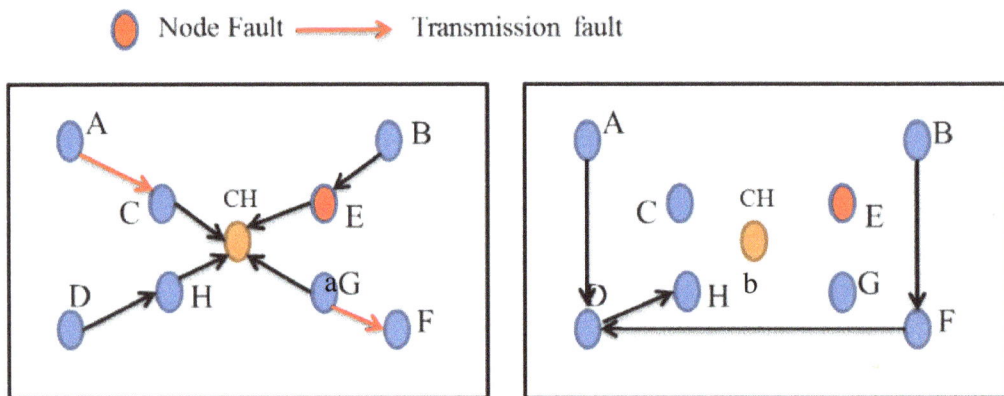

Figure4: Fault recovery policy in FTMRS technique

In the FTMRS technique if any transmission fault or node fault occurs in the network, then source nodes A, B, D or F does not receive any acknowledge message. Therefore, they are sending their data through duplicate path. In Figure4.a A to C node transmission fault occurs, hence'A' retransmitsits sensing data to D node. Similarly, when node E fails, node 'B'transmitsits data to 'F' node. However, node F does not gate any acknowledge message from G; hence it retransmits its data to node D (Figure4.b). In the same way cluster head transmits their data to the base station in FTMRS technique.

4.2. Node Hardware Fault Detection
In FTMRS scheme,the node collects data from nearest neighbour and transmits to the cluster head by a shortest path. If a node is not receiving any data from its neighbouring node for a period of time then the node sendsa health message to neighbour node and waits for replay message. If all neighbour nodes replay with respect to that health message, then node decides a transitions fault occurs in previous transmission. On the other hand if a node does not receive any replay message against health message then the node decidesits receiver circuit is faulty. However, if any one of the neighbour node is not replaying against health message then the node decidesthat the transmitter circuit of that neighbouring node is faulty. Then it informsthis to all other neighbour nodes. Sensor circuit fault is detected by the node itself by comparing its sensed data with data that he has been received from neighbour node. The comparison technique used here is explained in detail in [11].If sensing information is less thanthe threshold value, then the node's sensor circuit is in active condition. If sensing information is grater then the threshold value, then the sensor circuit is faulty. The FTMRS fault detection algorithm is described below.

Algorithm 1: Fault Detection Algorithm

Input: Insert all nodes into S (array of nodes)
Output: Check nodes hardware condition and find out fault nodes

---

```
1   WHILE S! =Null DO
2 WHILE network is alive DO
3 FOR each node DO
4          IF node receive data from neighbour THEN
5              Receiving data transmitted to shortest path
6          ELSE
7              Send health message all neighbour nodes
8          IF receive replay all neighbour nodes THEN
```

9	Transmission fault occurs in previous transmission
10	**ELSE IF** not receives replay from communication node **THEN**
11	Communication node is dead.
12	Inform to all neighbour node.
13	**ELSEIF** not receive replay from all neighbour **THEN**
14	nodes receiver circuit fault
15	**END IF**
16	**END IF**
17	**IF** Neighbour node data <= threshold value **THEN**
18	Sensor circuit node is good
19	**ELSE**
20	Sensor circuit of comparison nodes is Faulty
21	Inform to cluster head
22	**END IF**
23	**IF** node battery reading < threshold value **THEN**
24	Battery fault occur
25	**ELSE**
26	Battery is good
27	**END IF**
28	**END FOR**
29	**END WHILE**
30	**END WHILE**

4.3 Fault Recovery in FTMRS

Depending on the hardware condition of the node they are categorize as, *Normal Node, Traffic Node, End node, Dead Node* (Table 1) [11]. The categorization helps improving the network lifetime and decreases the percentage of dead node in the network.

Table 1

Categorization of nodes with respect to different hardware circuit failure

Node category	Microcontroller	Sensor circuit	Transmitter circuit	Receiver circuit	Battery /Power
Normal Node	Non Faulty	Non Faulty	Non Faulty	Non Faulty	Non Faulty
Traffic node	Non Faulty	Faulty	Non Faulty	Non Faulty	Non Faulty
End node	Non Faulty	Non Faulty	Non Faulty	Faulty	Non Faulty
Dead Node	Faulty	Faulty	Faulty	Faulty	Faulty

The FTMRS scheme reuses the faulty sensor node depending on the node's fault condition. If the sensor circuit is faulty, then sensor node is used as a *traffic node*. On the other hand if the node's receiver circuit is faulty, then this node works as an *end node*. If a node detects its transmitter circuit, microcontroller circuit, or battery is faulty, then the node is declared as a *dead node*. In FTMRS the faulty node recovery algorithm is shown below.

Algorithm 2: Faulty Node Recovery Algorithm

Input: Sensor nodes hardware condition
Output: According to hardware fault condition nodes responsibility distributed

---

1**IF** node detected sensor circuit fault **THEN**
2   Declare itself as traffic node and inform all other neighbour nodes
3**ELSE IF** receiver circuit fault occur **THEN**
4   Declare as end node and inform all other neighbour nodes
5**ELSE** transmitter circuit or microcontroller or batter fault occur **THEN**
6   Declare it as dead node by the cluster head and inform other cluster member nodes
7   The cluster head activate a neighbour standby node to replace the dead node.
8**END IF**

4.4 Traffic Management Scheme

FTMRS technique deals with data traffic congestion in the network. In FTMRS, every node maintains a time interval between two different data packet transmission in the same path. A sensor node when receives a new data from other node, they first checks shortest path condition for data transmission. If the shortest path is non-faulty, and it is currently not in use, then received data is transmitted through that shortest path. However, if shortest path is in use then they transmit new data through backup path as shown in the flow diagram (Figure 2). The received data transmission technique follows the fast come fast serve (FCFS) policy.

4.5 Energy Efficient Routing Methodology

In FTMRS technique every node maintains a time slot for data transmission through each path.

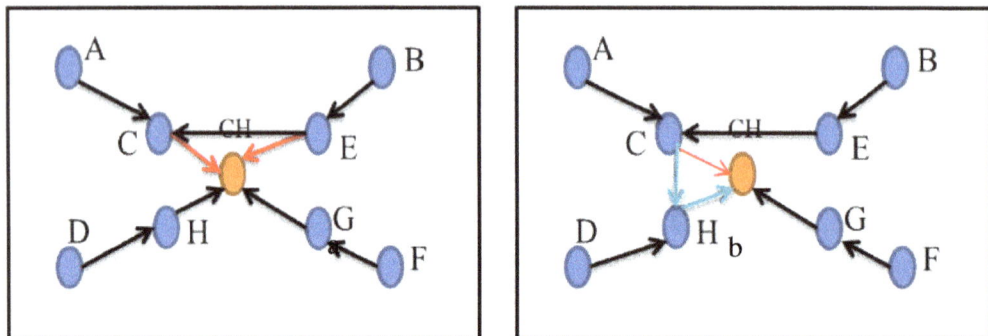

Figure5 : Data Traffic management in FTMRS technique

In Figure5.a when node 'E' transmits data to cluster head then transmission fault occurs. For this reason 'E' transmit data to neighbour node 'C'. When 'C' receivesE's messagethenitsends the data via available shortest path, as is shown in Figure5b, via node 'H'. In our proposed scheme, instead of initial multipath data propagation as in [13], [14], it sends the data through a single shortest path. However, if data transmission faults occur in that path then it will send the data through alternative backup path. Therefore, the energy wastage for multipath data propagation can be saved in FTMRS.Transmission energy loss of a sensor node is T_E[12].

$$T_E = (\alpha_1 + \alpha_2 * r^n) * \beta \tag{1}$$

Where α_1 [J/bit] is the energy loss per bit by the transmitter electronics circuit, and α_2[J/bit/m^4]is the dissipated energy in the transmitter op-amp. Transmission range is$r$[m]. The parameter n is power index for the channel path loss of the antenna. β[bit]is the message size which is transmitted by each node.Receiving energy loss of a node is R_E[J/bit].

$$R_E = (\alpha_3) * \beta \tag{2}$$

Where, α_3[J/bit] is energy per bit which is consumed by the receiver's electronics circuit used by the node. L_i message size which is received by each sensor node.

Lemma 1: The energy loss in FTMRS is less than multipath fault tolerant technique.

Proof:A single data communication energy loss is $E_{TR} = (T_E + R_E)$. In multipath data transmission communication energy loss is $ME_P = \sum^n E_{TR}$. Where n is the number of duplicate data transmission path in multipath data routing. In FTMRS scheme n value is 1 on the other hand multipath fault tolerance techniques n is always grater then 1. Therefore, energy conservation of FTMRS is grater then to other multipath fault tolerant techniques [13], [14].

The performance analysis of FTMRS technique is discussed next.

5. PERFORMANCE OF FTMRS

In this section, we present the result obtained from simulating different scenarios under different network sizes, different percentage of nodes faults and transmission faults. In order to evaluate the performance of FTMRS, four traditional metrics of WSN have been considered.(i) *Global Energy of Network:*this is the sum of residual energy of each node in the network. We calculate this value at each round of data transmission. (ii) *Average delay:*Average latency from the moment of data transmitted from source node to base station. (iii) *Average packet delivery ratio*: Number of packet transmitted to the source node and number of packet receive at the destination node.(iv) *Average dissipated Energy:* Total energy loss of the network and total number of sensor nodes ratio. The simulation parameters are taken from [12], [15], [16]. The table 2 shows parameters values which are used in simulation.

Table 2: Simulation parameters

Parameters	values
Number of node	1000-5000
Data Packet Size β	800bit
Initial Energy	0.5J
Energy consumed in the transmitter circuit $\propto_1$	50 nJ/bit
Energy consumed in the amplifier circuit $\propto_2$	10pJ/bit

Table 3: EEFTMR global energy loss in 0% and 40% node failure

Number of rounds	Global Energy (Joules/round)							
	In 0% node fault				In 40% nodes fault			
	Network size (300 m× 300m)							
	Deployed Node=1000	Deployed Node=1500	Deployed Node=2000	Deployed Node=2500	Deployed Node=1000	Deployed Node=1500	Deployed Node=2000	Deployed Node=2500
0	500 (J)	750(J)	1000(J)	1250(J)	500(J)	750(J)	1000(J)	1250(J)
100	458.35	696.43	937.50	1180.6	441.90	681.25	926.55	1170.60

200	416.7	642.86	875.00	1111.20	383.80	612.50	853.10	1091.20
300	375.05	589.29	812.50	1041.80	325.70	543.75	779.65	1011.80
400	333.40	535.72	750.00	972.40	267.60	475.00	706.20	932.40
500	291.75	482.15	687.50	903.00	209.50	406.25	632.00	853.00
600	250.10	428.58	625.00	833.60	151.40	337.50	554.30	774.60
700	204.45	374.01	562.50	764.20	93.00	268.75	485.85	694.20
800	166.8	321.44	598.00	694.80	32.20*	200.00	412.40	614.80
900	125.15	267.87	437.50	625.40	0.00+	131.25	338.95	535.40
1000	83.5	213.30	375.00	556.00	-	62.50*	265.50	456.23
1100	41.85*	160.73	312.50	486.60	-	0.00+	192.05	376.60
1200	0.00+	107.16	250.00	417.20	-	-	118.60	297.20
1300	-	53.59*	187.50	347.80	-	-	45.15*	217.80
1400	-	0.00+	124.00	278.40	-	-	0.00+	138.40
1500	-	-	62.50*	208.67	-	-	-	59.00*
1600	-	-	0.00+	139.60	-	-	-	0.00+
1700	-	-	-	70.00*	-	-	-	-

Table 3 depicts the global energy loss of WSN with different network size. In FTMRS, global energy lose increase when network size increase with no node failure. The life span of WSN decreases with 40% node failure, because of node fault detection and multipath data transition.

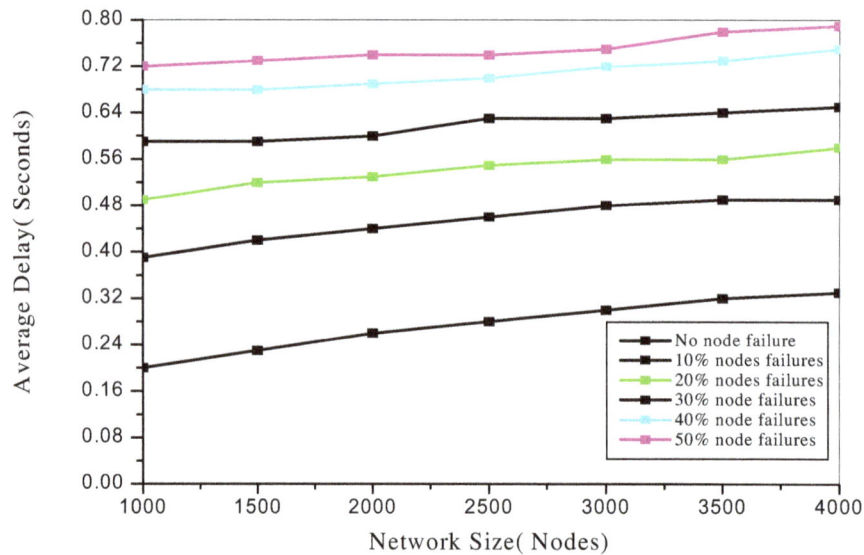

Figure6: Average delay in different percentage of nodes failures

Figure 6 shows the average packet transmission delay from sensor nodes to base station in different networks size. In FTMRS technique, data delivery time increase very slowly when node faults occurs. The Figure 7 shows the average packet delivery ratio from sender to base station. In the FTMRS technique, number of packet receives percentage in base station with respect to source node data transmission is very high. If any packet loss by nodes fault and path fault then backup path transmit duplicate data to cluster head as well as base station.

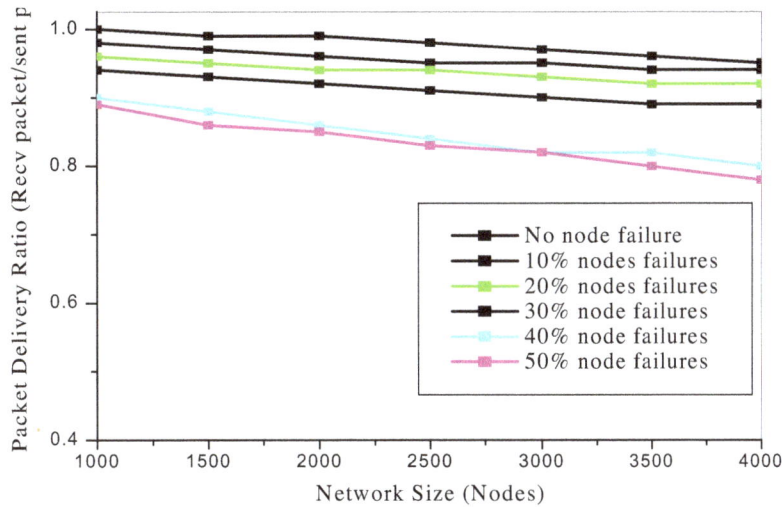

Figure 7: Average packet delivery in different percentage of nodes failures.

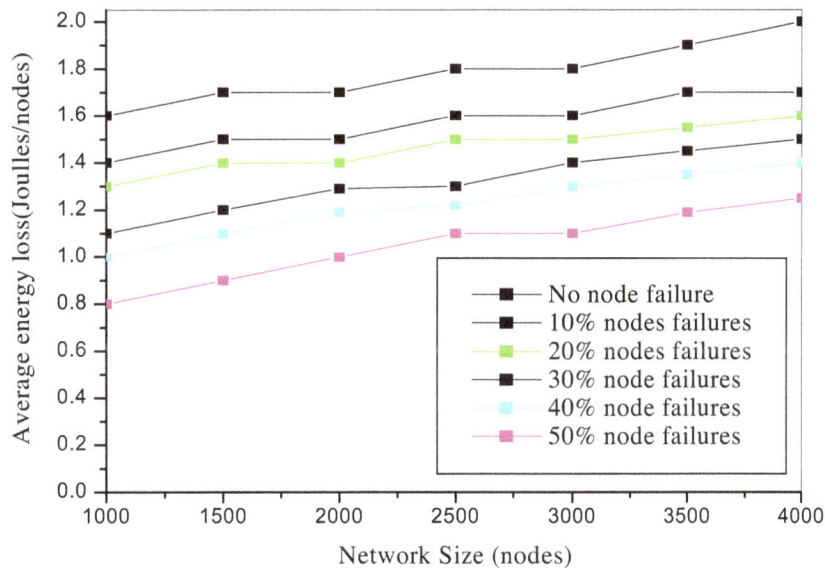

Figure8: Average dissipated energy in different percentage of nodes failures.

In FTMRS technique, energy loss rate of every node in different network size with different percentage of nodes failures is shown in Figure 8. When number of nodes fault percentage is low then energy loss of the networks is high because in this time maximum data is delivering to base station. If the number of node fault is increased then energy is loss of the network is decreased because in this condition data delivery to base station decreased.

Figure 9 shows the through put of the sensor nodes with respect to main routing path failures. In FTMRS technique throughput of sensor nodes is 49% batter in comparison to the fault-tolerant routing protocol for high failure rate wireless sensor networks(ENFT- AODV) [13] technique and 70% batter, compared to ad-hoc on-demand distance vector(AODV)[17]techniques.In the case of AODV technique, the throughput of the sensor nodes decreased rapidly when number of main path increases, because in this technique one main path have been failed then no other path is exited for retransmission of faulty data . ENFT-AODV technique used one backup path to improve retransmission of packet. If backup path fails then there is no way to transmit data to destination node. On the other hand FTMRS technique uses two backup paths for improvement of failure recovery.

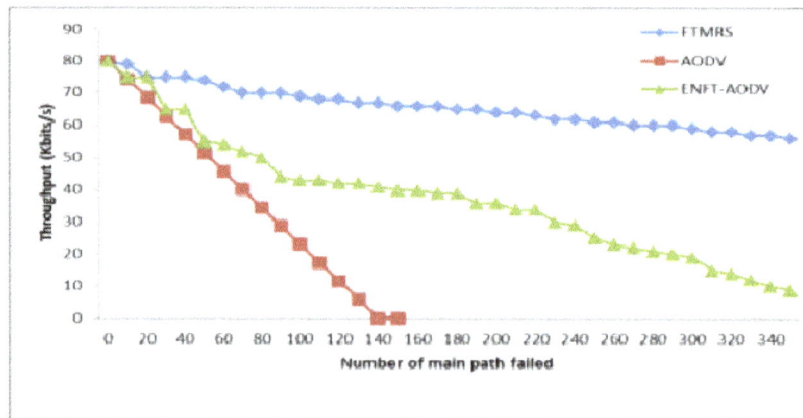

Figure 9: Throughput with respect to main routing path fault

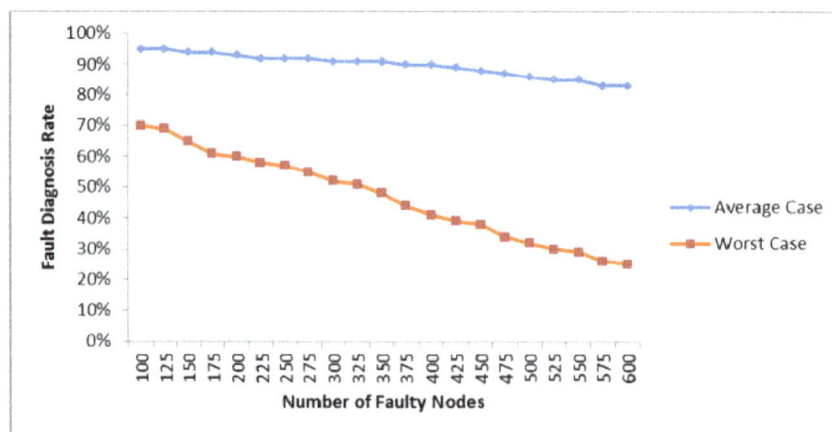

Figure 10: Successful diagnosis rate

Figure10 indicates the variation of the fault diagnosis rate with the number of faulty nodes. The fault diagnosis rate manifests the number of faulty node detected in each iteration. In average case if 100number of nodes are faulty,then approximately 95% of the faulty nodes are identified (95 out of 100), whereas approximately 70% of faulty nodes are detected (70 out of 100) in the worst case. In worst case high node failure in network leads to low fault diagnosis rate.

6.   CONCLUSIONS

In this paper, we present FTMRS as a fault tolerant multipath routing scheme for energy efficient WSN. The FTMRS technique recovers node fault and transmission fault and transmits data in energy efficient manner. In FTMRS technique, fault tolerant percentage is very high compare to other fault tolerant techniques. Data routing time in FTMRS is very fast and energy aware even at high percentage of nodes fault. The FTMRSalso proposes a faulty node recovery scheme that effectively reuses or replace the faulty node. The simulation results establish that the proposed routing give better monitoring of the nodes that effectively leads to an energy efficient maximally fault tolerant in sensor network.

In future we would like to improve and analyze the time complexity of the proposed algorithm. Moreover, the performance in worst case scenario improved by efficient detection of faulty nodes.

ACKNOWLEDGMENTS

This work is supported by the Council of Scientific and Industrial Research (CSIR) Human resource Development group (Extramural Research Division).

REFERENCES

[1] I.F. Akyildiz, W. Su, Y. Sankarasubramaniam, and E. Cayirci, "A Survey on sensor networks," *IEEE communications Magazine*, vol. 40. no. 8, pp. 102-114, August 2002.

[2] L. Paradis and Q. Han, " A survey of fault Management in wireless Sensor Networks," *Journal of Network and System Management*, vol. 15, no. 2, pp. 170-190, June, 2007.

[3] M. Yu, H. Mokhar, and M. Merabti, "A survey on Fault Management in wireless sensor networks,"*IEEE Wireless Communications*, vol. 14, no. 6, pp. 13 – 19, December 2007.

[4] Kim M, Jeong E, Bang Y.-C, Hwang S, Kim B, "Multipath energy-aware routing protocol in wireless sensor networks" , in *Proc. 5th international conference on networked sensing systems*, 2008, Pages 127-130.

[5] Li S, Wu Z, "Node-disjoint parallel multi-path routing in wireless sensor networks," *in Proc. of the second international conference on embedded software and systems (IEEE)*, 2006, pp. 432-437.

[6] Yang Y, Zhong C, Sun Y, Yang J. "Network coding based reliable disjoint and braided multipath routing for sensor networks, *Journal of Network and computer Applications*, vol. 33, no. 4,pp. 422 – 432, July 2010.

[7] Y. Challal, A. Ouadjaout, N. Lasla, M. Bagaa, A.Hadjidj, "Secure and efficient disjoint multipath construction for fault tolerant routing in wireless sensor networks,"*Journal of Network and Computer Applications*, vol. 34, no. 4,pp. 1380 – 1397,July 2011.

[8] W. L. Lee, A.D., R. Cardell-Oliver, "WinMS: Wireless Sensor Network-Management System, An Adaptive Policy-Based Management for Wireless Sensor Networks, " *School of Computer Science & Software Engineering, Univ. of Western Australia, tech. rep.* UWA-CSSE-06-001, 2006 2006.

[9] S. Chessa and P. Santi, "Crash fault identification in wireless sensor networks", *Computer Communications*, vol. 25, no. 14,pp. 1273 – 1282, 1 September 2002

[10] G. Gupta and M. Younis, "Fault tolerant clustering of wireless sensor networks," in *Proc. Wireless Communications and Networking (WCNC 2003)*,March 2003, pp. 1579-1584.

[11] Indajit Banerjee, PrasenjitChanak, Biplab Kumar Sikdar, HafizurRahaman, "DFDNM: Distributed fault detection and node management scheme in wireless sensor network," *Springer Link International Conference on Advances in Computing and Communications (ACC-2011)*, 22-23 July 2011, pp. 68-81.

[12] Indrajit Banerjee, PrasenjitChanak, HafijurRahaman "CCABC: Cyclic Cellular Automata Based Clustering for Energy Conservation in Sensor Network," *International Journal of wireless & Mobil Networks (IJWMN)*, vol. 3, no 4, pp. 39-44, 2011.

[13] Che-Aron. Z, Al-Khateeb, W.F.M., Anwar.F, "ENFAT-AODV: The Fault-Tolerant Routing Protocol for High Failure Rate Wireless Sensor Networks", in *Proc. 2nd International Conference on future computer and communication (ICFCC)*, 2010, pp.467 -471

[14] AzzedineBoukerche, Richard Werner NelemPazzi, Regina Borges Araujo, "Fault- tolerant wireless sensor network routing protocols for the supervision of context-aware physical environmental", *Journal of Parallel and Distributed Computing*, vol. 66, no. 4,pp. 586 – 599, April 2006.

[15] FarshadSafaei, HamedMahzoon, Mohammad SadeghTalebi, "A simple priority-based scheme for Delay-sensitive data transmission Over wireless sensor networks,"*International Journal of Wireless & Mobile Networks (IJWMN)*,vol. 4, no. 1, pp. 165-181, February 2012

[16] AbrarAlajlan, Benjamin Dasari, ZyadNossire, KhaledElleithy and VarunPande, "Topology management in wireless sensor Networks: multi-state algorithms," *International Journal of Wireless & Mobile Networks (IJWMN)*, vol. 4, no. 6, pp. 17-26, December 2012

[17] A. Wheeler, "Commercial Application of wireless Sensor Networks using ZigBee," *IEEE Communications Magazine*, vol. 45 , no. 4 , pp. 70 - 77 , April 2007.

5

A NOVEL QUERY DRIVEN POWER-BALANCED ROUTING PROTOCOL FOR WSN

Ayan Kumar Das[1] and Dr. Rituparna Chaki[2]

[1] Department of Information Technology, Calcutta Institute of Engineering and Management, Kolkata, India,
`ayandas24114057@yahoo.co.in`

[2]Department of Computer Science & Engineering, West Bengal University of Technology, Kolkata, India,
`rituchaki@gmail.com`

ABSTRACT

Wireless sensor networks consist of hundreds or thousands of small sensors with have limited resources. The energy resource is the most challenging one to be maintained. The major reason for power drain is the communication between sensor nodes. Most of the routing algorithms for sensor networks focus on finding energy efficient paths to prolong the lifetime of the networks. In this paper, a novel routing algorithm has been proposed to detect the source of an event in the network. This algorithm also maintains a balance between the powers of different nodes in the network, so that the longevity of the overall network may increase.

KEYWORDS

update sensing, event path, query node, Infinite loop.

1. INTRODUCTION

Sensor networks are among the fastest growing technologies that have the potential of changing our lives drastically. These collaborative, dynamic and distributed computing and communicating systems will be self organizing. They will have capabilities of distributing a task among themselves for efficient computation. A Wireless Sensor Network (WSN) contains hundreds or thousands of sensor nodes. These sensors have the ability to communicate either among each other or directly to an external base-station (BS). A greater number of sensors allows for sensing over larger geographical regions with greater accuracy.

Despite the innumerable applications of WSNs, these networks have several restrictions, e.g., limited energy supply, limited computing power, and limited bandwidth of the wireless links connecting sensor nodes. One of the main design goals of WSNs is to carry out data communication while trying to prolong the lifetime of the network and prevent connectivity degradation by employing aggressive energy management techniques. The design of routing protocols in WSNs is influenced by the above challenging factors. These factors must be overcome before efficient communication can be achieved in WSNs. The computation involving sensed data and communication of the same between the sensor nodes cause very high power consumption. This coupled with the inability to recharge once a node is deployed makes energy consumption which is the most important factor to determine the life of a sensor network, because usually sensor nodes are driven by battery and have very low energy resources. This can be achieved by having energy awareness in every aspect of design and operation of the network. The common routing techniques are based on power awareness, agent based, location based etc. To prolong the lifetime of sensor networks, most routing algorithm for sensor networks focus on finding energy efficient paths. As a result, the sensors on the efficient paths are depleted very early, and consequently the sensor networks become incapable of monitoring events from some of the parts of target areas.

It has been observed that most of the previous routing techniques do not maintain any information about the nodes which have been already traversed. Thus a query packet moving towards the event path randomly selects neighboring nodes without considering whether that node has already been traversed. This often leads to the infinite loop problem, leading to increased delay in routing. This paper proposes a query driven routing protocol with optimal power balancing techniques aimed at removing the infinite loop problem.

The remaining part of this paper is organized as follows: Section 2 deals with the review of state of the art routing topologies, section 3 gives a description of the proposed methodology, section 4 contains the simulation reports and section 5 is the concluding part.

2. REVIEW WORKS OF SENSOR NETWORKS

In this section, we survey the state-of-the-art routing protocols for WSNs. In general, routing in WSNs is categorized into power aware routing, query driven routing, zone based routing, agent based routing and cluster based routing.

2.1. Power Aware Routing Algorithm

This type of routing technique focuses on the effect of power efficient routing on the lifetime of multi hop wireless sensor networks (WSNs) [5]. All the nodes in the WSNs are divided into tiers. The nodes belonging to the highest tier are very important, because these nodes imply higher power consumption than that of nodes of any other tier. Obviously the batteries of nodes of 1st tier depletes sooner than these of any other tier. As soon as the nodes of 1st tier dies, the network becomes disconnected.

Reliable Energy Aware Routing (REAR) [8] is a distributed, on-demand, reactive routing protocol and is used to provide a reliable transmission environment for data packet delivery. To provide a reliable transmission environment to reduce retransmissions caused by unstable paths REAR introduces local node selection, path reservation and path request broadcasting delay. This algorithm efficiently utilizes the limited energy and available memory resources of sensor nodes. REAR takes precaution against errors, instead of finding a solution for the errors. The Simulation experiments show that REAR outperforms traditional schemes by establishing an energy-sufficient path from the sink to the source with special path request flooding, and also by distributing the traffic load more evenly in the network.

An approximation algorithm called max-min zPmin [27] has developed with a good empirical competitive ratio. This algorithm combines the benefits of selecting the path with the minimum power consumption and the path that maximizes the minimal residual power in the nodes of the network.

Minimum Cost Forwarding Algorithm (MCFA) [14] exploits the fact that the direction of routing is always known, that is, towards the fixed external base-station. Hence, a sensor node need not have a unique ID nor maintain a routing table. Instead, each node maintains the least cost estimate from itself to the base-station. Each message to be forwarded by the sensor node is broadcast to its neighbors. When a node receives the message, it checks if it is on the least cost path between the source sensor node and the base-station. If this is the case, it re-broadcasts the message to its neighbors. This process repeats until the base-station is reached.

GRAdient Broadcast [24] describes a technique of building a cost field toward a particular sensor node, and after that reliably routing query across a limited size mesh toward that sensor node. Overhead comes for a network flood to set up the cost field, but queries can be routed along an interleaved set of short paths, and can thus be delivered very cheaply and reliably. GRAB was not designed specifically to support in network processing but significantly influenced the work presented in its use of event-centric routing state in the network.

Directed Diffusion and Geo-Routing [19][21][22] provides a scheme for doing a limited flood for a query toward the event, and then setting up reverse gradients in order to send data back along the best route. Though Diffusion results in high quality paths, but an initial flood of the query for exploration will be the requirement. One of its primary contributions is an architecture that names data and that is intended to support in network processing. Rumor routing is intended to work in conjunction with diffusion, bringing innovation from GRAB and GOSSIP routing to this context.

Constrained Anisotropic Diffusion Routing (CADR) [25] aims to be a general form of directed diffusion. The key idea is to query sensors and route data in the network such that the information gain is maximized while latency and bandwidth are minimized. CADR diffuses queries by using a set of information criteria to select which sensors can get the data. This is achieved by activating only the sensors that are close to a particular event and dynamically adjusting data routes. The main difference from directed diffusion is the consideration of information gain in addition to the communication cost. In CADR, each node evaluates an information/cost objective and routes data based on the local information/cost gradient and end-user requirements. Estimation theory was used to model information utility measure.

COUGAR [26] is a data-centric protocol which views the network as a huge distributed database system. The key idea is to use declarative queries in order to abstract query processing from the network layer functions such as selection of relevant sensors and so on. COUGAR utilizes in-network data aggregation to obtain more energy savings. The abstraction is supported through an additional query layer that lies between the network and application layers. COUGAR incorporates architecture for the sensor database system where sensor nodes select a leader node to perform aggregation and transmit the data to the BS. The BS is responsible for generating a query plan, which specifies the necessary information about the data flow and in network computation for the incoming query and send it to the relevant nodes. The query plan also describes how to select a leader for the query. The architecture provides in-network computation ability that can provide energy efficiency in situations when the generated data is huge. COUGAR provided network-layer independent methods for data query. However, COUGAR has some drawbacks. First, the addition of query layer on each sensor node may add an extra overhead in terms of energy consumption and memory storage. Second, to obtain successful in-network data computation, synchronization among nodes is required (not all data are received at the same time from incoming sources) before sending the data to the leader node. Third, the leader nodes should be dynamically maintained to prevent them from being hot-spots.

2.2. Query Driven Routing

In Information-Driven Sensor Querying (IDSQ) [25], the querying node can determine which node can provide the most useful information with the additional advantage of balancing the energy cost. However, IDSQ does not specifically define how the query and the information are routed between sensors and the BS. Therefore, IDSQ can be seen as a complementary optimization procedure. Simulation results showed that these approaches are more energy-efficient than directed diffusion where queries are diffused in an isotropic fashion and reaching nearest neighbors first.

Rumor routing [18] is a variation of directed diffusion and is mainly intended for applications where geographic routing is not feasible. In general, directed diffusion uses flooding to inject the query to the entire network when there is no geographic criterion to diffuse tasks. However, in some cases there is only a little amount of data requested from the nodes and thus the use of flooding is unnecessary. An alternative approach is to flood the events if the number of events is small and the number of queries is large. The key idea is to route the queries to the nodes that have observed a particular event rather than flooding the entire network to retrieve information about the occurring events. In order to flood events through the network, the rumor routing algorithm employs long-lived packets, called agents. When a node detects an event, it adds such event to its local table, called events table, and generates an agent. Agents travel the network in order to propagate information about local events to distant nodes. When a node generates a query for an event, the nodes that know the route, may respond to the query by inspecting its event table. Hence, there is no need to flood the whole network, which reduces the communication cost. On the other hand, rumor routing maintains only one path between source and destination as opposed to directed diffusion where data can be routed through multiple paths at low rates. Simulation results showed that rumor routing can achieve significant energy savings when compared to event flooding and can also handle node's failure. However, rumor routing performs well only when the number of events is small. For a large number of events, the cost of maintaining agents and event-tables in each node becomes infeasible if there is not enough interest in these events from the BS. Moreover, the overhead associated with rumor routing is controlled by different parameters used in the algorithm such as time-to-live (TTL) pertaining to queries and agents. Since the nodes become aware of events through the event agents, the heuristic for defining the route of an event agent highly affects the performance of next hop selection in rumor routing.

2.3. Zone Based Routing

The zone-based routing [27] algorithm relies on max-min zPmin. It is scalable for large scale networks and is used to optimize the lifetime of the network. Zone-base routing is a hierarchical approach where the area covered by the sensor network is divided into a small number of zones. Each zone has many nodes and thus a lot of redundancy occurs in routing a message through it. To send a message across the entire area it finds a global path from zone to zone and gives each zone control over how to route the message within itself. A local path for the message is computes within each zone so as to not decrease the power level of the zone too much

2.4. Agent based Routing Algorithm

Apart from power aware routing some algorithms modeled after agent behavior. Agents traverse the network encoding the quality of the path they are traveled, and leave it the encoded path as state in the nodes [23]. In dealing with failure these algorithms are very effective, as there is always some amount of exploration, and especially around previously good solutions. However, due to the large number of nodes, the number of ant agents required to achieve good results tends to be very large, making them difficult to apply in sensor networks.

Ant-Colony Based Routing Algorithm (ARA) [17] used distance vector routing. Route discovery in ARA is done by broadcasting Forward Ants (FANT). ARA implements the packet sequencing mechanism to prevent packet loops. In ARA, destinations respond to multiple FANT packets received from the same node, thus supporting multi path routing.

MONSOON [1] [2] [3] proposed an evolutionary multi objective adaptation framework, in biologically inspired application architecture, called BiSNET/e. There are two types of software components- agents and middleware platforms. In MONSOON each application is implemented as a decentralized group of software agents, like a bee colony consists of bees.

Agents generally collect all the sensing data and/or detect an event on individual nodes, and also carry sensing data to the base station. Here each agent decides their behavior based on logical energy. It does not represent the amount of physical battery in a node, as it is a logical concept.

2.5. Cluster Based Routing Algorithm

In clustered network, nodes are clustered in any form of hierarchical structure. The advantage of cluster based approaches are improving routing efficiency, scalability, supporting QOS and saving power consumption in sensor nodes. Clustering transforms a physical network into a virtual network of interconnected clusters.

Low Energy Adaptive Clustering Hierarchy (LEACH) [20] is a cluster-based protocol, which includes distributed cluster formation. LEACH randomly selects a few sensor nodes as cluster heads (CHs) and rotates this role to evenly distribute the energy load among the sensors in the network. In LEACH, the cluster head (CH) nodes compress data arriving from nodes that belong to the respective cluster, and send an aggregated packet to the base station in order to reduce the amount of information that must be transmitted to the base station. LEACH uses a TDMA/CDMA MAC to reduce inter-cluster and intra-cluster collisions. However, data collection is centralized and is performed periodically. Therefore, this protocol is most appropriate when there is a need for constant monitoring by the sensor network. A user may not need all the data immediately. Hence, periodic data transmissions are unnecessary which may drain the limited energy of the sensor nodes. After a given interval of time, a randomized rotation of the role of the CH is conducted so that uniform energy dissipation in the sensor network is obtained. The authors showed that only 5% of the nodes need to act as cluster heads.

3. PROPOSED WORK

3.1. Basic Methodology

From the review of recent routing topologies, it is observed that many of the existing techniques fail to take care of rapid power loss by some nodes while other nodes remain unused, retaining the full power level. There is a possibility of cut-off of some parts of the network due to this un-balanced use of power. Also, many of the routing algorithms suffer from infinite loop problem, causing a packet to return to the point of origin after a while, instead of reaching the destination. This section describes a new routing protocol with aimed at reducing these problems. The network is modeled as a set of densely distributed wireless sensor nodes. Each node maintains a list of its neighbors, as well as an event table, with forwarding information of all the events in its knowledge. A node adds an event to its event table after encounter it. Then the node will propagate that event information to its neighbors and then to neighbors of neighbors and so on. This continues up to a certain distance, as it is unnecessary and also power wasting to inform the event to all the nodes of the network. The path which has the information about the event is known as event path. Any node from the network can generate a query and search for the event path. If it found that then it can reach to the event source node along with the event path.

However a problem may arise if we don't maintain any status checking option for the already traversed nodes. As a result the query while moving towards the event path randomly selects neighboring nodes without considering whether the next node has already been checked. This may lead to traversal or checking of the same node more than once. As a result lot of energy and time is wasted and may lead to the procedure getting stuck in an infinite loop. That is why the algorithm adds a checked/unchecked status to all the nodes to identify the previously visited node so that the infinite loop may be avoided.

A problem still exists as in all the cases the next hop node is chosen randomly without considering the available power of the node. Thus, the node with the least available power may be selected randomly and repeatedly as the node to be traversed next. This will result in the reduction of power of a node, which already has the smallest power while keeping the power of other nodes (with higher power than the node already selected as the next hop) constant. As a result of which, there exists a high chance of exhausting the total power of a particular node(s), keeping the power of other nodes almost unaffected. To overcome this disadvantage the algorithm maintains a data structure, containing the power content of all the nodes and depending on the power of these nodes the next hop is selected. The algorithm selects the node with the maximum power as the next hop. As a result of which, the power level within the network remain balanced.

Definitions:

Definition 1: Power required for transmission is directly proportional to the size of the packet that is send through the network and also the distance traveled by the packet, i.e., $P \alpha s*d$

Thus, we can write $P=\theta*s*d$ (1)

Where, s= size of the packet that is send through the network

    d= distance traveled by the packet, and θ is a constant.

Definition 2: The total consumed ENERGY(E)during the transmission of a packet can be defined as the sum of power required to transmit and receive the packet, i.e.,

    $E = \sum$(Power required to transmit the packet, Power required to receive the packet) .

    = P1+P2

    = $(\theta*s1*d1) + (\theta*s2*d2)$ (2)

Where s, s1 and s2 are the packet size whose value will be supplied by the user, d1 is the distance between the sending node and the current node and d2 is the distance between the current node and the receiving node.

3.2. Data Dictionary

Table 1. Variables list

Variable name	Description
N	Total Number of nodes
A	An array consists of the connections between the nodes
$C[a_0, a_1, a_2,..a_n]$	Event path nodes are stored in this array
Power[node_id,energy]	An array consists of initial power of each node
Visit[node_id]	Visited nodes are stored in this array
$P_i[$ node_id, status]	Neighbors of node i are stored in this array along with their status.
Next_node	next node chosen during the traversal
Dist	the distance between the last and second last visited nodes.
Tot	total distance traveled during one simulation
Node_max_pow[ ]	The neighboring nodes with the maximum power are stored in this array

3.3. Description

Step 1. Begin.

Step 2. Read size (for the packet)

Step 3: Find the neighboring nodes of source node (m) and store them in the array 'p'.

Step 4. Repeat the following statements n times (where n is the number of neighbor

    nodes of source m) —

    If status i = visited then   delete i[th] node from Visit[node_id]

Step 5. If all the neighboring nodes are visited previously, then Choose any one randomly.

    Else Find the node among (not visited) nodes with the maximum power and store in the array 'node_max_power'.

Step 6. If more than one node exists with maximum power content, then select any of them randomly as the next hop and store it as 'next_node'.

Step 7. Use equation 2 to deduce remaining energy level

Step 8. If the chosen next hop node falls on the event path then— calculate the distance traversed by the query to reach the event source and also calculate the average power reduction of each node for simultaneous queries.
Else
    Make the next hop node as the source node(m) and continue the process from step 3.
Step 9. End.

3.4. Case Study

Consider the following wireless sensor network as shown in Figure 1 formed with 15 nodes. Let the node n12 generates the query and sends the query packet to node n10 or n11 which one has the maximum power. After that the packet will be moved to node n8 and then from n8 moves to n9 then to n6, from n6 to n7 and ultimately returns to the already traversed node n8 thereby creating an infinite loop and will never get the event path.

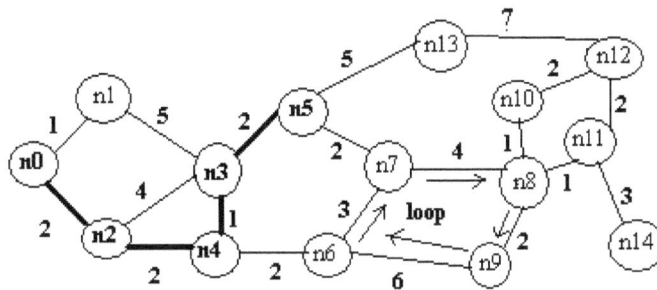

Figure 1. Infinite loop is formed

In order to overcome the above said limitation the proposed algorithm maintains a proper status checking to select the next hop node to avoid the infinite loop as shown in figure 2.

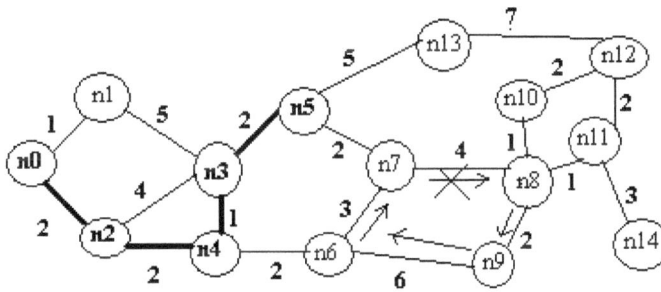

Figure 2. Infinite loop is not formed

Here since a track of all the traversed nodes were taken the packet which has started its journey from node n8 will not go back to same node n8 again after traversing node n7 as was the case in the previous example, instead it will move to node n5 as n6 and n8 are both already traversed nodes. Now node n5 is a node of event path and thus the packet will be able to reach at event source.

Through the network multiple simultaneous queries can be generated. For example if we send a query from node n12, then it will check the power of neighbor nodes n10 and n11 and select the node which contents of maximum power. If the selected node is n11 then it will check the power of n8 and n14 and select the node of maximum power and so on. When it will encounter any node of event path then it will move to the event source directly along with the event path.

4. SIMULATION RESULT

To analyze the performance of the algorithm multiple queries have been sent and after power consumption the power of all the nodes has measured and a graph of power vs node has drawn. It is also compared with the Rumor Routing algorithm. The parameter list is given below—

Table 2. Parameter list

Parameters	Description
Network size	15 nodes
Initial energy	50J per node
MAC Protocol	IEEE 802.15.4
Power consumption	Equivalent to packet size and distance
Number of queries	At least 6

According to the 'Rumor Routing Algorithm' where next hop node is choosing randomly, after sending 6 queries from node 15 the graph is shown below, where series1 denotes the initial power of the node, series2-series6 denotes the power of the node after each query is generated, and series7 denotes the average power of the nodes.

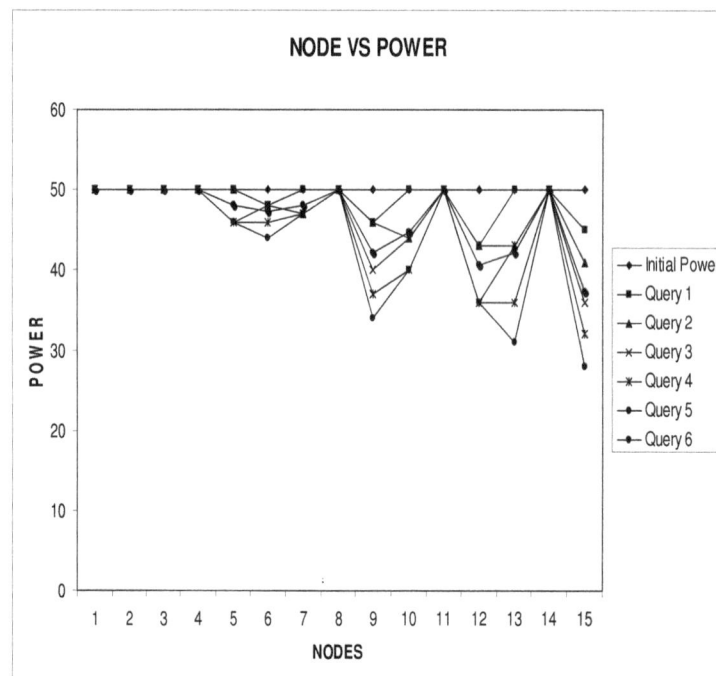

Figure 3. Node vs. Power graph for random neighbor selection

Now it is proposed that the next hop will be the node with the maximum power. When next hop is chosen based on the power of the neighboring nodes, the following graph is obtained, where series1 denotes the initial power of the nodes, series2-series6 denotes the power of the nodes after each query is generated, and series7 denotes the average power of the nodes.

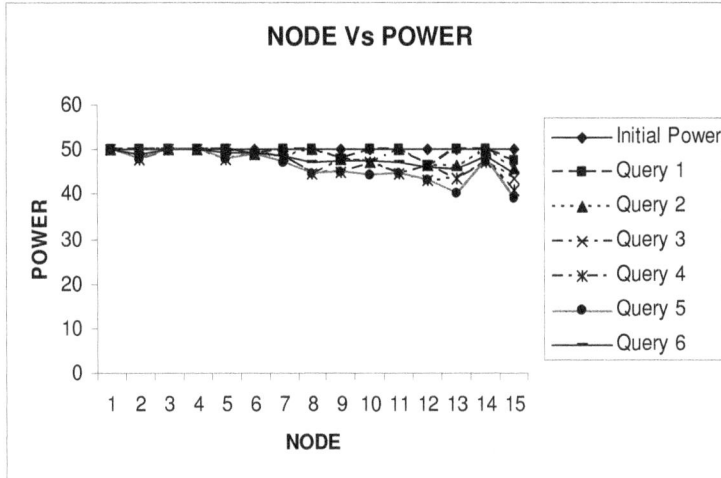

Figure 4. Node vs. Power graph for neighbor selection based on maximum power

Now, the average power of the nodes obtained in both are plotted in a graph, where series1 represents the average power of the node when the nodes are chosen randomly that is for Rumor Routing algorithm, series2 represents the power of the nodes when the next hop is chosen based on the maximum power of the neighboring nodes.

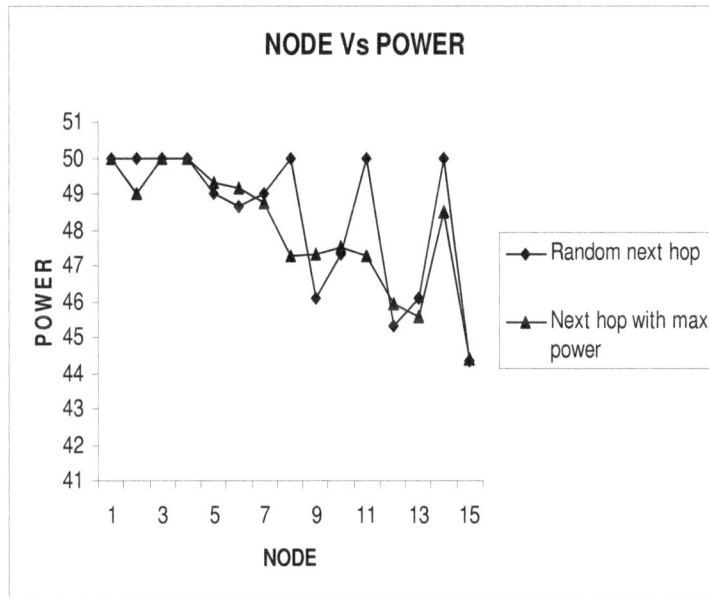

Figure 5. Node vs. Power graph by taking average power deduction for each node for random neighbor selection and neighbor selection based on maximum power

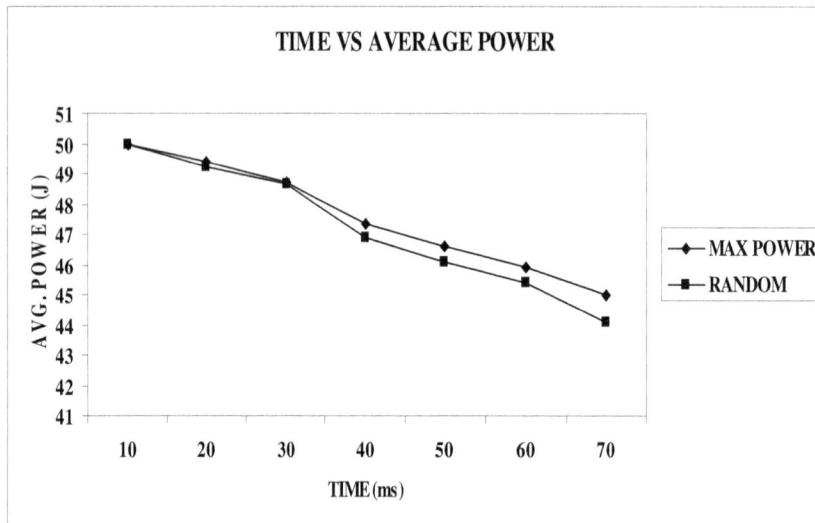

Figure 6. Average Power vs. Time graph for random neighbor selection and neighbor selection based on maximum power

In the above graph it is shown that when the time increases the average power deduction of the network for neighbor selection based on maximum power is less than that of random neighbor selection. Thus the network will be more stable if we choose the neighbor nodes based on maximum power contents and also the longevity of the network will be increased.

5. CONCLUSION

Sensor networks aim to achieve energy balance as well as energy efficiency. Till date, the energy-constrained nature of sensors poses the biggest problem in efficient management of sensor networks. The current state of the art is that for most of the power saving algorithms the cost factor is neglected, as they are not sending the packets in optimal path to balance the energy level between all the nodes in the network. This paper proposes an energy efficient technique to find out the source of the event, at the same time it selects the next hop node among the neighbor nodes which contain the maximum power. Thus there is a balance in power consumption for every node in the network. The simulation result shows the proposed algorithm increases the lifetime of every node in the network and thus it increases the longevity of the overall network.

6. REFERENCES

[1]   PruetBoonma, and Junichi Suzuki, "MONSOON: A Co-evolutionary Multiobjective Adaptation Framework for Dynamic Wireless Sensor Networks", In Proceedings of the 41st Hawaii International Conference on System Science, 2008.

[2]   PruetBoonma, and Junichi Suzuki, "Exploring self-star Properties in Cognitive Sensor Networking", In Proc of IEEE/SCS International Symposium on Performance Evaluation of Computer and Telecommunication Systems (SPECTS), Edinburgh, UK, 2008.

[3]   PruetBoonma, and Junichi Suzuki, "BiSNET: A biologically inspired middleware architechture for self managing wireless sensor networks", In Computer Networks, Vol.5.1, NO.16, pp. 4599-4616, 2007.

[4]   Wei-Ming Chen, Chung-Sheng Li, Fu-Yu Chiang, Han-Chieh Chao "Jumping Ant Routing Algorithm for sensor networks", published by Elsvier B.V. in 2007.

[5]   Suyoung Yoon, RudraDutta and Mihail L. Sichitiu,"Power Aware Routing Algorithm for Wireless Sensor Networks" in IEEE 2007.

[6]   Siva Kumar. D and Bhuvaneswaran.R.S, "Proposal on Multi agent Ants based Routing Algorithm for Mobile Adhoc Networks", in IJCSNS International Journal of Computer Science and Network Security, Vol-7, No. 6, June,2007.

[7]   M.Dorigo, "Ant Colony Optimization", University of Pretoria Etd, du Plessis J, 2006

[8]   HossamHassaanein and Jing Luo "Reliable Energy Aware Routing in Wireless Sensor Networks" in IEEE in 2006.

[9]   V.Laxmi, Lavina Jain, M.S.Gaur, "Ant Colony Optimization based Routing on NS-2" in the preceedings of International Conference On Wireless Communication and Sensor Networks,WCSN 2006

[10]   Ace Abrenica, RezanAlmojulea, Roger Dalupang, RhodelleMagnayon (2005). "The Development of a ZigBee Wireless Home Area Network Utilizing Solar   Energy as an Alternative Power Source". University of Manitoba.

[11]   PaymanArabshahi, Andrew Gray, IoannisKassabalidis, Arindam Das, "Adaptive Routing in Wireless Communication Network using Swarm Intelligence", Jet Propulsion Laboratory and University of Washington.,2004

[12]   KwangMongSim and Weng Hong Sun, Member IEEE, "Ant Colony Optimization or Routing and Load-Balancing: Survey and New Direction, part-A: System and Humans, Vol-33,No. 5 Sep,2003.

[13]   Sundaram Rajagopalam, Chien-Chung Shen, "A Routing Suite for Mobile Ad hoc Networks using Swarm Intelligence", Department of Computer and Information Sciences, University of Delaware, Newark,2003

[14]   ] F. Ye, A. Chen, S. Liu, L. Zhang, \A scalable solution to minimum cost forwarding in large sensor networks", Proceedings of the tenth International Conference on Computer Communications and Networks (ICCCN), pp. 304-309, 2001.

[15]   D. Gay, P. Levis, R. von Behren, M. Welsh, E. Brewer,   and D. Culler. The nesC language: A holistic approach to networked embedded systems. In SIGPLAN Conference on Programming Language Design and Implementation (PLDI'03), June 2003.

[16]   Dai, F., Wu. J. "Distributed dominant pruning in ad-hoc networks". In: Proceedings of ICC 2003.

[17]   M. Gunes, U. Sorges, I. Bouazizi, "ARA- the ant colony based routing algorithm for MANET",ICPP Proc of the 2002

[18]   David Braginsky and Deborah Estrin "Rumor Routing Algorithm for Sensor Networks" in WSNA '02, September 28, 2002.

[19]   Yu, Y. Govindan, R. and Estrin, D. Geographical and Energy Aware Routing: A Recursive Data Dissemination Protocol for Wireless Sensor Networks. UCLA Computer Science Department Technical Report UCLA/CSD-TR-01-0023, 2001.

[20]   W.R.Heinzelman,   A.Chandrakasan,   and   H.Balakrishnan,   "Energy   Efficient Communication Protocol for Wireless Microsensor Networks", In Proceedings of the 33rd Hawaii International Conference on System Sciences, 2000.

[21]   Intanagonwiwat, C. Govindan R. and Estrin, D. Directed Diffusion: A Scalable and Robust Communication Paradigm for Sensor Networks. In Proceedings of the sixth Annual International Conference on Mobile Computing and Networks (MobiCOM2000), 2000.

[22]   Karp, B. and Kung, H.T. GPSR: Greedy perimeter stateless routing for wireless networks, In Proceedings of the ACM/IEEE International Conference on Mobile Computing and Networking, pages 243-254, boston, Mass., USA, 2000.

[23]   Subramanian, D. Druschel, P. Chen, J. Ants and Reinforcement Learning: A Case Study in Routing in Dynamic Data Networks. In Proceedings of IJCAI-1997.

[24]   ]. C. Schurgers and M.B. Srivastava, "Energy efficient routing in wireless sensor networks", in the MILCOM Proceedings.

[25]   ] M. Chu, H. Haussecker, and F. Zhao, "Scalable Information-Driven Sensor Querying and Routing for ad hoc Heterogeneous Sensor Networks," The International Journal of High Performance Computing Applications, Vol. 16, No. 3, August 2002.

[26]   Y. Yao and J. Gehrke, "The cougar approach to in-network query processing in sensor networks", in SIGMOD Record, September 2002.

[27]   Qun Li, Javed Aslam, Daniela Rus, "Hierarchical Power aware Routing in Sensor Networks", In DIMACS Workshop on Pervasive Networking, Rutgers University, May 21, 2001.

6

ZRP with WTLS Key Management Technique to Secure Transport and Network Layers in Mobile Adhoc Networks

Dr.G.Padmavathi[1], Dr.P.Subashini[2], and Ms.D.Devi Aruna[3]

[1]Professor and Head, Department of Computer Science,
Avinashiligam University for Women, Coimbatore – 641 043
ganapathi.padmavathi@gmail.com

[2]Associate Professor, Department of Computer Science,
Avinashiligam University for Women, Coimbatore – 641 043
mail.p.subashini@gmail.com

[3]Project fellow, Department of Computer Science,
Avinashiligam University for Women, Coimbatore – 641 043
deviaruna2007@gmail.com

ABSTRACT

A mobile ad hoc network (MANETs) is a self-organizing network that consists of mobile nodes that are connected through wireless media. A number of unique features, such as lack of infrastructural or central administrative supports, dynamic network topologies, open communication channels, and limited device capabilities and bandwidths, have made secure, reliable and efficient routing operations in MANET a challenging task. The ultimate goal of the security solutions for MANET is to provide security services, such as authentication, confidentiality, integrity, anonymity, and availability to mobile users. To achieve the goals, the security solution need for entire protocol stack. The primary focus of this work is to provide transport layer security for authentication, securing end-to-end communications through data encryption. It also handles delay and packet loss. The MANET transport layer protocols provide end-to-end connection, reliable packet delivery, flow control and congestion control. The proposed model combines Zone Routing Protocol(ZRP) with Wireless Transport Layer Security(WTLS) provides authentication, privacy and integrity of packets in both routing and transport layers of MANET and also to defend against Denial of Service(DoS) attack.ZRP with WTLS is found to be a good security solution even with its known security problems. The simulation is done using network simulator qualnet 5.0 for different number of mobile nodes. The proposed model has shown improved results in terms of Average throughput, Average end to end delay, Average packet delivery ratio and Average jitter.

KEYWORDS

MANET,WTLS,ZRP,Denial of Service attack.

I. INTRODUCTION

Mobile ad hoc networks (MANETs) have received marvelous attention because of their self-maintenance capabilities. While early research effort assumed a friendly environment and paying attention on problems such as multihop routing and wireless channel access, security has become a main concern in order to provide protected communication between nodes in a potentially hostile environment. Although security has extensive been an active research topic in wireline networks, the unique characteristics of MANETs present a new set of nontrivial challenges to security design. These challenges include shared wireless medium, stringent resource constraints, open network architecture and highly dynamic network topology. So, the existing security solutions for wired networks do not directly apply to the MANET domain.

The vital goal of the security solutions for MANETs is to provide security services, such as confidentiality, integrity, authentication, anonymity, and availability, to mobile users. To achieve the goals, the security solution need for complete protocol stack. DoS attacks can be launched against any layer in the network protocol stack particularly transport layer which is a challenging one to defend against. In this type of attack, an attacker attempts to prevent legitimate and authorized users from the services offered by the network Table 1 describes the security issues in each layer. The proposed model combines hybrid routing protocol ZRP with WTLS to defend against   DoS attack and it also   provides *authentication, privacy and integrity* of packets in both routing and transport layer of MANET. The primary focus of this work is to provide transport layer security for authentication, securing end-to-end communications through data encryption, handling delays, packet loss and so on. The MANET transport layer protocols provide end-to-end connection, congestion control, reliable packet delivery and flow control.

Table 1: Layer wise Security Challenges

Layer	Security issues
Application layer	Detecting and preventing viruses, worms, malicious codes, and application abuses
Transport layer	Authenticating and securing end-to-end communications through data encryption
Network layer	Protecting the ad hoc routing and forwarding protocols
Link layer	Protecting the wireless MAC protocol and providing link-layer security support
Physical layer	Preventing signal jamming denial-of-service attacks

The paper is organized in such a way that Chapter 2 discusses Review of Literature, Chapter 3 discusses proposed method, Chapter 4 discusses Experimental evaluation and Chapter 5 gives the conclusion

II. REVIEW OF LITERATURE

This chapter briefly describes Denial of Service attacks for MANET and related work.

1. Denial of Service attack

An attacker attempts to stop authorized and legitimate users from the services obtainable by the network. A denial of service (DoS) attack can be carried out in many ways. The typical way is to flood packets to any centralized resource present in the network so that the resource is no longer accessible to nodes in the network, as a result of which the network no longer function in the manner in which it is designed to operate. This may lead to a failure in the delivery of certain services to the end users. DoS attacks can be launched against any layer in the network protocol stack. On the physical and MAC layers, an adversary could employ jamming signals which disrupt the on-going transmissions on the wireless channel. On the network layer, an adversary could take part in the routing process and exploit the routing protocol to disrupt the normal functioning of the network. For example, an adversary node could contribute in a session but simply drop a certain number of packets, which may lead to degradation in the QoS being offered by the network. On the higher layers, an adversary could bring down serious services such as the key management service. For example, consider the following: In figure1 assume a shortest path that exists from **S** to **X** and **C** and **X** cannot hear each other, that nodes **B** and **C** cannot hear each other, and that **M** is a malicious node attempting a denial of service attack. Suppose **S** wishes to communicate with **X** and that **S** has an unexpired route to **X** in its route cache. **S** transmits a data packet towards **X** with the source route **S** --> **A** --> **B** --> **M** --> **C** --> **D** --> **X** contained in the packet's header. When **M** receives the packet, it can alter the source route in the packet's header, such as deleting **D** from the source route. Consequently, when **C** receives the altered packet, it attempts to forward the packet to **X**. Since **X** cannot hear **C**, the transmission is unsuccessful [6][7][9].

$$S \leftrightarrow A \leftrightarrow B \leftrightarrow M \leftrightarrow C \leftrightarrow D \leftrightarrow X$$

Figure 1. Denial of Service attack

2. Related Work

The following list of papers show the relative work carried out for MANET attacks and the possible solutions.

1) Wormhole Attack Detection in Wireless Sensor Networks: This paper discusses the nature of wormhole attack and existing methods of defending mechanism and then proposes round trip time (RTT) and neighbor numbers based wormhole detection mechanism [14].

2) Enhanced Intrusion Detection System for Discovering Malicious Nodes in Mobile Ad Hoc Networks: The main characteristic of the proposed system is its capability to discover malicious nodes which can partition the network by falsely reporting other nodes as misbehaving and then it proceeds to protect the network [16].

3) A Distributed Security Scheme for Ad Hoc Networks: It discusses the DoS attack like flooding using AODV protocol and concludes with an direct enhancement to make the limit-parameters adaptive in nature. [13].

4) A Secure Routing Protocol against Byzantine Attacks for MANETs in Adversarial Environments: This considers an integrated protocol called secure routing against collusion (SRAC), in which a node makes a routing decision based on its trust of its neighboring nodes [15].

5) Detecting Network Intrusions via Sampling: A Game Theoretic Approach: This paper discusses the problem of detecting an intruding packet in a communication network [12].

The majority of the related study covers only few network layer attacks,In the proposed approach, attempts to identify transport layer attacks and it provides authentication and secure end-to-end communication.

III. PROPOSED METHOD

This chapter briefly describes proposed method combines Zone Based Routing protocol (ZRP) and transport Layer security in Mobile Adhoc Networks.

Routing protocols can be classified mainly into three types proactive, reactive and hybrid routing protocols. Proactive routing protocols maintain routing information all the time and always update the routes by broadcasting update messages. However, reactive routing is started only if there is a demand to reach another node. Reactive protocols acquire routing information only when it is actually needed. Hybrid protocols combine the advantages of proactive and of reactive routing. The widely used hybrid routing protocol Zone Based Routing protocol (ZRP) is taken for the proposed work. It is considered to be the most suited one for ad hoc networks [2][3]. A brief description of the ZRP routing protocol is given below.

1. Zone Routing Protocol (ZRP)

Zone Routing Protocol (ZRP) uses both a proactive and a reactive routing .ZRP was first introduced by Haas in 1997. ZRP is proposed to decrease the reactive routing protocols latency caused by route discovery and to reduce the proactive routing protocols control overhead. ZRP defines a zone around each node consisting of its k-neighborhood (e. g. k=3). In ZRP, the distance and a node, all nodes within hop distance from node belong to the routing zone of node. It is formed by two sub-protocols, a proactive routing protocol: Intra-zone Routing Protocol (IARP), is used inside routing zones and a reactive routing protocol: Inter-zone Routing Protocol (IERP), is used between routing zones, respectively. A route to a destination within the local zone can be established from the proactively cached routing table of the source by IARP; therefore, if the source and destination is in the same zone, the packet can be delivered immediately.

Route discovery happens reactively when routes beyond the local zone. The source node sends a route requests to its border nodes, containing its own address, the destination address and a unique sequence number. Border nodes are nodes which are exactly the maximum number of hops to the defined local zone away from the source. The border nodes check their local zone for the destination. If the requested node is not a member of this local zone, the node adds its own address to the route request packet and forwards the packet to its border nodes. If the destination is a member of the local zone of the node, it sends a route reply on the reverse path back to the source. The source node uses the path saved in the route reply packet to send data packets to the destination [5][10].

Advantages: Provides scalability.

Disadvantages: Routing security in mobile Adhoc networks.

2. Transport Layer security in Mobile Adhoc Networks

The MANET transport layer protocols provide end-to-end connection, reliable packet delivery, flow control and congestion control. The security issues associated to transport layer are handling delays, authentication, end-to-end Communications through data encryption, and packet loss. The nodes in a MANET are also susceptible to the Denial of Service (DoS) attacks. The wide use of mobile communication has created an important demand for value-added services. WAP (Wireless Application Protocol) is a framework for developing applications to run over wireless networks. WAP is developed by WAP Forum. WTLS (Wireless Transport Layer Security) is the security protocol of the WAP protocol suite. WTLS operates over the transport layer and provides end-to-end security, where one end is WAP gateway and the other end is the mobile client. WAP gateway acts as a proxy of the mobile client to access an application server hosted anywhere on the Internet. The communication beyond the WAP gateway is conducted using the regular Internet (TCP/IP) protocol suite. A set of handshake messages is exchanged in order to set up a secure environment between the server (WAP gateway) and mobile client. Cryptographic algorithms, keys and related parameters are negotiated during the handshake. Once the handshake messages are exchanged and session key is generated, all WTLS and upper layer protocol messages can be exchanged in encrypted form. In this way, confidentiality and integrity are provided. Authentication is an optional service in WTLS. Authentication is provided if the parties provide digital certificates during the handshake. Certificates are digital identities that contain public-keys to be used during the key exchange. Certificates are issued by trusted Certification Authorities (CA) with a digital signature on the certificate content. Validation of a certificate means the legitimacy of the enclosed public-key. A party, who does not have a certificate, should use an unapproved public-key. Therefore, that party cannot be authenticated. Authentication, certificate validation, and session key exchange use asymmetric public-key cryptosystems that require computation-intensive processes, and are therefore slow. Speed is inversely proportional to the key size used in public-key cryptosystems. Since the processing power of mobile clients is limited, relatively smaller keys are selected for WTLS. Furthermore, data transfer rate is also limited in mobile communication environment and using smaller keys would help to save bandwidth [1][2].

Public-key cryptosystems in WTLS

Public-key cryptosystem operations use two different keys: public-key and private-key. Public-key operations are for signature verification and encryption. Private-key operations are for signature issuance and decryption. Public-key cryptosystems are used in the WTLS handshake for key exchange and certificate verification purposes. Authentication is mechanically provided when key exchange is performed using certified keys. WTLS supports two public-key cryptosystems: ECC (Elliptic Curve Cryptography) and RSA (Rivest- Shamir-Adleman).Public-key cryptosystems is used for key exchange and certificate verification .If RSA is to be used for key exchange, If ECC is to be used, ECDH (Elliptic Curve Diffie-Hellman) key exchange method is employed. Regular DH (Diffie-Hellman) [6] method is proposed as another key exchange mechanism in WTLS standard. Anonymous handshakes are vulnerable to man-in-the-middle-attacks, where an adversary impersonates both parties. Therefore, we do not consider anonymous handshakes as secure methods and do not include them in our performance evaluation. Besides DH, WTLS also propose anonymous versions of RSA and ECDH methods that we disregard as well. Certificate verification is a public-key operation. Both RSA and ECC can be used. If RSA is to be used, its verification feature is employed. If ECC is to be used, ECDSA (Elliptic Curve Digital Signature Algorithm) is employed. [3].

Key exchange suites of WTLS

WTLS supports numerous alternative key exchange suites. However, only two of them offer an acceptable level of security:

RSA and ECDH_ECDSA key exchange suites.

1. ECDH_ECDSA: ECDSA is used for certificate verification.

2. RSA: RSA cryptosystem is used for both key exchange and certificate verification

IV. EXPERIMENTATION AND EVALUATION

Qualnet5.0 network simulator is used for experimentation. Mobility scenarios are generated using a Random waypoint model by varying 10 to 50 nodes moving in a terrain area of 1500m x 1500m. The image of the network as it appears in Qualnet 5.0 is presented in Figure-2. The simulation parameters are summarized in Table 2.

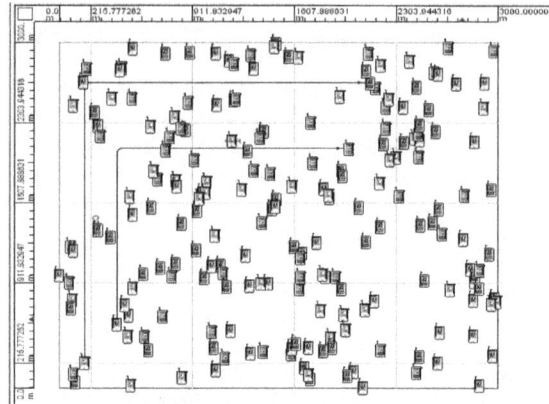

Figure2. The image of the network as it appears in Qualnet 5.0

Table2. Simulation Parameters

Parameter	Value
Simulator	Qualnet 5.0
Simulation time	100 s
Number of nodes	50
Traffic Model	CBR
Pause time	2 (s)
Maximum mobility	60 m/s
No. of sources	15
Terrain area	1500m x 1500m
Transmission Range	250m

The simulation is done to investigate the performance of the network with various parameters. The metrics used to evaluate the performance are:

1) Average packet delivery ratio
2) Average end-to-end delay
3) Average delay jitter
4) Average throughput

Average packet delivery ratio: The packet delivery ratio (PDR) of a receiver is defined as the ratio of the number of data packets actually received over the number of data packets transmitted by the senders.

Average end-to-end delay: The end-to-end delay of a packet is defined as the packet takes a time to travel from the source to the destination. The average end-to-end delay is the average of the end-to-end delays taken over all the received packets Eqn (1) is used to find the end to end delay of the packet.

$$delay = \frac{1}{nbx} \sum_{i \in x} \sum_{iey} \frac{delay_j}{nby} \quad ---- (1)$$

x: is the set of destination nodes that received data packets.

nbx: is the number of receiver nodes

y: is the set of packets received by node i as the final destination.

Average delay jitter: Delay jitter is the variation (difference) of the inter-arrival times between the two successive packets received. Each receiver calculates the average per-source delay jitter from the received packets originated from the same source. The receiver then takes the average over all the sources to obtain the average per-receiver delay jitter.

Average throughput: The throughput of a receiver (per-receiver throughput) is defined as the ratio of the number of bits received over the time difference between the first and the last received packets. The average throughput is the average of the per-receiver throughputs taken over all the receivers. Eqn (2) is used to find the throughput of the packet.

$$Throuhput(\%) = \frac{\text{Re}\,ceivedpackets}{Sentpackets} * 100 \quad ---(2)$$

Performance comparison of routing protocol ZRP and WTLS for ZRP routing protocol with Denial of Service attack

The different parameters are considered for evaluation. Average packet delivery ratio, Average throughput, should be higher and Average end-to-end delay, Average delay jitter must be lower.Figure 3 shows that Average packet delivery ratio is higher in WTLS with ZRP with Denial of Service attack compared to ZRP. Figure 4 shows that Throughput is higher in WTLS with ZRP with Denial of Service attack compared to ZRP. Figure 5 shows that Average Jitter is lower in WTLS with ZRP with Denial of Service attack compared to ZRP. Figure 6 shows that End to End Delay is lower in WTLS with ZRP with Denial of Service attack compared to ZRP.

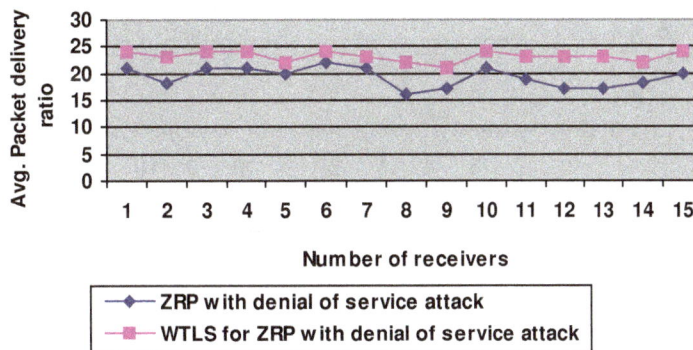

Figure 3. Comparison of Average packet delivery ratio of ZRP and ZRP for WTLS with Denial of Service attack

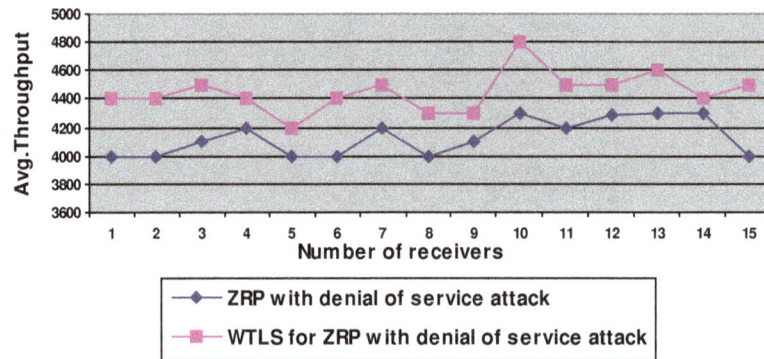

Figure 4. Comparison of Throughput of ZRP and ZRP for WTLS with Denial of Service attack

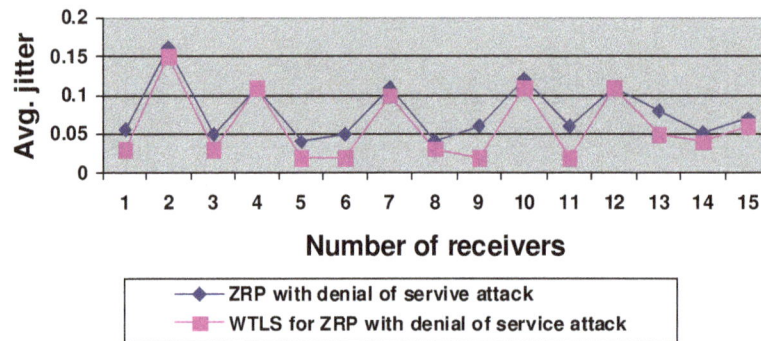

Figure 5: Comparison of Average Jitter of ZRP and ZRP for WTLS with Denial of Service attack

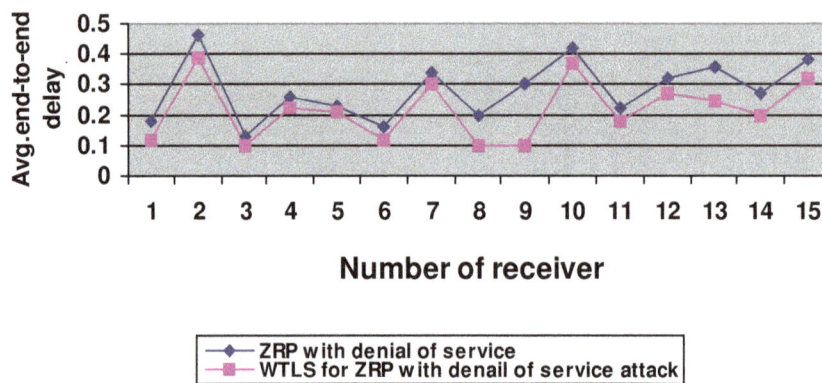

Figure 6: Comparison of End to End delay of ZRP and ZRP for WTLS with Denial of Service attack

From the simulation results it is observed that proposed model is robust against denial of service attacks and it also provides authentication, securing end-to-end communications through data encryption, handling delays, packet loss in routing and transport layer of MANET.

V. CONCLUSION

A mobile ad hoc network (MANET) is a self-organizing network consisting of mobile nodes that are connected through wireless media. A number of unique features, such as lack of infrastructural or central administrative supports, dynamic network topologies, open communication channels, and limited device capabilities and bandwidths, have made secure, reliable and efficient routing operations in MANET a challenging task. The ultimate goal of the security solutions for MANET is to provide security services, such as authentication, confidentiality, integrity, anonymity, and availability, to mobile users. To achieve this goal, the security solution need for whole protocol stack. The main focus of this work is to provide transport layer security for authentication, securing end-to-end communications through data encryption, packet loss and handling delays, The MANET transport layer protocols provides end-to-end connection, reliable packet delivery, flow control and congestion control. The proposed model combines hybrid routing protocol ZRP with WTLS to defend against Denial of Service(DoS) attack and it also provides authentication, privacy and integrity of packets in both routing and transport layers of MANET.

ACKNOWLEDGMENT

The authors would like to thank the University Grants Commission (UGC) for supporting this Major Research project (MRP).

References

1. K. Sundresses, V. Anantharaman, H. Y. Hsieh, and R. Sivakumar. ATP," **A Reliable Transport Protocol for Ad Hoc Networks**". In Proceedings of ACM MOBIHOC 2003, pp. 64-75, June 2003.

2. Kahraman, Gokhan," *An Investigation of WAP Transaction Protocol Performance for Packet Radio Network's,* Master Thesis, Electrical and Electronics Engineering, Graduate School of Natural and Applied Sciences, The Middle East Technical University, Ankara, Turkey, April 2002

3. The WAP Forum, "**Wireless Transaction Protocol**", Version 10-Jul-2001, http://www.wapforum.org

4. B. Aerobic, R. Curtmola, H. Rubens, D. Holmer, and C. Nita-Rotaru, "*On the survivability of routing protocols in ad hoc wireless networks*," IEEE, 2005.

5. Z. J. Haas, M. Perlman, "*The Performance of Query Control Schemes of Zonal Routing Protocol*", *IEEE Trans. on Networking*, vol. 9, no. 4, pp. 427-438(2001).

6. M.K. Denko, "*A Localized Architecture for Detecting Denial of Service (DoS) Attacks in Wireless Ad Hoc Networks*", In Proc. IFIP INTELLCOMM'05, Montreal, Canada.

7. Aad, J.P, Hubaux, and E.W. Knightly, "*Denial of Service Resilience in Ad Hoc Networks*", ACM MOBICOM 2004, Philadelphia, PA, USA.

8. V. Gupta, S. Krishnamurthy, and M. Faloutsos," *Denial of Service Attacks at the MAC Layer in Wireless Ad Hoc Networks*". In Proc. of MILCOM, 2002.

9. A. Habib, M. H. Hafeeda, and B. Bhargava, "*Detecting Service Violation and DoS Attacks*", In Proc. of Network and Distributed System Security Symposium (NDSS), 2003.

10. *Zone Routing Protocol (ZRP) for Ad Hoc Networks*", *IETF* Internet Draft, Version 4, July 2002.

International Journal of Wireless & Mobile Networks (IJWMN) Vol. 4, No. 1, February 2012

11. Prince Samar, Marc Pearlman and Zygmunt Haas, *"Independent Zone Routing: An Adaptive Hybrid Routing Framework for Wireless Networks"*, *IEEE/ACM* Transactions on Networking, 12, No. 4, August 2004. pp: 599.

12. Murali Kodialam T. V. Lakshman, *"Detecting Network Intrusions via Sampling: A Game Theoretic Approach"*, IEEE INFOCOM, 2003.

13.   Dhaval Gada, Rajat Gogri, Punit Rathod, Zalak Dedhia and Nirali Mody Sugata Sanyal, Ajith Abraham, *"A Distributed Security Scheme for Ad Hoc Networks"*, ACM Publications, Vol-11, Issue 1, 2004, pp. 5 – 5.

14.Zawtun and Aung Htein Maw, *"Wormhole attack detection in wireless sensor networks"*, World Academy of Science, Engineering and Technology, 46, 2008.

15.Ming Yu; Mengchu Zhou; Wei Su, *"A Secure Routing Protocol Against Byzantine Attacks for MANETs in Adversarial Environments"*, IEEE Transactions on Vehicular Technology Vol-58, Issue 1, Jan. 2009 , pp.449 – 460.

16. Nasser, N.; Yunfeng Chen, *"Enhanced Intrusion Detection System for Discovering Malicious Nodes in Mobile Ad Hoc Networks"*, IEEE International Conference on Communications, ICC apos; Vol-07 , Issue 24-28 June 2007 , pp.1154 – 1159.

17. Oscar F. Gonzalez, God win Ansa, Michael Howarth, and George Pavlou, *"Detection and Accusation of Packet Forwarding Misbehavior in Mobile Ad-Hoc networks",* Journal of Internet Engineering, vol-2, 2008, pp.1.

7

MOBILE NODE LOCALIZATION IN CELLULAR NETWORKS

Yasir Malik[1], Kishwer Abdul Khaliq[2], Bessam Abdulrazak[1], Usman Tariq[3]

[1]Department of Computer Science, University of Sherbrooke, Quebec, Canada
`yasir.malik, bessam.abdulrazak[@usherbrooke.ca]`
[2]Center of Research in.Networks and Telecom (CoReNeT), Mohammad Ali Jinnah
University, Islamabad, Pakistan
`kishibutt@gmail.com`
[3]Department of Information Systems, College of Computer and Information Sciences,
Al-Imam Mohammed Ibn Saud Islamic University, Riyadh, Saudi Arabia
`usman@usmantariq.org`

ABSTRACT

*Location information is the major component in location based applications. This information is used in different safety and service oriented applications to provide users with services according to their Geo-location. There are many approaches to locate mobile nodes in indoor and outdoor environments. In this paper, we are interested in outdoor localization particularly in cellular networks of mobile nodes and presented a localization method based on cell and user location information. Our localization method is based on hello message delay (sending and receiving time) and coordinate information of **Base Transceiver Station (BTSs)**. To validate our method across cellular network, we implemented and simulated our method in two scenarios i.e. maintaining database of base stations in centralize and distributed system. Simulation results show the effectiveness of our approach and its implementation applicability in telecommunication systems.*

KEYWORDS

Cellular network, Mobile computing, Location base service, Network algorithm.

1. INTRODUCTION

Recent immense growth in wireless networks and related technologies allows its user to be mobile and still get access to information they need. This roaming freedom with the seamless mobility between neighbouring base stations facilitates its users to communicate anywhere [8]. While the user is mobile it is very important for service providers to know the physical location of its users to provide services according to their location. For instance with the latest regulation by Federal Communications Commission (FCC)[1], it is required by all network providers to implement the E911 service[2] which will help to get the exact physical location of users when the 911 service is requested. Consequently the physical location data of the user is very important input for **L**ocation **B**ase Services (LBS).

The process of estimating the physical location of a wireless device is called localization. The core of the process lies in getting the location of the mobile device. There have been different mechanisms to find the location of mobile nodes, however these mechanisms are not good enough to support the requirements of LBS in technologies like GSM and UMTS. Global Positioning System (GPS) is widely used for the location information to provide services with respect to physical location of user. There are many mobile devices which are equipped with GPS and they work with networks such as GSM, UMTS etc, however these solutions leads to

---

[1] http://www.fcc.gov/

[2] http://www.fcc.gov/ 911/enhanced/

increase in cost, battery consumption, etc. [4] and often are not suitable for urban area. In this paper we provide a solution without using the GPS system, our solution is based on GSM/3G network and does not require any special hardware. The location information is collected with the existing telecom infrastructure which makes it easier for the network operator to use the same network to locate nodes in network, and for users to use their devices without needing any special hardware upgrades. Our approach of node localization is based on hello message delay (sending and receiving time) and coordinates information of BTSs, and hence locates the node location across the cellular network. We have tested our approach in two scenarios i.e. centralized and distributed databases on each BTS and BSS. The rest of the paper is organized as follows. The next section briefly summarizes the state of the art focusing on localization in cellular networks. Section 3 presents our localization approach in both scenarios along with respective algorithms. Results are presented in section 4 and we conclude the paper in section 5.

2. RELATED WORK

Several models and methods have been presented for location-based services systems in cellular indoor and outdoor networks. In this section, we review some preventative work that addresses localization in different cellular networks and related technologies present in the domain. Sinha and Das presented a localization method where mobile node in a cellular network sends a special distress signal to covering base station which computes the localization coordinates of mobile node with the help of adjacent base station and detailed road map [7]. Kiran and colleagues proposed a localization system that finds the mobile node within a cell based on the cell-id, signal strength and hello packet delay [6]. The approximate location of mobile node is found by using the signal strength which is received by the neighbour-receivers. Andreas Hartl in [5] presented a lightweight solution that communicates the cell information to web services. This solution is provider-independent and easily extensible. Authors in [9] focused on the localization problem in out of coverage and non GPS equipped devices in UMTS networks and proposed to use a cooperative localization method based on ETSI/3GPP LCS architecture that enable devices to estimate their position by performing power measurements on signals emitted by mobile phones with satellite navigation receivers and known-position. Similar efforts have been presented in [3] where authors proposed to utilize additional information obtained from short-range links and later combine the time difference of arrival (TDOA) and received signal strength (RSS) in their simulation using advanced data fusion techniques for node localization. In another effort, authors extended the Kalman Filter to merge the time difference of arrival and the received signal strength retrieved from the long and short range [2]. Authors in [1] presented a lookup table correlation technique that applies multiple positioning and locating techniques to be used with advance propagation model in conjunction with Kalman predictive filtering for node localization. Authors in [10] presented a zero-length technique based on received signal strength to compute node localization. This allows a less detailed path loss model to use without significant impact to the location estimation. For a comprehensive reading about localization techniques readers may refer to [8].

3. PROPOSED LOCALIZATION MECHANISM

To find the exact location of the mobile node in cellular network, our approach relies on time of sending and response of 'hello' messages, and also requires to maintain the database for all BTS. The main idea is that the Mobile Node (MN) sends a query to nearest BTS for location, that servicing BTS generates a hello message to the MN and MN respond to the BTS. As the same time servicing base station also communicates to neighbour BTS for MN location, on the basis of control messages exchange and time difference of sending and receiving these messages, the MN location is calculated. For this purpose we design an algorithm that finds exact location of mobile node in a cellular network which is tested and validated with two scenarios i.e. distributed and centralized databases on base stations. In the next sections, we describe our method for both scenarios. Our solution is based on the following assumptions.

- We need to maintain a database on:
  - o Each BTS about the location of base station (for distributed system scenario).
  - o Each BSC about the location of all base stations present in BSS (for centralized system scenario).
- Channels are kept reserved at each base station for lookup services.
- The mobile remains stationary during the whole process.

A.   DISTRIBUTED DATA BASE APPROACH (DDBA)

In the first approach we consider a cellular network where cells are arbitrary shaped and need to maintain a data base on each base station about its adjacent base stations. The data base contains the coordinates of the adjacent base stations. The numbers of base stations are fewer; in case of hexagonal shape it will be maximum six. The serving base station is known as the master base station. The mobile node sends a request for lookup services (hello packet) to nearest base station. The corresponding base station (i.e. Master BTS) receives the request and tracks the mobile node M by sending a message and mobile node M acknowledges to corresponding BTS. Then the Euclidean distance between master base station and mobile node is calculated using equation 1. Then Master base station sends messages to two adjacent base stations known as slave base stations. The slave base stations locate the mobile node and acknowledge to master base station. The distance is calculated using distance equations 2 and 3. The master base station BTS calculates the coordinates of the mobile node and the end result is sent to the BSC where lookup services are implemented. The system flow chart illustrating communication between BTS, mobile node and BSC and messages used for communication is shown in **Figure 1**. The communication sequence between Master BTS and M is shown in Figure 2a, and the communication sequence between Master BTS, Slave BTS and M is shown in Figure 2b.

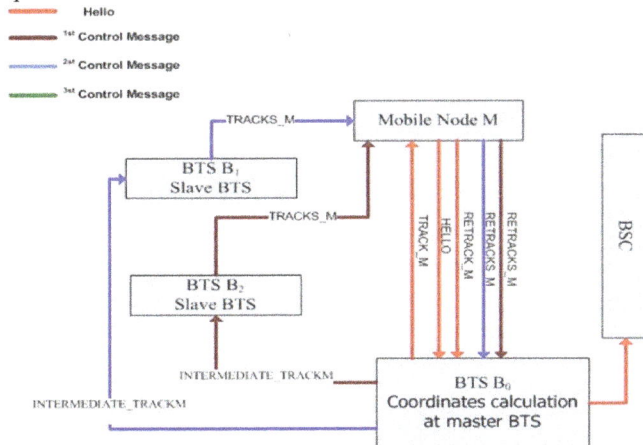

Figure 1.System Diagram of DDBA

The coordinates are calculated when the mobile user sends hello message to the servicing base station, called Master base station, the Master base station sends tracking message to the mobile node M at time T_1 and mobile node M acknowledges to the Master base station at time T'_1 as shown in the Figure 2a. The Euclidean distance between mobile node M and Master BTS is calculated using equation 1.

$$T'_1 - T_1 = \left(\frac{2d_1}{c}\right) [7] \qquad (1)$$

Here T'_1 and T_1 are the time stamp value of message sending and receiving respectively, d_1 is the distance of master BTS to the mobile Node M and c is the velocity of light. To calculate d_1

the master base station sends tracking message to adjacent BTS to track mobile node M. After receiving tracking message, the mobile node M acknowledged to Master BTS, and the distance is calculated with equations 2 and 3.

$$T'_1 - T_1 = (d_{01} + d_1 + d_2)/c \text{ [7]} \qquad (2)$$

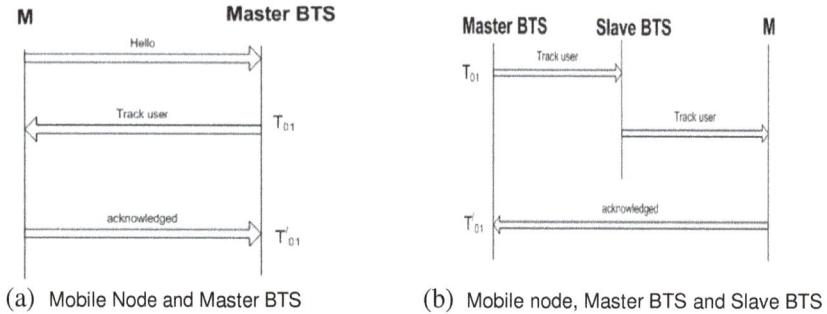

(a) Mobile Node and Master BTS   (b) Mobile node, Master BTS and Slave BTS

Figure 2 .Communication Sequence Diagram.

Where

d_{01}=distance between B_0 and B_1
d_1= distance between B_0 and the mobile M
d_2= distance between B_1 and the mobile M

$$T'_1 - T_1 = (d_{02} + d_3 + d_1)/c \text{ [7]} \qquad (3)$$

Where

T'_1 and T_1 is the sending and receiving times
d_{02} = distance between B_0 and B_2
d_3 = distance between B_2 and the mobile M

The computation results of $d1$, $d2$ and $d3$ in equations 1, 2 and 3 are later used to calculate the exact location of the mobile node M using equations 4, 5 and 6 as used in [7],

$$(x - x_0)^2 + (y - y_0)^2 = d1^2 \qquad (4)$$
$$(x - x_1)^2 + (y - y_1)^2 = d2^2 \qquad (5)$$
$$(x - x_2)^2 + (y - y_2)^2 = d3^2 \qquad (6)$$

Where (x_0, y_0), (x_1, y_1), (x_2, y_2) are the geographical coordinates of three base stations B_0, B_1, B_2. From equations 4, 5 and 6, we obtain the linear equations 7 and 8.

$$a_1 x + b_1 y + c_1 = 0 \qquad (7)$$
$$a_2 x + b_2 y + c_2 = 0 \qquad (8)$$

By solving equations 7 and 8 we obtain the value of (x, y) that provides the geographical position of mobile Node M. The proposed method is designed for cellular network where a BSC controls many BTSs; each BTS provides services to mobile users. Here we assume a model where a BSC controls many base stations, a mobile node sends a request for lookup services, such as a list of nearest hotels, by querying to serving base station. The servicing base station, first finds the exact location of mobile node M and then sends the information about mobile location to BSC where BSC responds to mobile node M for the requested query. The localization method procedural flow for the model is shown in Figure 3; each step is labelled and the description of each is provided below.

Step 1: The mobile user M initiates the process by sending hello packet to the nearest BTS.

Step 2: On receiving hello packet from mobile user M, B_0 sends out the TRACK M message to mobile user M appending B_0 identity master-ID at the end of the message received from mobile user M. As the retransmission starts, B_0 notes its local time instant t_0.

Step 3: As soon as a mobile node *M* detects that the type of the message it is receiving is TRACK M, receives the whole packet, checks the mobile-ID field, and if it is its own ID, it retransmits the message modifying it into a RETRACK *M* message, otherwise it simply ignores the packet. B_0, on receiving RETARCK M packet, notes its receiving time $t'0$.

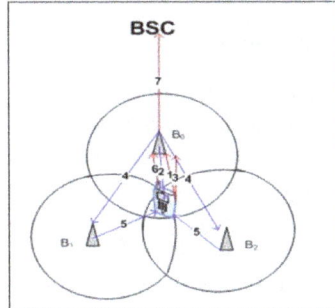

Figure 3. Node Localization in DDBA Scenario

Step 4: Simply ignores the packet. B_0, on receiving RETARCK M packet, notes it B_o sends an INTERMEDIATE TRACKM packet to nearest base stations that are called slave base stations (lets say B_1 and B_2), appending the master-ID and the slave-ID at the end of the hello packet. B_o also notes the local time t_{01} at the start of sending the packet.

Step 5: When a base station detects that the type of the message it is receiving is an INTERMEDIATE TRACK M message, it receives the whole packet and modifies it into TRACK M message and sends this packet to mobile station *M*.

Algorithm 1 Distributed Data Base Approach(DDBA)

1: ⟨ **Variable** ⟩
2: *D1, D2: Distance of BTS B0 from BTS B1 and BTS B2 respectively.*
3: *d0, d1, d2: Distance between mobile node M and BTS B0, BTS B1, BTS B2 respectively.*
4: *equation1, equation2 and equation3: Equation of circle*
5: *matrix[r][c]: Resultant linear equations in matrix.*
6: *Count: contain the number of communication*
7: *timeDiff:= T2-T1 where T1 is communication starting time and T2 is the ending time.*
8: ⟨ **Procedure** ⟩
9: ***Mobile_BTS_HELLO()***:
10: ***BTS_To_Mobile_TRACK_M()***:
11: ***Mobile_To_BTS_RETRACK_M()***:
12: *computeDistance (count, timeDiff)*:
13: ***BTS_To_Slave_BTS_INTERMEDIATE_TRACK_M()***:
14: ***Slave_BTS_To_Mobile_TRACKS_M()***:
15: ***Mobile_To_BTS_RETRACKS_M()***:
16: ***Elimination (matrix[r][c])***:
17: mobileCoordinates (d0, d1, d2):
18: ⟨ **Main Algorithm** ⟩
19: If*(Mobile_BTS_HELLO(msgType, msgLength, MobileID, data))*
20: *Number of communication Count=1*
21: *Communication starting time T1:*
22: *BTS_To_Mobile_TRACK_M(msgType, msgLength, MobileID, data, masterID, masterTimeStamp)*:
23: *Mobile_To_BTS_RETRACK_M(msgType, msgLength, MobileID, data, masterID)*:
24: *Communication ending time T2:*
25: *timeDiff =T2-T1*:
26: *computeDistance (count, timeDiff)*:
27: for *i ←1 to 2*
28: *Count++*:
29: *Communication starting time T1:*
30: *BTS_To_Slave_BTS_INTERMEDIATE_TRACK_M(msgType, msgLength, MobileID, data, masterID, slaveID)*:
31: *Slave_BTS_To_Mobile_TRACKS_M(msgType, msgLength, MobileID, data, masterID, slaveID)*:
32: *Mobile_To_BTS_RETRACKS_M(msgType, msgLength, MobileID, data, masterID, slaveID)*:
33: *Communication ending time T2:*
34: *timeDiff =T2-T1*:
35: **endfor**
36: *computeDistance(count, timeDiff)*:
37: *mobileCoordinates(d0, d1, d2)*:
38: **endif**

Step 6: Mobile station receives TRACK M message, modifies it to RETRACK M and sends it to master base station B_0.

Step 7: B_0 calculate the distance of mobile node using equations and then sends the calculated distance to BSC.

B.   CENTRALIZED DATA BASE APPROACH (CDBA)

In the Centralized Data Base Approach (CDBA) approach, we have the same cell environment as in DDBA However, in this scenario, we need to maintain a data base only on BSC about the coordinates of base stations. The mobile node sends a request for lookup services (hello packet) to nearest base station. The corresponding base station receives the request and forwards the request to the BSC, which tracks the mobile node M through servicing BTS by sending a message and mobile node M acknowledges to BSC. Then the distance between BSC and mobile node is calculated using the same equations as in DDBA approach. BSC sends messages to two adjacent base stations known as slave base stations. The slave base stations locate the mobile node and acknowledge to BSC. The BTS calculate the coordinates of the mobile node where lookup services are implemented. The system diagram illustrating communication between BTS, mobile node and BSC and messages used for communication is shown in Figure 4

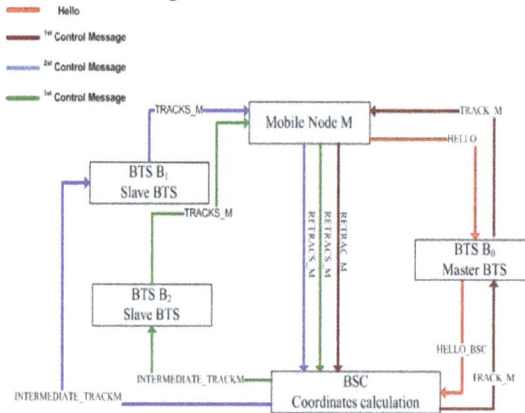

Figure 4. System Diagram of CDBA

When a mobile user sends hello message to servicing base station, the corresponding base station forwards the hello message to BSC and sends tracking message to mobile node M at time T_1 and acknowledges to the Master base station at time T'_1 (Figure 5).

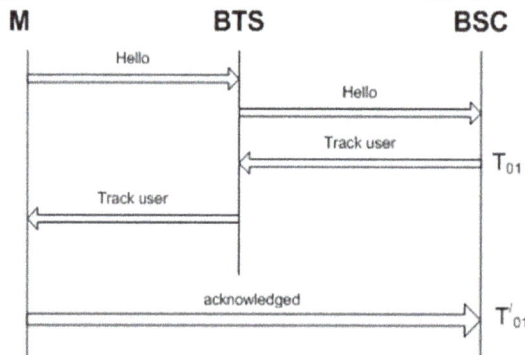

Figure 5. Communications Sequence Diagram (Mobile Node, BTS and BSC)

The Euclidean distance between mobile node M and Master BTS is calculated using equation 9.

$$T''_1 - T'_1 = (d_1 + d_{01})/c \qquad (9)$$

Where T'_1 and T_1 are the initial and ending time of message sending and receiving, d_1 is the distance of servicing base station BTS to the mobile Node M, d_{01} is the distance of BSC to serving base station and c is the velocity of light. Then BSC sends tracking message to adjacent BTS to track mobile node M. After receiving tracking message, the mobile node M acknowledged to Master BTS, the distance is calculated using coordinate equation. The computation of d_1, d_2 and d_3 in the equations 9, 2 and 3 is performed, and then these computations are used in the computation of the exact location of the mobile node M using

---

Algorithm 2 Centralized Data Base Approach (CDBA)

1: ⟨ **Variable** ⟩
2: *D0, D1, D2: Distance between BSC and BTS B0, BTS B1 and BTS B2 respectively.*
3: *d0, d1, d2: Distance between mobile node M and BSC via BTS B0, BTS B1, and BTS B2 respectively.*
4: *equation1, equation2 and equation3: Equation of circle*
5: *matrix[r][c]: Resultant linear equations in matrix.*
6: *Count: contain the number of communication*
7: *timeDiff: = T2-T1 where T1 is communication starting time and T2 is the ending time.*
8: ⟨ **Procedure** ⟩
9: *Mobile_BTS_HELLO():*
10: *BTS_To_BSC_HELLO_BSC():*
11: *BSC_To_BTS_TRACK_M():*
12: *BTS_To_Mobile_TRACK_M():*
13: *Mobile_To_BSC_RETRACK_M():*
14: *computeDistance(count, timeDiff):*
15: *BTS_To_BSC_INTERMEDIATE_TRACKM():*
16: *BTS_To_Mobile_TRACKS_M():*
17: *Mobile_To_BTS_RETRACKS_M():*
18: *Elimination(matrix[r][c]):*
19: *mobileCoordinates(d0, d1, d2):*
20: ⟨ **Main Algorithm** ⟩
21: **IF** *(Mobile_BTS_HELLO(msgType, msgLength, MobileID, data))*
22: *BTS_To_BSC_HELLO_BSC(msgType, msgLength, MobileID, data, S_slaveID):*
23: *Number of communication Count=1*
24: *Communication starting time T1;*
25: *BSC_To_BTS_TRACK_M(msgType, msgLength, MobileID, data, S_slaveID, flag, BSCID);*
26: *BTS_To_Mobile_TRACK_M(msgType, msgLength, MobileID, data, S_slaveID, flag, BSCID);*
27: *Mobile_To_BTS_RETRACK_M(msgType, msgLength, MobileID, data, S_slaveID, flag, BSCID);*
28: *Communication ending time T2;*
29: *timeDiff =T2-T1;*
30: *computeDistance(count,timeDiff);*
31: *for i ← 1 to 2*
32: *Count++;*
33: *Communication starting time T1;*
34: *BTS_To_BSC_INTERMEDIATE_TRACKM(msgType, msgLength, MobileID, data, flag, S_slaveID, BSCID, slaveID);*
35: *BTS_To_Mobile_TRACKS_M(msgType, msgLength, MobileID, data, flag, S_slaveID, BSCID, slaveID);*
36: *Mobile_To_BSC_RETRACKS_M(msgType, msgLength, MobileID, data, flag, S_slaveID, BSCID, slaveID);*
37: *Communication ending time T2;*
38: *timeDiff=T2-T1;*
39: **endfor**
40: *computeDistance(count, timeDiff);*
41: *MobileCoordinates(d0, d1, d2)*
42: **endif**

---

equations 4, 5, 6, 7 and 8. Assuming a model where a BSC controls many base stations, a mobile node sends a request for lookup services, such as a list of nearest hotels, by querying to servicing base station. The BSC finds the exact location of mobile node M and then responds to mobile node M for the requested query. The topology for mobile nodes M coordinates calculation is shown in the Figure 6; each step is labelled and their description is given below.

Step 1: The mobile user M initiates the process by sending hello packet to the nearest BTS, let's say B_0. The Base Station B_0, on receiving the message packet, forwards that message to the BSC by appending its own ID.

Step 2: On receiving hello packet from mobile user M, BSC sends TRACK M message to mobile user M through B_0 appending B_0 master-ID at the end of the message received from mobile user M. As the retransmission starts, BSC notes its local time instant t_0.

Step 3: BSC sends an INTERMEDIATE TRACKM packet to nearest base station that are called slave base station (in this case, named B_1 and B_2), appending the master-ID and the slave-ID at the end of the hello packet. B_0 also notes the local time t_{01} at the start of sending the packet.

Step 4: B_1 and B_2 each send TRACKS M message to mobile node M and note their receiving and retransmitting times (i.e. T_1, T_2).

Step 5:   As soon as a mobile detects that the type of the message it is receiving is TRACK M (or TRACKS M), it immediately notes its local time stamp, receives the whole packet, checks the mobile-ID field and if it is its own ID. The mobile retransmits the message modifying it into a RETRACK M (or RETRACKS M) message and sends to BSC; otherwise, it simply ignores the packet. Then BSC calculates the distance of mobile node.

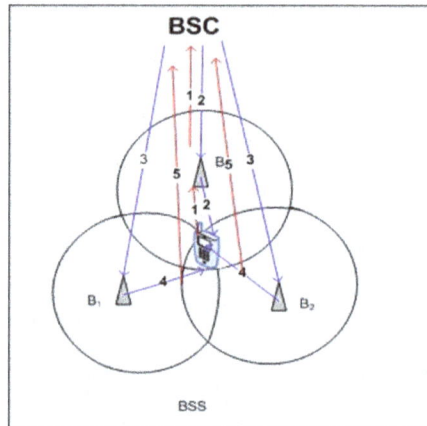

Figure 6. Node Localization in CDBA Scenario

4. Simulation Results

To validate the performance of proposed localization method, we implemented the algorithm in NS-2[3]. There are three base stations and mobile nodes. Mobile nodes initiate the request for services they need in their vicinity. There are three control messages in both scenarios, and nine communication messages in DDBA and eleven in CDBA. The complexity of localization algorithm is calculated with the sum of hello packet initialization and the node distance calculation from nearest BTS. Suppose our mobile node M_1 is in the BSC whose ID is 111. M_1 establishes a connection to base station BTS_1 by sending a hello packet, this base station is named BTS1 (coordinates (1, 2)). When BTS1 communicates with its two nearest base stations (called slave base stations) named BTS2 (coordinates (4, 6)) and BTS3 (coordinates (9, 8)), for the mobile node's location. BTS1 sends all data to ID 111, which replies with the mobile node's coordinates (for this example M1 (coordinates (0.922827, 7.43964)) and distance of the mobile node (e.g. 5.129 km). The average system time spent on localization of ten mobile nodes is 26.8 seconds as illustrated in Figure 7a; in CDBA, the average system time for same number of nodes is 32.6 seconds, as illustrated in Figure 7b. Figure 7a shows algorithm complexity for CDBA approach for 10 nodes where database maintenance on each base station is required, therefore, the query response takes less time. The query processing time in the CDBA approach (where database is only maintained at the server) is shown in Figure 7b and, to resolve query, 11 communication messages are required. The CDBA approach reduces the cost of database maintenance at each base station with little delay and two additional communication messages.

---

[3] http://isi.edu/nsnam/ns/

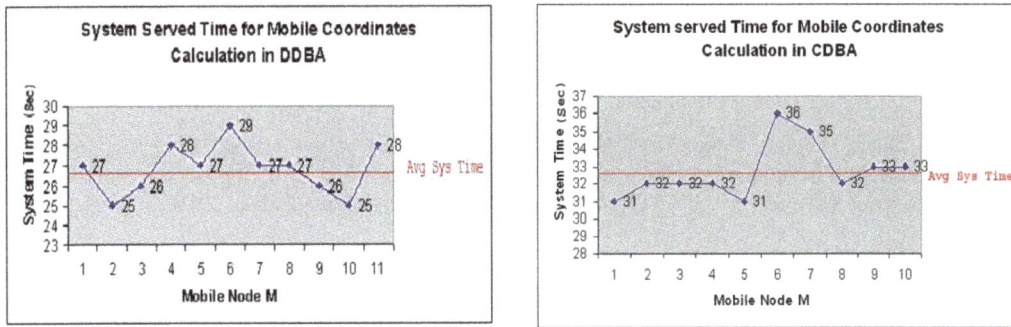

A. DISTRIBUTED DATA BASE APPROACH (DDBA)   B. CENTRALIZED DATA BASE APPROACH (CDBA)

Figure 7. Algorithm Complexity in (DDBA) and (CDBA)

5. Conclusion

Location base services (LBS) are developed using the information specific to a location. With the immense increase in the use of mobile phones, it would be a real advantage for a cellular company to provide LBS to their consumers. This has become the hottest issue today and many mobile companies are trying to find different ways to implement LBS in GSM network. To provide LBS, it is important to find the exact location of mobile node in cellular network. In this paper we presented the localization methods and simulated the in two scenarios. . Our approach of node localization is based on hello message delay (sending and receiving time) and coordinates information of BTSs, and hence locates the node location across the cellular network. The method is evaluated in two different scenarios. In the first scenario (DDBA) where the coordinates of neighboring BTS of serving BTS are maintained on each BTS, while in the second scenario, the centralized data base approach (CDBA), the coordinates of all BTS are maintained in each BSC. The promising benefit of this approach is that user doesn't have to carry special devices, There is not special hardware upgrade for the service providers.

REFERENCES

[1]. Marco Anisetti, Claudio A Ardagna, Valerio Bellandi, Ernesto Damiani, and Salvatore Reale. Advanced localization of mobile terminal in cellular network. *International Journal of Communications, Network and System Sciences.* 1(1):95–103, 2008.

[2]. Lhom Edouard, Frattasi Simone, Figueiras Joao, and Schwefel Hans-Peter. Enhancement of localization accuracy in cellular networks via cooperative ad-hoc links. *In Proceedings of the 3rd international conference on Mobile technology, applications & systems, Mobility* USA, 2006. ACM.

[3]. Mayorga Carlos Leonel Flores, Della Rosa Francescantonio,Wardana Satya Ardhy, Gianluca Simone, Raynal Marie Claire Naima, Joao Figueiras, and Simone Frattasi. *Cooperative Positioning Techniques for Mobile Localization in 4G Cellular Networks*, pages 39–44. IEEE, 2007.

[4]. Sayed A H, Tarighat A, and Khajehnouri N. Network-based wireless location: challenges faced in developing techniques for accurate wireless location information. *Signal Processing Magazine, IEEE*, 22(4):24–40, 2005.

[5]. A Hartl. A provider-independent, proactive service for location sensing in cellular networks. In *GTGKVS Fachgesprch (Online Proceedings)*, 2005.

[6]. S Kiran, M Bhoolakshmi, and G Varaprasad. Algorithm for finding the mobile phone in a cellular network. *International Journal of Computer Science and Network Security*, 7(10):306–310, 2007.

[7]. Sinha Koushik and Das Nabanita. Exact location identification in a mobile computing network. *In Proceedings of the 2000 International Workshop on Parallel Processing, ICPP '00*, pages 551–, Washington, DC, USA, 2000. IEEE Computer Society.

[8].   Santosh Pandey and Prathima Agrawal. A survey on localization techniques for wireless networks. *Journal of the Chinese Institute of Engineers*, 29(7):1125–1148, 2006.

[9]. Francesca Lo Piccolo. A new cooperative localization method for UMTS cellular networks. *In Proceedings of the Global Communications Conference, 2008. GLOBECOM* USA, pp 2383–2387

[10]. Qing Zhang, Chuan Heng Foh, Boon-Chong Seet, and Alvis Cheuk M Fong. Applying springrelaxation technique in cellular network localization. *In 2010 IEEE Wireless Communications and Networking Conference, WCNC 2010,* Proceedings, Australia, pages 1–6, 2010.

8

PROVIDE A MODEL FOR HANDOVER TECHNOLOGY IN WIRELESS NETWORKS

Abbas Asosheh[1], Nafise Karimi[2] and Hourieh Khodkari[3]

[1]Faculty of Technical Engineering, Tarbiat Modares University, Tehran, Iran
asosheh@modares.ac.ir
[2] Faculty of Technical Engineering, Tarbiat Modares University, Tehran, Iran
n.karimi@modares.ac.ir
[3] Faculty of Technical Engineering, Tarbiat Modares University, Tehran, Iran
khodkari@gmail.com

ABSTRACT

Fast Handovers for the MIPv6 (FMIPv6) has been proposed to reduce the Handover latency, in the IETF. It could not find the acceptable reduction, so led to more efforts to improve it and however the creation of multiple Handover methods in the literature.

A stable connection is very important in mobile services so the mobility of device would not cause any interruption in network services and thus mobility management plays a very important role. Mobile IPv6 has become a general solution for supporting mobility between different networks on the internet which a flawless connection needs to be managed properly.

In order to select the appropriate method‹ in this paper, all the proposed methods have been classified according to the identified performance metrics. Call blocking probability, Handover blocking probability, Probability of an unnecessary handover, Duration of interruption *and delay, as the most* important Handover algorithm *performance metrics are* introduced.

The AHP method will be deployed to weight the metrics in a sample topology according to the selected sound application. Then the TOPSIS method will be employed to find the appropriate Handover algorithm.

KEYWORDS

Handover, Handover Performance Metrics, FMIPv6, AHP Method, TOPSIS Method.

1   Introduction

IPv6 is a next generation network protocol, which was standardized to take the place of current protocols. This protocol will become the infrastructure of the next generation internet and in comparison with IPv4, it has improved dramatically in these areas: security, dynamism, convergence, scalability and was standardized in 1990s by IETF.[1] Integrated management in next generation network provides management functions for NGN resources and maintains connections between management plans themselves and other NGN renounces or services.[1] MIPv6 is seen as the de facto standard for mobility management in next generation networks (NGN) with IPv6 nodes.[2]

A management framework is needed in order to improve the costumer service satisfaction and simultaneously decrease the operator expenses using new technology, business models and new functional methods. One of the available services included in next generation networks is the possibility of communication between different devices and connections among fixed networks and mobile ones or wired and wireless networks. Such service requires a secure and reliable environment and to gain more efficient results it must be used with a proper management framework.[1]

The handover process happens when the MN(Mobile Node) moves from one access medium to another, and it should accomplish three operations: movement detection, new CoA(Care-of Address) configuration, and BU(Binding Update).[3] To make a MN stay connected to the Internet regardless of its location, mobile IPv6 is proposed as the next generation wireless Internet protocol. This is achieved primarily through using CoA to indicate the location of the MN. Although the Mobile IPv6 protocol has many promising characteristics and presents an elegant mechanism to support mobility, it has an inherent drawback. That is, during a handover process, there is a short period that the mobile node is unable to send or receive packets because of link switching delay and IP protocol operations.[4] This handover delay is intolerable for most applications. Proposed methods, mostly with study on most effective parameters in improving the QoS(Quality of Service) , including improve delay ,jitter and packet lost parameters are trying to improve the performance of Handover. But regardless of categories, in different conditions, the proposed methods will not enough performance, and a pretreatment is necessary to distribute the criteria in various classes having the same characteristics e.g. delay and jitter.[5]

A stable connection is very important in a mobile network so the mobility of device would not cause any interruption in network services. It shows the importance of the mobility management role. To determine the parameters that affect the performance of handover, classification of existing methods is required. It is also necessary to determine handling handover procedures. After identifying the parameters that can affect the efficiency of handover, choosing the appropriate algorithm can be done by using Multi-Criteria Decision Making Methods.

When looking on a handover from an architectural point of view there are two different types, vertical and horizontal. The horizontal handover is a handover between base stations belonging to the same type of network technology while the vertical handover is made between base stations attached to different network technologies.[6] MIH framework is a standard being developed by IEEE802.21 which proposes to enable handover between heterogeneous networks.[7]

From the perspective of geographical, mobility management solutions are divided in to two categories: macro-mobility and micro-mobility solutions. The mobility between two network domains known as macro-mobility and between the subnets in a domain known as micro-mobility. Several micro protocols have been proposed, which include HAWAII (Handover-Aware Wireless Access Internet Infrastructure)[8], CIP(Cellular IP)[9], HMIP (Hierarchical MIP)[10], IDMP (Intra-Domain Mobility Management).[11]

Due to the time of connection to new access point and its better management, three types of Handover are defined. In the hard handover scheme the MN changes its point of attachment with a short interruption of service. The old link is released and a new one created at the new BSs. The time the system needs to set up the path is referred to as the network response time. If the old radio link is broken up before the network completes the setup, the connection is dropped even if there are channels available in the cell.[12] Therefore this method is called brake before make.[13]

The seamless handover is based on the concept of changing between cells using the old and the new connection simultaneously with only one of them being active. Data is broadcast via both links. The old link stays active as long as the new path is activated. In comparison to the hard handover the seamless approach is more reliable since the old link is release after a new one has been established. However the utilization of two links during the handover phase degrades the number of available channels, which has a negative impact on the number of users that can be carried.[12]

The soft handover allows a transient phase during which multiple links can be used for communication simultaneously with all of them being active - which has the advantage that if one link fails the MN can communicate using the remaining links -. Soft handover can be used to extend the time that is available to make a handover decision without any loss of QoS. This allows reduction of the service interruption to a minimum when changing between cells. However in addition to limiting the efficient use of the frequency spectrum, this results in high data overhead since packets are transmitted on all links.[12]

When looking on a handover from layer point of view there are different types, The sub network layer, network layer, transport layer, session layer and application layer, that the SCTP(Stream Control Transmission Protocol), SLM(Session Layer Mobility Management) and SIP(Session Initiation Protocol) Handover procedures are examples of transport, sessions and application layer, respectively.[14],[15]

In the literature, handover performance metrics in order to select handover algorithm is as follows: Call blocking probability, Handover blocking probability, Handover probability, Call dropping probability, Probability of an unnecessary handover, Rate of handover, Duration of and Delay.[16],[17]

A number of procedures for handling handoffs have been proposed in the literature. A common handoff priority scheme is one in which a specified number of channels is set aside for the exclusive use of handoffs. The number to be set aside can be made adaptable with traffic intensity to satisfy a given handoff dropping/blocking probability combination. This priority strategy is often termed a guard-channel approach. Another procedure proposed in the literature is one in which neighboring cells send each other periodically an indication of their channel utilization. By predicting ahead, a given cell can determine the chance of a newly admitted call being denied service in a neighboring cell if it is subsequently handed off. If that probability turns out to be above a given threshold, it is better to deny service to the new call in the first place. Calculations indicate that this strategy provides an improvement over the guard-channel scheme, but it does require periodic communication between cells. Other simple scheme is that of buffering handoff calls up to some maximum time if no channel is initially available. The handoff dropping probability does of course reduce as a result, at the cost of a delay in continuing service. If this delay is not too high, it may be acceptable to the participants in an ongoing call.[18] In this paper the Guard-the channel scheme has been studied.

In the related work session, examples of algorithms in the literature have been studied. In the next session, the proposed methodology has been introduced. Then in **implementation and evaluation** Session, performance metrics for these algorithms are calculated and optimal algorithm has been found between them.

2   Background and related Works

Handover algorithms are classified from different view. To reduce the handover latency, two categories of protocols have been proposed. One focuses on the change in network architecture such as HMIP and IDMP. The other focuses on the mechanism to reduce latency by MN and AR(Access Router) themselves, hence change in design, such as fast handover. In this paper, Examples of each class of handover is considered so that change in design or architecture is evident.

2.1   Change in design

In design change process, the characteristics of MIPv6 (Mobile Internet Protocol version 6) are implemented to improve the efficiency parameters. Some important protocols as fast handover enhanced fast handover and seamless MIPv6 will be discussed as follows.

2.1.1   Fast Handover Protocol

The protocol enables an MN to do movement detection and create nCoA(New CoA), by providing the new access point and the associated   subnet prefix information when the MN is still connected to its   current subnet[19]. Unlike in FMIPv6 algorithms in MIPv6, L2 handover should be done before L3 handover. Handover in layer 2 includes: channel scanning, association and authentication.[20]

In FMIPv6 to prevent the packet loss, a bidirectional tunnel between PAR and NAR is established. the binding updates to the HA and CN(Correspondent Node) are performed after the time point when the MN is IP-capable on the new subnet link.[3] Because of this, the MN communicates with the CN directly via the NAR, before completing the BU, using this tunnel in a very late time. Figure 1 shows the messages exchanged during FMIPv6.

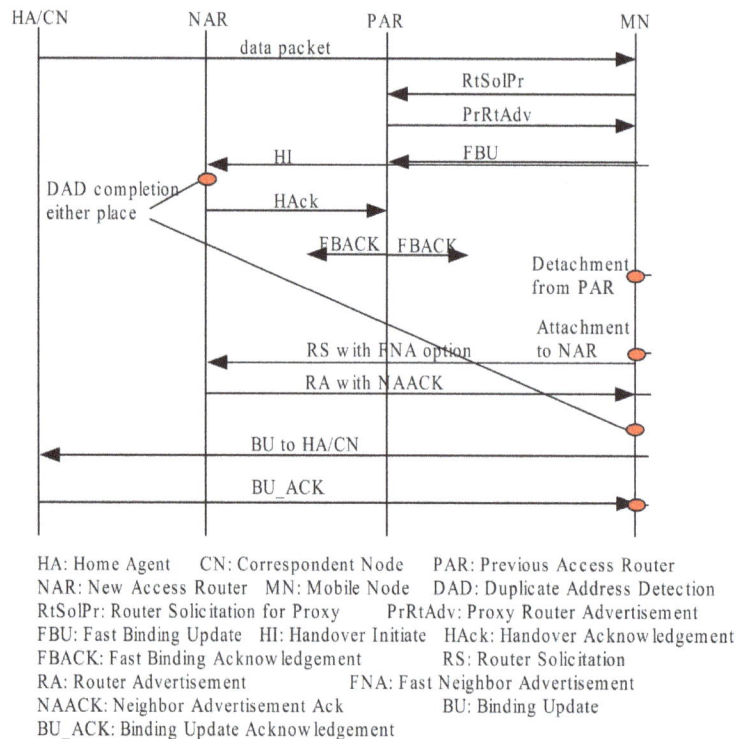

HA: Home Agent   CN: Correspondent Node   PAR: Previous Access Router
NAR: New Access Router   MN: Mobile Node   DAD: Duplicate Address Detection
RtSolPr: Router Solicitation for Proxy   PrRtAdv: Proxy Router Advertisement
FBU: Fast Binding Update HI: Handover Initiate HAck: Handover Acknowledgement
FBACK: Fast Binding Acknowledgement   RS: Router Solicitation
RA: Router Advertisement   FNA: Fast Neighbor Advertisement
NAACK: Neighbor Advertisement Ack   BU: Binding Update
BU_ACK: Binding Update Acknowledgement

Figure 1: messages exchanged during FMIPv6[3]

2.1.2   Enhanced Fast Handover Protocol

In EFMIPv6, LI has stated that, unlike the FMIPv6 the nCoA generation and DAD procedure can be performed before handover starts. At the same time, that when nCoA is informed to PAR, the handover to the new access point will definitely happen. Therefore, It is known that the binding update to the HA/CN can be performed at the time point when the new CoA is known by PAR. Also It has allowed that new AR construct a new CoA, perform DAD for the

MN and store this new CoA to the nCoA table when anticipating that a handover for an MN is about to happen. At the same time, to reduce the registration latency in the binding update, the binding update to the HA/CN will be performed after the PAR knows the nCoA.[2] To describe the optimized scheme clearly, the detailed timing graph for the enhanced scheme is provided in Figure 2.

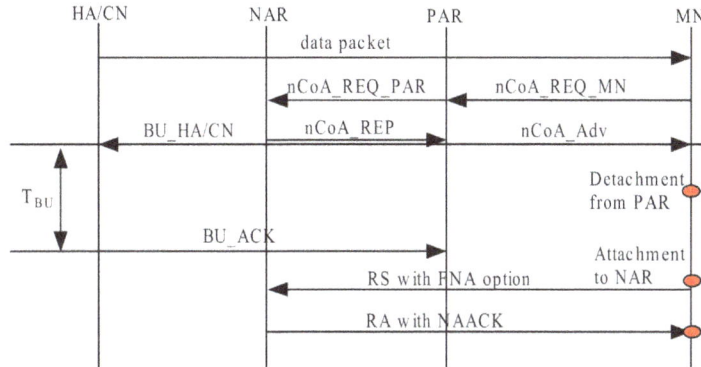

Fig.2: messages exchanged during EFMIPv6[3]

2.1.3 Seamless Mobile IPv6 Protocol

SMIPv6 makes use of users' mobility patterns to predict the cell where the next handover will occur. Based on this knowledge, the protocol updates all its CNs with its new address before leaving its current network and entering a new one. Furthermore, using layer 2 information, SMIPv6 is able to predict the exact time the handover will occur. Using its mobility pattern, a mobile node will send update messages to its correspondent nodes only when a change of network is in sight. Normally, these updates occur at regular intervals. SMIPv6's mobility management model is divided into two components: a mobility pattern learning module implemented in each mobile node and a mobility management protocol executed by all entities in the network. The L3 handover is performed upon the reception of a layer 2 trigger. The trigger contains identification information about the new access point. Based on this identifier, a mobile node can verify if this AP(Access Point) is part of its mobility profile. If it is, the NCoA based on the sub network's prefix is created without waiting for the RAs to be sent by the AR. Upon the completion of the address creation phase, the MN sends BUs containing its NCoA to all its CNs as well as to its HA. Then in this algorithm delay of RtSolPr and PrRtAdv messages exchange and delay of BU are deleted. Fig. 3 shows the messages exchanged during SMIPv6 handover.[21]

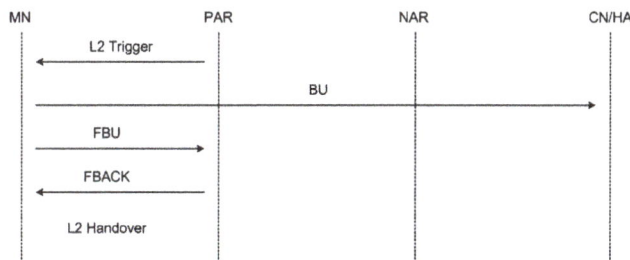

Figure 3 : messages exchanged during SMIPv6[21]

2.2   Change in architecture

In architecture change process, one or more entities to improve performance are added to the existing architecture. For example in HMIPv6, one or more MAP(Mobility Anchor Point) are added to the network architecture or in[2] functional network entity, called the handover coordinator (HC), to the IP core to be shared and utilized by the internetworking heterogeneous wireless networks (i.e. both source and target networks) in a PMIPv6 micro-mobility domain.

2.2.1   Hierarchical MIPv6 protocol

This method is design for handover delay problem when the HA or CN is located geographically far away from the MN and when a mobile node moves in a small coverage area (micro-mobility).[10] Authenticating binding updates requires approximately 1.5 round-trip times between the mobile node and each correspondent node. In addition, one round-trip time is needed to update the Home Agent; this can be done simultaneously while updating correspondent nodes. For these reasons a new Mobile IPv6 node, called the Mobility Anchor Point, is used and can be located at any level in a hierarchical network of routers, including the AR. The MAP will limit the amount of Mobile IPv6 signaling outside the local domain The introduction of the MAP provides a solution to the issues outlined earlier in the following way:

- The mobile node sends Binding Updates to the local MAP rather than the HA (which is typically further away) and CNs.

- Only one Binding Update message needs to be transmitted by the MN before traffic from the HA and all CNs is re-routed to its new location. This is independent of the number of CNs that the MN is communicating with.[7]

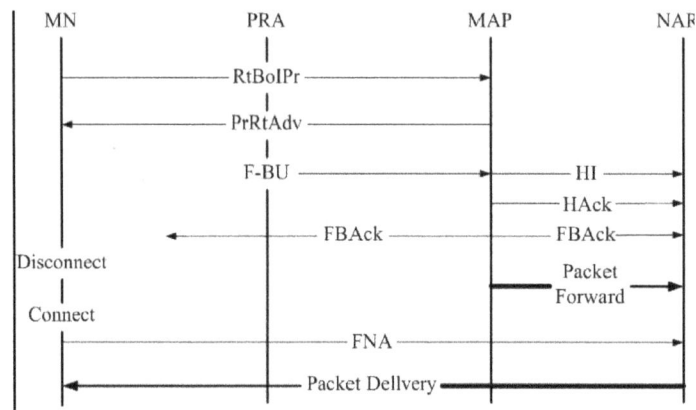

Figure 4 : messages exchanged during HMIPv6 [22]

3   Proposed methodology

The proposed methodology to choose the best and proper protocol in different situations includes four steps. It should be noted that voice packet as an example, is used in data analyzing.

3.1   Determine handover class

In the first step the class of studied handover algorithms should be determines. In the proposed methodology, handover algorithms occurred in the network-layer that can be run horizontally are compared. Determining the time of connection to the new access point, is important to determining the number of channels used in the algorithm. Determining the geographic scope for the studied algorithms is important to feasibility of change in design or architecture.

3.2   The performance metrics calculation

After determine the class of each algorithm, in the second step, according to the topology used in Figure 5, the delay of each step should be calculated. Processing delay of a node n, is assumed equal to T. All delays on wired links hold value f except for link (N1, N2) which holds value F. This link represents both local and global mobility and in HMIPv6 study, determine domain. Each radio Link will have a delay equal to d. L2 Handover delays hold a value equal to h. It is necessary to note that, except processing delay and propagation delay, other delays are ignored. But other delay scan be easily calculated or based on Cisco recommends[23], using worst case in design. Due to the importance of DAD delay, in proposed methodology, this delay is calculated separately. This delay in the worst case that referred in MIPv6 reference algorithm[24], is intended $D= 1\ s$ and is add to total signaling delay of handover algorithm that is not adjusted or deleted on them. It can be seen the calculating details of performance parameters for the mentioned protocols in the following. To become more transparent, the results of the MIPv6 reference algorithm also have been studied.

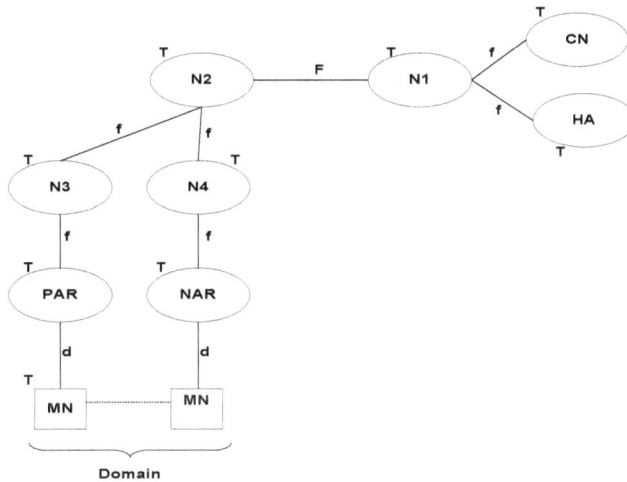

Fig.5: Proposed topology for evaluating handover algorithms performance

3.2.1   Case 1: MIPv6 handover

Fig. 6 shows messages exchanged during an MIPv6 handover. Table 1 points out the chronological details of messages exchanged as well as the analytical delay found for each event. The last packet through the PAR was received at t = T. The first packet through the NAR was received at $t = 36T+22f+6d+h+2F$. Hence, the total handover delay amounts to: $t = 35T+22f+6d+h+2F$

From the moment where the MN initiates the handover to when the CN sends its packets to the new NCoA, packets sent to the previous CoA are lost. The exact number of packets lost can be calculated using the following formula: $(35T+22f+6d+h+2F)*\ Throughput$.

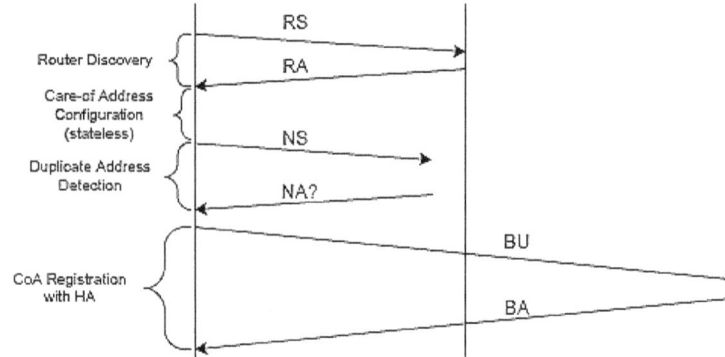

Figure 6 : MIPv6 signaling[25]

The signalization latency starts precisely when the mobile node receives the RA and ends when the BU is received by the MN's correspondent node. Thus, the total value of the signalization delay is equal to: $t = 30T + 19f + 5d + h + F + D$.

Table 1:. Chronological details of an MIPv6 handover

Delay	Event	Time
T	L2 Trigger	$t = 0$
$6T+4f+d$	RS	$t = T$
$6T+4f+d$	RA	$t = 6T+4f+d$
$6T+4f+d$	NS	$t = 12T+8f+2d$
$6T+4f+d$	NA	$t = 18T+12f+3d$
H	L2 Handover	$t = 24T+16f+4d$
$3f+F+d+6T$	BUs sent to HA/CN	$t = 24T+16f+4d+h$
$3f+F+d+6T$	Packets sent by CNs@NCOA	$t = 30T+19f+5d+h+F$
	Packets sent by CNs are received	$t = 36T+22f+6d+h+2F$

3.2.2   Case 2: FMIPv6 handover

Figure 1 shows the messages exchanged during an FMIPv6 handover. Table 2 points out the chronological details of messages exchanged as well as the analytical latency found for each event. The last packet through the PAR was received at $t = 4d + 8f + 18T$. The first packet through the NAR was received at $t = max\ t = max\ (6d + 8f + h + 22T,\ 12f + 4d + 23T)$. Hence, the total handover delay is given by: $Max\ (2d + h + 4T,\ 4 + 5T)$

No packets are lost since the PAR starts rerouting packets toward the NAR before proceeding with the handover .All packets received in the meantime, that is, before the L2 handover is performed, are stored in a buffer thus ensuring that no packets are lost. Following the reception of the FNA, all packets are sent to the MN. Although packet losses are null, the signalization delay is quite high. The L2 trigger is only received by the MN at time and the CN and HA receive their respective BUs at $t = 11f + 7d + F + h + 28T$. Thus, the signalization delay is equal to: $14f + 6d + 2F + h + 30T + D$

3.2.3   Case 3: SMIPv6 handover

Figure 3 shows the messages exchanged during an SMIPv6 handover. Table 3 presents the chronological details of messages exchanged as well as the analytical delay found for each event. The last packet going through the PAR is received at $t = 2d + 5T$. The first packet passing through the PAR is received at t = min = *min(2d + 4f + 9T, 2d + 6f +2F + 12T)*. Hence, the handover delay is equal to: *4f + 4T*

There are no packets lost since the PAR reroutes packets through the NAR before performing the actual handover. Indeed, the MN joins the new network before packets sent by the CNs or rerouted by the PAR reach the new network. the first rerouted packet arrive at *4f + 3d + 6T* and that the MN joins the new *network at 2d + h + 5T* Thus, if we subtract the time the rerouted packets arrive from the time the MN reaches its new network, we get 4f _ h + T, a positive value since h is near 0 (L2 handover delay) and T is relatively small. The signalization delay equal to: 3f + F + d + 5T

Table 2: Chronological details of a FMIPv6 handover

Delay	Event	Time
T	L2 Trigger	*t = 0*
d+2T	RtSolPr	*t = T*
d+2T	PrRtAdv	*t=d+2T*
d+2T	FBU	*t =2d + 4T*
4f + 5T	HI	*t =3d + 6T*
4f + 5T	HACK	*t =3d +4f + 11T*
d+2T	FBACK	*t =3d +8f + 16T*
4f + 5T	Packets are rerouted through	*t =3d +8f + 16T*
h	L2 Handover	*t =4d +8f + 18T*
d+2T	FNA	*t =4d +8f + h + 18T*
d+2T	FNA -ACK	*t =5d +8f + h + 20T*
3f + F + d + 6T	BUs sent to HA/CN	*t =6d +8f + h + 22T*
d+2T	PAR sends packets to MN	*t = max (5d +8f + h + 20T, 12f +3d + 21T)*
	Packets are received by MN	*t = max (6d +8f + h + 22T, 12f +4d + 23T)*
	BUs are received by CNs	*11f +7d + F + h + 28T*
	Bus-ACK are received by MNs	*t =14f +8d + 2F + h +*

Table 3: Chronological details of a SMIPv6 handover

Delay	Event	Time
T	L2 Trigger	*t = 0*
d + 2T	FBU	*t = T*
3f + F + d + 6T	BU	*t = T*
d + 2T	FBACK	*t = d + 3T*
4f + d + 6T	Rerouting of packets	*t = d + 3T*
h	L2 Handover	*t =2d + 5T*
3f + F + d + 6T	Packets sent by CNs@NCOA	*t = d +3f + F + 6T*
	Rerouted packets are received	*t = 4f +2d + 9T*
	Packets sent by CNs are received	*t = 6f +2F +2d + 12T*

3.2.4 Case 4: EFMIPv6 handover

Figure 2 shows the messages exchanged during an EFMIPv6 handover. Table 4 presents the chronological details of messages exchanged as well as the analytical delay found for each event. Like as fast handover No packets are lost, then the handover delay is equal to: *max (3f+h+F+5T,4f+3T).*The signalization delay equal to:*3d+11f+21T+F+h*

3.2.5   Case 5: HMIPv6 handover

Figure 4 shows the messages exchanged during a HMIPv6 handover. Table 5 presents the chronological details of messages exchanged as well as the analytical delay found for each event. Like as fast handover No packets are lost, then the handover delay is equal to: *max (2d + h + 6T, 2f +d+ 5T).* The signalization delay equal to: *10f +3d + h + 19T+D*

Table 4: Chronological details of a EFMIPv6 handover

Delay	Event	Time
T	L2 Trigger	$t = 0$
$d + 2T$	nCoA-REQ-MN	$t = T$
$4f + 5T$	nCoA-REQ- PAR	$t = d + 2T$
$4f + 5T$	nCoA-REP	$t =d+4f+7T$
$3f + F + 5T$	BUs sent to HA/CN	$t =d+8f+12T$
$d+2T$	nCoA-Adv	$t =d+8f+12T$
$4f + 5T$	Packets are rerouted through PAR	$t =d+8f+12T$
$3f + F + 5T$	BU_ACK	$t =d+11f+17T+F$
H	L2 Handover	$t =d+11f+17T+F$
$d+2T$	FNA	$t =d+11f+17T+F+h$
	Rerouted packets are received	$t =2d+12f+17T$
	Packets are received by MN	$t=max(2d+11f+19T+F+h, 2d+12f+17T)$
$d+2T$	NAACK	$t =2d+11f+19T+F+h$
	NAACKs are received by MN	$t =3d+11f+21T+F+h$

Table 5: Chronological details of a FHMIPv6 handover

Delay	Event	Time
T	L2 Trigger	$t = 0$
$d+2T$	RtSolPr	$t = T$
$d+2T$	PrRtAdv	$t = d + 2T$
$d+2T$	FBU	$t =2d + 4T$
$4f+5T$	HI	$t =3d + 6T$
$4f+5T$	HACK	$t =3d +4f + 11T$
$d+2T$	FBACK	$t =3d +8f + 16T$
$4f + 5T$	Packets are rerouted through PAR	$t =3d +8f + 16T$
h	L2 Handover	$t =4d +8f + 18T$
$d+2T$	FNA	$t =4d +8f + h + 18T$
$d+2T$	FNA -ACK	$t =5d +8f + h + 20T$
$2f +d+T$	BUs sent to MAP	$t =6d +8f + h + 22T$
	PAR sends packets to MN	$t = max (5d +8f + h + 20T, 12f +3d + 21T)$
	Packets are received by MN	$t = max (6d +8f + h + 22T, 12f +4d + 23T)$
	BUs are received by MAP	$t =10f +7d + h + 23T$
	Bus-ACK are received by MN	$t =12f +8d + h + 24T$

After calculating values of Packet loss ‹Handover Delay and Signaling Delay, using available formulas,[18] we can calculate Call blocking and Handover blocking probability.

3.3   Weighting the metrics based on AHP algorithm

Performance metrics in each method is obtained, the weight of these metrics should be allocated, till can use these metrics in MCDM methods. AHP, fuzzy AHP, fuzzy TOPSIS, TOPSIS methods respectively, are as most efficient MCDM Compensatory methods.[26]

The work of selecting the appropriate handover method in the literature[27, 28], AHP technique as a method of weighting the quantitative and qualitative criteria are considered.

3.4   The appropriate method according to TOPSIS

According to the literature[5] in the Fourth step, using TOPSIS algorithm among the various available handover methods, appropriate method is selected.

4   Implementation and evaluation

In this section the performance of FMIPv6 ‹EFMIPv6 ‹SMIPv6 and HMIPv6 will be evaluated according to the described methods in the previous section.

4.1   Handover class

Class of each method determine in table 6. In SMIPv6 protocol, to preparation and installation mobility pattern learning module on each node and planning and implementation of the mobility management protocol to the project cost will be added. In HMIPv6 protocol, to add a MAP, the cost will be added to the project.

Table 6: Classifying studied algorithms

Change in design/ architecture	support micro/macro mobility	Hard/Soft handover*	Algorithm Class
-------------------------	macro mobilitysupport	Hard	MIPv6
Change in design	support macro mobility	Soft	FMIPv6
Change in design	support macro mobility	Soft	EFMIPv6
Change in design	support macro mobility	Soft	**SMIPv6
Change in architecture	support micro mobility	Soft	***HMIPv6

*The algorithms are implemented as soft, only half of the channels are available.

4.2   The performance metrics calculation

To calculating performance metrics, the following conditions are considered:

Speed of mobile node: 60 km/h, average call holding time is 300 sec and cell radius is r = 10 km. There are ten channels in each cell that three channels are considered as guard channels. Using the above values, can be calculate Call blocking probability and Handover blocking probability. Also, we have:

Propagation speed on the wireless link is equal to $2*10^8$ m/s. Propagation speed on the wired link is equal to $3*10^8$ m/s. the length of wireless link d= 500 m, the length of f wired link is f=35m and the length of F wired link is F= 2 km. Then the propagation delays on above link are 2.5 μsec ‹0.12 μsec and 6.7 μsec respectively. Given ADPCM, G.726as a coder, the processing delay at each node in best case is equal to T= 2.5 ms .[23] The cost for reference algorithm is considered as 1000. The results of the algorithms are given in Table 7 and 8.

Table7: The results of evaluating algorithms as parametric

Algorithm	Packet lost	Handover Delay	Call blocking probabilit	Handover blocking probabilit	Signaling Delay	Price
MIPv6	$35T+22f+6$ $d+h+2F$	$12T+6f+2d+h+2F$	$1.82*10^-3$	$6.74*10^-11$	$30T+19f+5d+h+2$ $F+D$	1000
FMIPv6	0	$max (2d + h +6T,$ $4f + 7T+d)$	0.56	$2.5*10^-5$	$14f +6d + 2F + h +$ $30T+D$	1000
SMIPv6	0	$4f + 4T$	0.56	$2.5*10^-5$	$3f + F + d + 5T+D$	1500
EFMIPv6	0	max $(3f+h+F+5T,4f+3$ $T)$	0.56	$2.5*10^-5$	$3d+11f+21T+F+h$	1000
HMIPv6	0	$max (2d + h + 6T,$ $2f +d+ 5T)$	0.56	$2.5*10^-5$	$10f +3d + h +$ $19T+D$	1500

Table 8: The results of evaluating algorithms as numerical

Algorithm	Packet lost	Handover Delay	Call blocking probability	Handover blocking probability	Signaling Delay	Price
MIPv6	0.0875310	0.0300191	$1.82*10^-3$	$6.74*10^-11$	1.0750281	1000
FMIPv6	0.0000000	0.0175029	0.56	$2.5*10^-5$	1.0750300	1000
SMIPv6	0.0000000	0.0100004	0.56	$2.5*10^-5$	1.0125095	1500
EFMIPv6	0.0000000	0.0125070	0.56	$2.5*10^-5$	0.0525155	1000
HMIPv6	0.0000000	0.015005	0.56	$2.5*10^-5$	1.0475087	1500

4.3   Weighting the metrics based on AHP

To weight to metrics, using experts' opinion. Finally, weight of each metric with respect to the output of the software is as follows:

```
Packet lost                     .546
Handover Delay                  .133
Call blocking probability       .050
Handover blocking probability   .145
Signaling Delay                 .095
Price                           .031
Inconsistency = 0.06
   with 0  missing judgments.
```

Fig.7: the weight of each metrics according to the expert choice software

4.4   The appropriate method according to TOPSIS

Decision matrix to select the optimal Handover algorithm, after calculating all the types of metrics shown in table 9.

Table 9: Decision matrix to select the optimal Handover algorithm

Algorithm	Packet lost	Handover Delay	Call blocking probability	Handover blocking probability	Signaling Delay	Price
MIPv6	0.00011854	0.00004912	$1.82*10^-3$	$6.74*10^-11$	1.00010318	1000
FMIPv6	0	0.000015	0.56	$2.5*10^-5$	1.00010508	1000
SMIPv6	0	0.00001048	0.56	$2.5*10^-5$	1.00002206	1500
EFMIPv6	0	0.00001956	0.56	$2.5*10^-5$	0.00006802	1000
HMIPv6	0	0.000015	0.56	$2.5*10^-5$	1.0000612	1500

Finally, the rating options are as follows:

1. EFMIPv6
2. SMIPv6
3. HMIPv6
4. FMIPv6
5. MIPv6

5   Conclusion

The handover process happens when the MN moves from one access medium to another, and it should accomplish three operations: movement detection, new CoA configuration, and BU. During handover period, the MN is unable to send or receive packets as usual. The length of this period which is called handover latency is very critical for the delay-sensitive and real-time services. To reduce the handover latency and increase its efficiency several methods have been proposed in the literature. In this paper, a methodology for choosing the appropriate algorithm between the existing methods is presented. It was clarified that BU and DAD signaling are critical points of handover algorithms then methods that try to improve this point, are successful in improving the overall effectiveness of Handover.

As expected, EFMIPv6 protocol is the best selection, because of eliminate DAD delay and reduce the delay of BU. The cost of the SMIPv6 algorithm is increased and the time required for BU signaling effectively reduced and time needed to exchange RtSolPr and PrRtAdv messages are deleted. Normally, in practical, Algorithms that have changed in design or architecture should be examined separately. In HMIPv6 algorithm, when the mobile node moves within a domain, If the change in topology in HMIPv6 and MAP or MAPs is/are adding, increase the cost of this algorithm should also be considered. Despite packet loss in the algorithms that use of hard handover, when traffic is low sensitivity to packet loss, weight of packet loss parameter in the AHP algorithm is reduced and due to the efficient use of bandwidth in these algorithms, their use is preferred. For evidence result, algorithms have been selected that, have obvious difference. But in methods that in which change in design or architecture are complex or similar, using proposed methodology is very effective.

Reference

1.   Asosheh.A, Khodkari.H., *"The integrated management of the challenges in next generation networks"*, in *Fourth National Conference on ICT*. 2010.

2.   Magagula, L.A., Chan, H.A., Falowo, O.E., *"Handover Coordinator for Improved Handover Performance in PMIPv6-Supported Heterogeneous Wireless Networks"*, in *Wireless Communications and Networking Conference (WCNC)*. 2010 Sydney, NSW p. 1 - 6

3.   Li, R., Li,J., Wu,K., Xiao,Y., Xie.J., *"An Enhanced Fast Handover with Low Latency for Mobile IPv6"*. IEEE Transactions on Wireless Communications, 2008. **7**(1): p. 334 - 342

4.   AI, M., CHEN,S., LI,W., SHI,Y., *"Design and Evaluation of A New Movement Detection and Address Configuration Method for supporting Seamless Mobile IPv6 Handover"*, in *International Conference on Wireless Communications, Networking and Mobile Computing*. 2007: Shanghai p. 1739 - 1744

5.   Mohamed, L., Leghris, C., Adib, A., *"A Hybrid Approach for Network Selection in Heterogeneous Multi-Access Environments"*, in *4th IFIP International Conference on New Technologies, Mobility and Security (NTMS)*. 2011. p. 1 - 5

6. Lax, M.C., X Dammander.A., "*WiMAX - A Study of Mobility and a MAC-layer Implementation in GloMoSim*", in *Computing Science*. 2006, Ume°a University: Sweden. p. 99.

7. Ma, C., Fallon, E., Qiao,Y., Lee, B.," *Optimizing Media Independent Handover Using Predictive Geographical Information for Vehicular Based Systems*", in *Fourth UKSim European Symposium on Computer Modeling and Simulation (EMS)*. 2010. p. 420 - 425

8. Ramjee, R., Varadhan, K., Salgarelli, L., Thuel, S.R., Wang,S.Y., La Porta, T. , "*HAWAII: A Domain-based Approach for Supporting Mobility in Wide-area Wireless Networks*". IEEE/ACM Transactions on Networking, 2009. **10**(3): p. 396 - 410

9. Campbell, A.T., Gomez, J., Kim, S., Valko, A.G., Wan,C-H., Turanyi, Z.R., "*Design, Implementation, and Evaluation of Cellular IP*". IEEE Personal Communications, 2000. **7**(4): p. 42 - 49

10. Soliman, H., Castelluccia,C. , El Malki,K., Bellier,L., "*RFC 4140 - Hierarchical Mobile IPv6 Mobility Management (HMIPv6)*". 2005, IETF.

11. Misra, A., Das, S., Dutta, A., McAuley, A., Das, S.K. ," *IDMP-based Fast Handoffs and Paging in IP-based 4G Mobile Networks*". IEEE Communications Magazine, 2002. **40**(3): p. 138 - 145

12. Bauer, C.I., Rees, S.J. , "*Classification of handover schemes within a cellular environment*", in *The 13th IEEE International Symposium on Personal, Indoor and Mobile Radio Communications*. 2002. p. 2199 - 2203.

13. Lin, Y.B., Pang,A.C., "*Comparing Soft and Hard Handoffs*". IEEE Transactions on Vehicular Technology, 2000. **49**(3): p. 792 - 798

14. Szabó, C.A., Szabó,S ., Bokor,L., "*Design Considerations of a Novel Media Streaming Architecture for Heterogeneous Access Environment*, in *workshop on Broadband wireless access for ubiquitous networking*". 2006: Alghero, Italy.

15. Eddy, W.M., "*At What Layer Does Mobility Belong?*" IEEE Communications Magazine, 2004. **42**(10): p. 155 - 159

16. Pollhi, G.P., "*trends in Handover Design*". IEEE Communications Magazine, 1996.

17. Poethi Boedhihartono, G.e.M., "*Evaluation of the guaranteed handover algorithm in satellite constellations requiring mutual visibility*". INTERNATIONAL JOURNAL OF SATELLITE COMMUNICATIONS AND NETWORKING, 2003: p. 163–182.

18. Schwartz, M., "*Mobile Wireless Communications*". 2005: Cambridge University Press.

19. Koodli, R., "*RFC4068 - Fast Handovers for Mobile IPv6*". 2005, Nokia Research Center.

20. An, Y.Y., Yae,B.H., Lee, K.W., Cho,Y.Z.,Jung,W.Y., "*Reduction of Handover Latency Using MIH Services in MIPv6*", in *20th International Conference on Advanced Information Networking and Applications*. 2006: Vienna, Austria p. 229-234.

21. Quintero, A., Pierre,S., Alaoui,L., "*A mobility management model based on users' mobility profiles for IPv6 networks*". Computer Communications, 2006. **30**(1): p. 66-80.

22.   Zheng, G., Wang,H., *"Optimization of Fast Handover in Hierarchical MobileIPv6 Networks"*, in *Second International Conference on Computer Modeling and Simulation*. 2010. p. 285-288.
23.   Cisco. *"Understanding Delay in Packet Voice Networks"*. 2008; Available from: http://kbase.cisco.com/paws/servlet/ViewFile/5125/delay-details.xml?convertPaths=1.
24.   Wei, A., Wei,G., Dupeyrat,G., *"Improving Mobile IPv6 handover and authentication in wireless network with E-HCF"*. International Journal of Network Management, 2009. **19**(6): p. 479–489.
25.   Xie, G., Chen,J., Zheng,H., Yang,J., Zhang,Y., *"Handover Latency of MIPv6 Implementation in Linux"*, in *Global Telecommunications Conference*. 2007: Washington, DC p. 1780 - 1785
26.   Vafaie, f., *"Measuring Efficiency of MADM methods with DEA"*, in *20th International Conference on Multiple Criteria Decision Making* 2009: china.
27.   Bi, Y., Huang,J., Iyer, P., Song, M., Song, J., *"An Integrated IP-layer Handover Solution for Next Generation IP-based Wireless Network"*, in *IEEE 60th Vehicular Technology Conference*. 2004. p. 3950 - 3954.
28.   Yang, S.F., Wu, J.S., Huang,H.H., *"A Vertical Media-Independent Handover Decision Algorithm across Wi-FiTM and WiMAX TM Networks"*, in *5th IFIP International Conference on Wireless and Optical Communications Networks*. 2008: Surabaya p. 1-5.

9

EFFICIENT MIXED MODE SUMMARY FOR MOBILE NETWORKS

Ahmed E. El-Din[1] and Rabie A. Ramadan[1]

[1] Computer Engineering Department, Cairo University
Cairo, Egypt

{Ahmed_ezzeldin@eng.cu.edu.eg, Rabie@rabieramadan.org}

ABSTRACT.

Cellular networks monitoring and management tasks are based on huge amounts of continuously collected data from network elements and devices. Log files are used to store this data, but it might need to accumulate millions of lines in one day. The standard name of this log is in GPEH format which stands for General Performance Event Handling. This log is usually recorded in a binary format (bin). Thus, efficient and fast compression technique is considered as one of the main aspects targeting the storage capabilities. On the other hand, based on our experience, we noticed that experts and network engineers are not interested in each log entry. In addition, this massive burst of entries can lose important information; especially those translated into performance abnormalities. Thus, summarizing log files would be beneficial in specifying the different problems on certain elements, the overall performance and the expected network future state. In this paper, we introduce an efficient compression algorithm based log frequent patterns. In addition, we propose a Mixed Mode Summary-based Lossless Compression Technique for Mobile Networks log files (MMSLC) as a mixed on-line and off-line compression modes based on the summary extracted from the frequent patterns. Our scheme exploits the strong correlation between the directly and consecutively recorded bin files for utilizing the online compression mode. On the other hand, it uses the famous "Apriori Algorithm" to extract the frequent patterns from the current file in offline mode. Our proposed scheme is proved to gain high compression ratios in fast speed as well as help in extracting beneficial information from the recorded data.

KEYWORDS. Logs, Compression, Frequent Patterns

1.INTRODUCTION

Cellular networks monitoring and management tasks are based on huge amounts of data that are continuously collected from elements and devices from all around the network [5][11]. Log files are considered the most famous storage file type for distributed systems like cellular networks. These files are collections of log entries and each entry contains information related to a specific event took place within the system. The importance of these files comes from the fact of using them to monitor the current status of the system, track its performance indicators, identifying frauds and problems to help in decision making in all aspects of the network [2][3].
Unfortunately, these files are recorded in a binary format (bin files) that needs huge effort to be parsed first, and then evaluated.

Initially, log files were used only for troubleshooting problems [8]. However, nowadays, it is used in many functions within most organizations and associations, such as system optimization and measuring network performance, recording users' actions and investigating malicious activities [2][3]. Unfortunately, these log files can be of order tens of gigabytes per day, and must be stored for a number of days as history. Thus, efficient and fast compression of these log files becomes very important issue facing the storage capabilities of the system. The essential problem falls not only in the storage but also in the speed of extracting meaningful information from this huge data in a reasonable time.

Traditional compression techniques such as Lempel-Ziv [6] and Huffman Encoding [7] exploit the structures at the level of consecutive bytes of data. Therefore, the drawback of applying those techniques to cellular network log files is that in order to access a single tuple, the entire data page must be accessed first. On the other hand, those techniques did not give importance to the dependencies and the correlations between the attributes of the same log entries or between those of different log files [7].

Several compression techniques developed further targeting log files problems, and falls into two categories which are : 1) Offline compression in which it compresses the log files after been totally stored and it is considered as the most famous compressing approach; 2) Online compression in which the log file is compressed on the fly. Although offline compression leads to high compression ratios, this technique needs to scan the log file multiple times before compression. At the same time, online compression methods, in most of the cases, produces lower compression ratios than the offline compression methods.

Data mining and knowledge discovery methods are considered promising tools for systems' operators to gain more out of their available data [5][9]. These methods build models representing the data and can be used further in the rapid decision making. For large and daily updated log files, it is difficult or almost impossible to define and maintain a priori knowledge about the system, as new installed network components trigger new events. Fortunately, there is still a possibility to use Meta information that characterizes different types of log entries and their combinations. "Frequent patterns (FP)" is considered the most famous type of Meta information [7][8].

FP capture the common value combinations that occur in the logs. This type of presentation is quite understandable for experts and can be used to create hierarchical views. These condensed representations can be extracted directly from highly correlated and/or dense data.

Experts and network operators are not interested in each log entry. At the same time, this massive burst of log entries can lose important information, especially those translated into performance abnormalities of certain element in the network. In addition, indeed, rare combinations can be extremely interesting for system monitors. Thus from the experts and monitors point of view, frequent patterns summary can help fast detection of many of the network problems and abnormalities in the components performance over its lifetime.

In this paper, we propose a new lossless compression scheme based on mining frequent patterns named Mixed Mode Summary-based Lossless Compression for Mobile Networks log files. Our scheme first uses the famous Apriori technique for mining frequent patterns, assigns unique codes according to their compression gain, and uses these codes in compressing the file. MMSLC exploits the high correlation between the consecutively recorded log files by introducing mixed online and offline compression modes. MMSLC uses the FPs extracted from previous log files in offline mode to compress the current log files in online mode. At the same time, it extracts the frequent patterns from the current log files in offline mode to be used in compressing the new log files.

Our offline compression achieves high compression ratio, provides fast summary of the log file holding the frequent patterns to be used in network monitoring, while being able to restore all details if needed. MMSLC works on the attribute/tupelo-level to exploits the semantic structures in the recorded data. The rest of this paper is organized as follows: Section 2 introduces the related work; GPEH log files structures and standards are elaborated in section 3; the frequent pattern generator algorithm is elaborated in section 4; MMSLC is introduced in section 5; MMSLC performance analysis is elaborated in section 6; finally this paper is concluded in section 7.

2   RELATED WORK

Jakub Swacha et. al. [1] described a lossless, automatic and fully reversible log file compression transformation. This transformation is not tuned for any particular type of logs, and is presented in variants starting from on-line compression variant to off-line compression one. They dealt with log files as plain text files in which every line corresponds to a single logged event. Thus, for the first variant, as the neighbouring lines are very similar in structure and content, this variant replaces tokens of a new line with references to the previous line, on byte level. However, single log often records events that may belong to more than one structural type. Thus, similar lines are not always blocked, but they are intermixed with lines differing in content or structure. The next variant searches the block, for each new line, for the line that returns the longest initial match is used as reference line. These variants are on-line schemes and addressed the local redundancy. However, this kind of compression affects the final compression ratio as it depends on the region around the line the algorithm and it searches for the reference. For off-line variant, it handles words frequency throughout the entire log. The dictionary has an upper limit and if reaches that, it became frozen. During the second pass, words within the dictionary are replaced with their respective indexes. The dictionary is sorted in descending order of frequencies; therefore frequent words have smaller index values than the rare ones. However, the limited size of dictionary may affect negatively on the final compression ratio as it freezes without adding higher frequent word. At the same time, compressing with respect to word frequencies does not give high compression ratio compared to that with respect to pattern frequencies.

Following the same concept, Rashmi Gupta et. at [4] proposed a multi-tiered log file compression solution, where each notion of redundancy is addressed by separated tier. The first tier handles the resemblance between neighbouring lines. The second tier handles the global repetitiveness of tokens and token formats. The third tier is general-purpose compressor which handles all the redundancy left. These tiers are optional and designed in several variants differing in required processing time and obtained compression ratio. However this compression scheme addresses the local redundancy and affects the final compression ratio as it depends on the region around the line the algorithm searches for the reference to compress the current line.

Kimmo et. at. [5] Presented a comprehensive log compression (CLC), method that uses frequent patterns and their condensed representations to identify repetitive information from large log files. A log file may be very large. During one day, a log file might accumulate millions of lines. Thus, for file evaluation of an expert, manual check is required. However, the most dangerous attacks are new and unseen for an enterprise defence system. CLC filters out frequently occurring events that hide other, unique or only a few times occurring events, without any prior domain knowledge. This separation makes it easier for a human observer to perceive and analyze large amounts of log data. However this scheme is used only in summarizing the frequent patterns in offline mode, and not used further in lossless compressing the corresponding log file.

These previously mentioned techniques and more others take into consideration that the log files are in a text format. However, the log files that we are dealing with are in a binary format. In order to deal with extracting such data, we had to deal with the recorded standard for each event. We stress on this part since we spend a lot of time working with different standards as well as their versions before reaching the step of compression and data mining. In the following, we briefly explain how the log files are recorded.

3   GPEH LOG FILES STRUCTURE AND RECORDING STANDARDS

Log files are used to record events that occured within the system. For cellular mobile and telecommunication, the events are related to every ascpet concerning the core network to the base stations ending with the cellular phones. Consecuently each log file can be of order tens of gigabytes per day, and must be stored for number of days as history. Each event within the log file follows certain format with respect to the different features recorded, the order and the size of each feature. Concerning the field of Telecommunication, these log files are in binary format. Thus for faster dealing with data, we parse each event and its corresponding messages into separate MySql database table. Finally we have database table for each unique GPEH event.

Taking GPEH Internal event "Admission Control Respose" as an example of the events recorded in log files. This event is captured whenever the admission control function responds to an admission request [15]. Table 1 shows the Structure of the event, the order of parameters, the number of bits of each parameter and the range of each parameter. Each single event recorded in the log file, has its own structure, parameters' order and sizes.

As mentioned before, the accumilated size of the log file is very huge. "Downlink Channelization Code Allocation" internal event for example accumilated nearly 1,622,900 rows (220 MB) for just 4 minites and "RRC Measurement Report" accumilated nearly 10GB for 15 minites and 68GB for 3 hours. Thus compressing these events is considered essential before storing as history.

Table 1. Internal Admission Control Event

Event ID 391 INTERNAL ADMISSION CONTROL RESPONSE		
EVENT_PARAM	PPS	Range
RESULT	3	0..3
FAILURE_REASON	8	0..78
…………..		
GBR_UL	15	0..16000

4 FREQUENT PATTERN GENERATOR ALGORITHM

In this section, we state the basic terminologies that will be used further in the paper then introduce frequent pattern generator algorithm that will be used for lossless compression, as follow:

4.1. Basic Terminologies:

We assume normal relational table T with m records. Each record can be viewed as n (attribute - attribute value) pairs and each pair is considered as an item. Each record contains a set of n items and the group of items in the record is called itemset. The Support of an itemset is the number of records in the table T where this itemset occurs. The itemset is called Frequent Pattern (FP), if the support count is greater than or equal to a user-defined threshold. These frequent patterns can be used in summarizing the GPEH log files and compressing the data as well.

Pattern Gain is a measure of the contribution of this pattern in the table T [7]. The higher the gain is, the larger area covered by this pattern and the better compression can be gained by using this pattern. The storage gain of the pattern P can be calculated as in equation (1):

$$Gain(P) = \begin{cases} |P| \cdot S(P), & if \; |P| > 1 \; and \; S(P) > 1 \\ 0, & if \; |P| = 1 \; or \; S(P) = 1 \end{cases} \quad (1)$$

where *|P|* is the number of individual items in the pattern and *S(P)* is the support of pattern P.

4.2.Frequent Pattern Generation:

Discovering all combinations of itemsets with support above the user-defined threshold requires multiple passes over the current data, and that is why Apriori algorithm is used [12][13][14]. The basic property used in Apriori algorithm and its extensions is that the large frequent itemsets is generated by joining smaller sized frequent itemsets, and removing the infrequent itemsets from the further search.

In each pass, the previously generated frequent itemsets are used, new items are added, and new candidate itemsets are generated. For the new pass, the support of these candidate itemsets is found and those with support higher than the threshold to be the seed for the next pass are selected. This process continues until no new itemsets are found in the data. The following pseudo code summarizes the algorithm:

1. The first pass on the data counts the frequency of each individual item to determine 1-itemsets and selects the frequent items as the next seeds.

2. The Kth pass consists of two phases.
(1) Use the frequent (K-1)-itemsets produced from the $(K-1)^{th}$ pass, add new 1-itemsets   to generate the candidate K itemsets.
(2) Scan the data and calculate the support of each candidate K itemsets.
(3) Use the frequent K itemsets in the next pass.

3. Stop when there are no frequent itemsets.

A	B	C
a_1	b_1	c_1
a_2	b_2	c_2
a_3	b_3	c_3
..	..	..
..	..	..

Figure 1: Simple GPEH log file.

For example, the relational table shown in Figure (1) has three columns; each column holds different values for three GPEH log file features. Using Apriori algorithm, the first scan gets the support and the gain for each individual attribute value in each column, as declared in figure 2. Using the user defined threshold, discard the frequent patterns with frequency below the specified threshold, this will keep faster compression by neglecting the infrequent patterns.

Using frequent patterns generated from the first pass, the different combinations of these features are generated as candidates for the next run, as shown in figure (3). During the second scan, the support and the gain for each itemset are calculated; the above threshold frequencies to be the candidates for the next run are also selected. The scheme stops when there is no an itemset above the threshold.

Finally, those frequent itemsets generated from the first run till the end are sorted in descending order with respect to their gain. Each frequent itemset is given a pattern in ascending order for big overall compression gain, as shown in Figure (4). These patterns are used to compress the main log file in lossless mode.

Finding all combinations of itemsets with support above the user defined threshold could be solved using breadth-first search. However, it is not that efficient and it is considered time consuming especially for large GPEH log files. Thus, our scheme, as will be shown later, will compress the files based on selected features. Offline compression needs several passes on the current log file to extract the frequent patterns and to assign codes. At the same time, using old codes in online compressing all the upcoming log files, may causes lose of new frequent patterns and then affects negatively on the decision making process. Thus our proposed scheme mixes the offline and online compression modes to keep fresh summary data for accurate decision making. Traditional compression schemes operate in offline fashion to avoid losing new patterns. Our scheme made use of the strong correlations of the consecutive recorded log files and the previously stored frequent-pattern look-up table to compress the current log files in on-line fashion.

Itemset	Width	Support	Gain
a1	1	5	5
a2	1	15	15
a3	1	18	18
...	...	...	...

⇨

Itemset
a2
a3

Itemset	Width	Support	Gain
b1	1	20	20
b2	1	8	8
b3	1	16	16
...	...	...	...

⇨

Itemset
b1
b3

Figure 2: First Pass.

Itemset	Width	Support	Gain
a2b1	2	11	22
a2b3	2	13	26
a3b1	2	0	0
a3b3	2	0	0
...	...	...	...

⇨

Itemset
a2b1
a2b3
....

Itemset	Width	Support	Gain
b1c1	2	15	30
b1c2	2	0	0
b3c1	2	0	0
b3c2	2	19	38
...	...	...	...

⇨

Itemset
b1c1
b3c2
....

Figure 3: Second Pass.

Itemset	Gain	Pattern
$a_2b_1c_1$	40	P_0
$a_3b_1c_1$	22	P_2
...	...	...
b_1c_1	12	P_{14}
...	...	...
c_1	2	P_{24}
a_3	2	P_{25}

Figure 4: Codes assignment.

5 Mixed Mode Summary-Based Lossless Compression for log files (MMSLC):

In this section, our lossless compression scheme (MMLC) is introduced. The aim is to extract the frequent patterns from GPEH selected features, and then use these patterns to compress the GPEH log files. Our proposed algorithm falls into three phases are as follow:

5.1 Step1: Frequent Patterns Mining:

Log files hold several attributes, and each attribute has wide range of the different values. Compressing using all these attributes leads to huge storage save with unbounded time limit. On the other hand, experts and network operators are in need of fast summary from the current log file for certain features, as will be shown later in performance analysis section. Thus, for efficient and fast compression, summarizing GPEH log file will be with respect to some selected features. This would be beneficial in specifying the different problems on certain element, performance, and the expected future state.

Our scheme, MMLC, applies the frequent pattern Generator algorithm on those selected features. As discussed in the previous section, the frequent pattern Generator algorithm makes offline summary from the different combinations of these selected features. This summary holds the patterns, their codes, their supports and gains.

5.2 Step2: Compression and decompression:

Our scheme assigns unique codes in the ascending order of the patterns' gains, as shown in figure 4. These codes are used for lossless compressing/decompressing lookup table for the GPEH log file. In addition, two other issues can lead to more compression gain:

1- There are some features with constant value, single value for the whole log file. MMSLC exploits these features, by replacing them with hints at the end of the compressed file, as shown in figure (5).

2- GPEH log entries are associated with time when the events are triggered. This time feature cannot be compressed for the wide variation of its values. However, some patterns are repeated number of times consequently for the same log file. MMSLU makes good use of this repetition by stating the start, the end time, and the frequency of the consecutive repetition, as shown in figure (6).

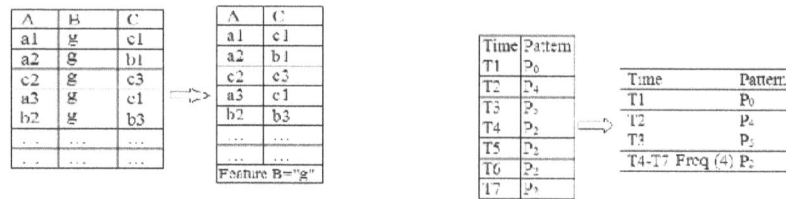

Figure 5: Compressing Uni-Valued Columns.   Figure 6: Compressing with respect to time.

5.3 Step 3: Mixed modes:

The compressed file with the used frequent pattern list represents the frequent and infrequent patterns along the scanned file. Although offline compression is the most widely used technique, this kind of compression delays the decision making process, that depends on the frequent and infrequent patterns obtained after scanning and compressing the log files. Thus, online compression can benefit decision makers for network monitoring, especially for huge amount of data reported by the telecommunication networks.

Our scheme proposed mixed mode between the offline frequent pattern mining process and the online compression using these patterns. MMSLC, mainly, assumes high correlation between the consecutive log files which is the case of cellular networks. Thus, the frequent patterns extracted from number of log files can be used to compress the direct next log files. As shown in figure (7), for the first history of log files, frequency patterns are extracted in offline mode. Then, the new log files are compressed in online mode using these patterns. During compression, MMSLC deals with these log files as new history and extracts new frequency patterns list in offline mode. Then, these patterns are used for compressing the new log files.

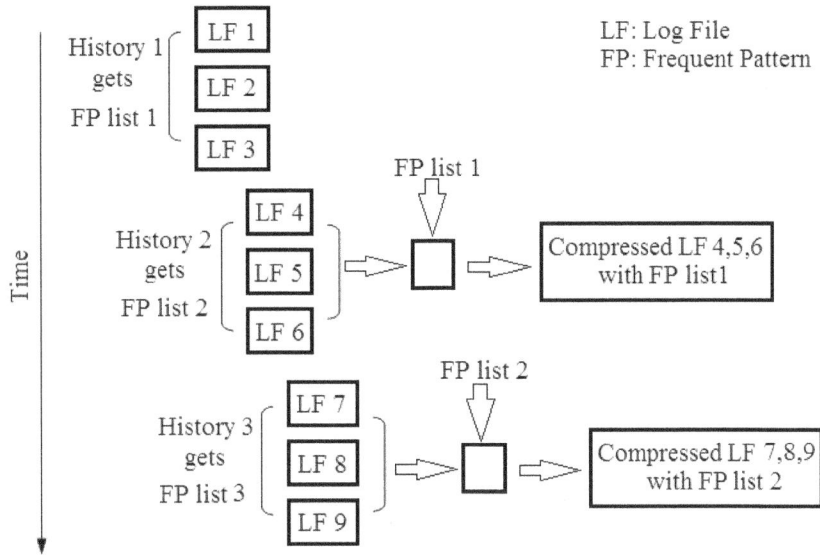

Figure 7: MMSLC Progress by Time.

6 Performance Analysis

In this section, we evaluate the performance of our proposed scheme, MMSLC. The scheme is implemented using java and the experiments are run on live recorded GPEH log files. Our algorithm runs on GPEH log files to compress and make summary with respect to certain parameter selected by experts. These parameters' frequencies are recorded in offline mode and the given patterns are presented according to their compression gain. Finally, these patterns are used to compress the log file. Our scheme was run on 3 different GPEH internal events as follows:

6.1   GPEH Events:

Our Scheme was run on GPEH "*System Block*" events to make summary on the blocked call problems recorded in the bin files. Experts usually collect the summary with respect to the different combinations of the cell id, the different source configuration and the different block reasons, instead of monitoring each record in the file that can leads to inefficient analysis of the system status. Figure 8 shows small part of this summary.

Figure 8: Part of the output Summary file from Block Calls internal event.

The second event was the "*Rab Establishment*" GPEH event, we collected summary with respect to the physical layer category, the source connection properties and the target configuration. The third event was the "*Soft handover Evaluation*" and we collected summary with respect to RNC module, the Event trigger and the Action. Experts can make use of the output summary for fast and accurate decision making, sample of the possible output summary of system block event are the source configuration shown in figure 9 and Cell ids shown in figure 10.

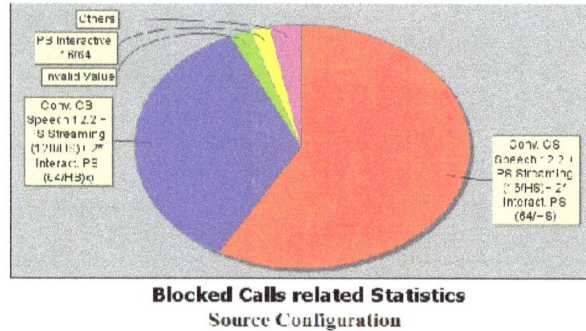

Blocked Calls related Statistics
Source Configuration

Figure 9: Source Configuration Summary representation.

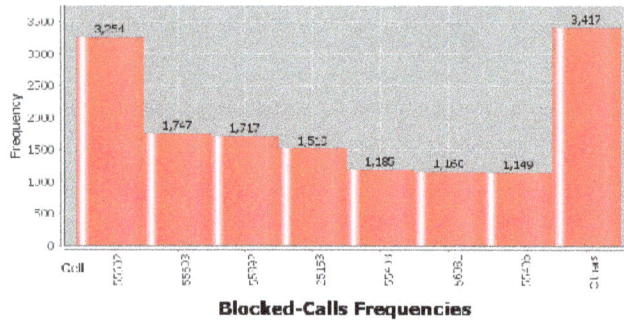

Blocked-Calls Frequencies

Figure 10: Cell ID Summary representation.

As stated before, in Step2, there are uni-valued features with constant value for the whole log file. MMLC exploits these features, states them in the summary file as shown in figure 11 and remove these columns from the compressed file. Also for the semi static valued columns, MMSLC gets the most frequent value in these fields, states them in the summary file as shown in figure 11. Thus for lossless compression, the log entries, with different values for those fields, will be written in the compressed file.

Figure 11: Static and Semi-Static Columns.

6.2   Memory and Time Analysis:

Figure 12 shows comparison between the main file, compressed file and the summary file in terms of the required storage memory. As shown, the compression gain is incomparable compared to the main file. These summaries are the most frequently used in telecommunications, instead of deep immerse in huge block of data.

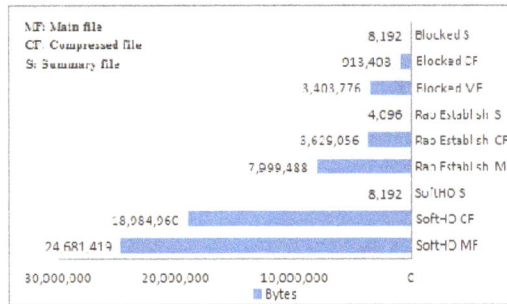

Figure 12: Memory needed to store the different types of files.

MMSLC allows faster online compression of log files using the offline summary of the previous *"Rab Establishment"* and *"Soft handover Execution"* events and this is very obvious in figure 13. This figure shows the time difference between offline compressing file after extracting its summary, which is the natural mode of most current compression techniques, and between online compressing the file based on offline saved summary on the three previously described GPEH events.

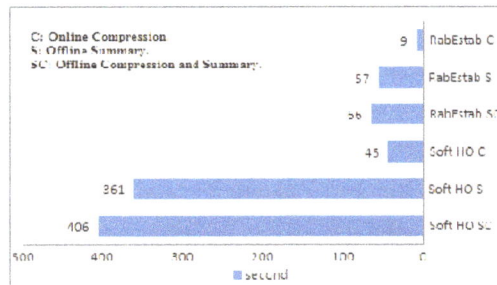

Figure 13: Time Compression.

6.3   GPEH Joint Events:

Our proposed scheme was applied on the join of three related GPEH events, named Internal_IMSI, Internal_SystemRelease and RRC_ConnectionRequest to collect statistics on the IMSI Roamers. IMSI Roamer is the mobile IMSI from outside the country and considered financial profit for the mobile system than the normal IMSI.

The collected statistics focus on displaying the number of Roamers connected to the current cells on Google Map and identifying the countries of these roamers.

As shown in figure 14, from the output summary of the events' join, this map shows the number of IMSI Roamers related to each cell with color indicators. These indicators divide the max number of Roamers into four groups, each indicates the importance of each cell. Green indicator means that this cell holds the number of roamers bigger than that of the yellow indicator. Thus for experts, cell with green indicator must have higher priority in maintenance and monitoring than that of the red indicator.

Figure 15 shows the same result in a different representation for experts; table shows the distribution of the roamers on the current active cells in the field.

Figure 14: Google Map showing the number of Roamers connected to each cell.

Cells	Roamers Count
15251	32
k33052	30
19937	17
10311	16
k32994	15
04803	14
k36833	14
15253	12
19933	12
01203	11
00133	11
01223	10
01462	10
n03251	9

Figure 15: The number of Roamers connected to each cell.

Country distribution of IMSI Roamers is shown with two different representations, table and pie chart in figure 16 and 17 respectively. This kind of analysis is very important for experts from financial point of view.

Country	Roamers Count
Saudi Arabia	196
United Arab Emirates	50
Palestine	33
Jordan	20
Malaysia	16
Kuwait	13
Netherlands	13
Republic of Macedonia	9
Turkey	9
Libya	9
Denmark	8
China	8
Greece	6
Oman	6
Mexico	3
South Africa	3
Austria	3
Iraq	2

Figure 16: Country distribution of the current Roamers.

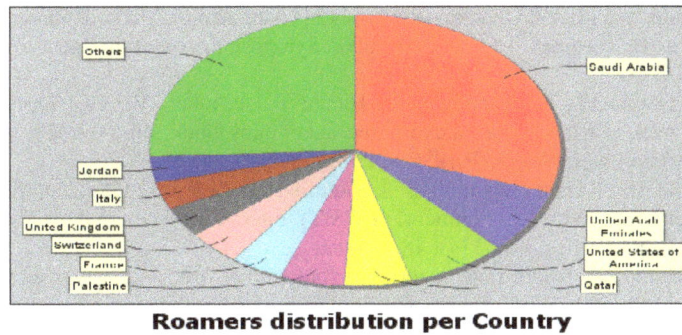

Roamers distribution per Country

Figure 17: Pie Chart displaying the Top IMSI Roamers Countries.

7 CONCLUSION

In this paper, we introduced a new lossless compression scheme based on mining frequent patterns named Mixed Mode Summary-based Lossless Compression for log files (MMSLC). Our scheme uses the famous *Apriori technique* for mining frequent patterns to be used in compressing the file. Massive burst of entries can lose important information; especially those translated into performance abnormalities. In addition, experts and network engineers are not interested in each log entry. MMSLC provides summarization for the current log files that would be beneficial in specifying the different problems on certain elements, the overall performance and the expected network future state. MMSLC exploits the high correlation between the consecutively recorded log files by introducing mixed online and offline compression modes. MMSLC uses the FPs extracted from previous log files in offline mode to compress the current log files in online mode. At the same time, it extracts the frequent patterns from the current log files in offline mode to be used in compressing the new log files. Our proposed scheme is proved to gain high compression ratios in fast speed as well as help in extracting beneficial information from the recorded data.

ACKNOWLEDGMENTS.

This paper is part of "TEMPO" project funded by NTRA Egypt, and we also would like to express our gratitude to all TEMPO Team for their help and effort to accomplish this work.

REFERENCES

1. Christian Borgelt: Recursion Pruning for the Apriori Algorithm. Workshop on Frequent Itemset Mining Implementations-FIMI (2004).
2. Christopher H., Simon J. Puglisi, Justin Z.: Relative Lempel-Ziv Factorization for Efficient Storage and Retrieval of Web Collections. Proceedings of the VLDB Endowment (PVLDB), 2011, Vol. 5, No. 3, pp. 265-273.
3. Ericsson: General Performance Event Handling RNC Description, 2010.
4. Haiming Huang: Lossless Semantic Compression for Relational Databases. B.E., Renmin University of China, Beijing, P.R.China, 1998.
5. K. Hatonen: Data mining for telecommunications network log analysis, University of Helsinki, 2009.
6. Karahoca, A.: Data Mining Via Cellular Neural Networks In The GSM Sector. The 8th IASTED International Conference Software Engineering and Applications, Cambridge, MA, pp. 19-24, 2004.
7. Kimmo H., Jean-fancois B., Mika K., Markus M., Cyrille M.: Comprehensive Log Compression with Frequent Patterns. Data Warehousing and Knowledge Discovery DaWak, pp. 360-370, 2003.
8. Mr.Raj Kumar Gupta, Ms.Rashmi Gupta: An Evaluation of Log File & Compression Mechanism. International Journal of Advanced Research In Computer and Communication Engineering, VOL 1, ISSUE 2, 2012.

9.   Piotr Gaqrysiak and Michal Okoniewski: Applying Data Mining Methods for Cellular Radio Network Planning. In Proceedings of the IIS'2000 Symposium on Intelligent Information Systems, Mieczyslaw, 2000.
10. Rakesh A., Heikki M., Ramakrishnan S., Hannu T., A. Inkeri V.: Fast Discovery of Association Rules. Advances in Knowledge dicovery and datamining book, pp. 307-328, American Association for Artificial Intelligence Menlo Park, CA, USA, 1996.
11. Rashmi Gupta, Raj Kumar Gupta, 2012. A Modified Efficient Log File Compression Mechanism for Digital Forensic in Web Environment. (IJCSIT) International Journal of Computer Science and Information Technologies, Vol. 3 (4), 4878-4882.
12. Risto Vaarandi: A breadth-first algorithm for mining frequent patterns from event logs. In Proceedings of the 2004 IFIP International Conference on Intelligence in Communication Systems.
13. Salvatore O., Paolo P., Raffaele P.: Enhancing the Apriori Algorithm for Frequent Set Counting . International conference Data Warehousing and Knowledge Discovery-DaWak, pp. 71-82, 2001.
14. Sebastian D., Szymon G.: Subatomic field processing for improved web log compression. International Confereonce on Modern Problems of Radio Engineering, Telecommunications and Computer Science, 2008.
15. Skibiński P., Swacha J.: Fast and efficient log file compression. CEUR Workshop Proceedings of 11th East-European Conference on Advances in Databases and Information Systems (ADBIS 2007).

10

PERFORMANCE EVALUATION OF DIFFERENT WIRELESS AD HOC ROUTING PROTOCOLS

Niranjan Kumar Ray[1] and Ashok Kumar Turuk[2]

Department of Computer Science and Engineering,
National Institute of Technology Rourkela, India
[1]rayniranjan@gmail.com, [2]akturuk@nitrkl.ac.in

ABSTRACT

One of the major challenges in wireless ad hoc network is the design of robust routing protocols. The routing protocols are designed basically to established correct and efficient paths between source and destination. In the recent years several routing protocols have been proposed in literature and many of them studied through extensive simulation at different network characteristics. In this paper we compare the performance of three most common routing protocols of mobile ad hoc network i.e. AODV, DSR and ZRP. Performance of these three routing protocols are analysed based on given set of parameters. We compare the packet delivery ratio, average end to end delay, average jitters and energy consumption behaviours of the AODV, DSR and ZRP.

KEYWORDS

Routing protocols, Wireless ad hoc network, Energy efficiency, Performance evaluation, Simulation, Mobility

1. INTRODUCTION

Mobile technology has strongly influenced our personal and professional lives in the recent time due to its applicability and versatility in different fields. Cellular phones, PDA, blue-tooth devices, and many hand-held computers enhanced our computing, communication skills and information accessing capabilities through their inherent advantages. Furthermore numerous traditional home appliances are blending with wireless technology, are extend the wireless communication to a fully pervasive computing environment. Mobile networking is emerging as one of the most pervasive computing technologies. There are typically two types of wireless networks: (i) infrastructure based, and (ii) infrastructure less. The first category requires access points or base stations to support communications, while the second type does not require such technology (or device) and commonly known as ad hoc network. Multi hop wireless networks in all their different form such as mobile ad hoc network (MANET) and vehicular ad hoc network (VANET), wireless sensor network(WSN), wireless mesh network(WMN), etc are coming under this category. In multi-hop technique destination nodes may be multiple hops away from the source node. This approach provides a number of advantages as compare to single-hop networking solution. Some of its advantages are (i) support for self configuration and adaption at low cost, (iii) support of load balancing for increasing network life, (iv) greater network flexibility, connectivity, etc[1], [2], [3]. However irrespective of these advantages it also suffered many challenges associated with unpredictable mobility, restricted battery capacity, routing, etc.[4]

The most fundamental problem of ad hoc network is how to deliver data packets efficiently among the mobile nodes. Since node's topology changes frequently that makes routing very problematic. Also low bandwidth, limited battery capacity and error prone medium adds further complexity in designing an efficient routing protocol. Estimation of end-to-end available bandwidth is one of the key parameter as resource allocation; capacity planning, end-to-end file sharing, etc depends on it. As nodes are battery operated and it is always a cumbersome task to recharge them in most of the environments, for which proper utilization of available power is the key issue for power aware routing protocols. Flooding based routing protocols rely on message forwarding by broadcasting the message among them. This mechanism consumes a major portion of battery power at node level. Coverage and connectivity are also other consideration in ad hoc routing protocols along with load balancing and flow control mechanism. Most of the algorithms form a tree structure when nodes do not have location information in its local cache. However a tree based topology does not perform well in terms of energy efficiency and scalability. So far the works done on routing in ad hoc networks categorised as: (i) reactive, (on-demand) (ii) proactive (table driven) and (iii) hybrid (mixed). Reactive routing protocols establish path based on the requirements. However proactive based works differently, in this approach each node keeps the path information in the routing table and establish the path based on the routing table information. Reactive protocols are considered as the most suitable for networks with higher mobility. In contrast proactive protocols are best fit to the static networks where node information does not changes frequently but they suffered with *loop to infinity problem* when a link fails. Hybrid routing protocols in other hand carry some features of on-demand protocols and some from table driven protocols. Regardless its advantages very few works have been focused on hybrid protocols.

The remaining structure of the paper described as follows. Functionalities and working procedures of AODV, DSR and ZRP were discussed in section 2. Section 3 demonstrates the performance evaluation, comparison and simulation results of AODV, DSR, and ZRP. Related literature study is presented briefly in section 4. Conclusion is made on section 5.

2. BACKGROUND

2.1. Ad-hoc On-Demand Distance Vector Routing (AODV)

Perkin *et al.* first presented AODV in [5] which then standardize in IETF RFC 3561 in 2003[6]. It is an improvement on the DSDV [7] algorithm and modified message broadcasting procedure to minimize the number of required broadcasts by creating routes on a demand basis instead of maintaining a complete list of routes as observed in the DSDV algorithm. AODV makes use of destination sequence numbers to ensure all routes are loop-free and contain the most recent route information. The destination sequence number is to be included along with any routes the destination sends to requesting nodes. Given the choice between two routes to a destination, a requesting node is required to select the one with the greater sequence number. The greater sequence number indicates that route is a fresh route as compared to its counterpart. Three message types defined by AODV are: *Route Requests* (RREQs), *Route Replies* (RREPs), and *Route Errors* (RERRs).

A path discovery process is initiated when a source node wishes to send a message to a destination node and does not have a valid route. The source broadcasts RREQ, recipients of RREQ forward the message if they don't have the path to the destination. The forwarding continues till message reach the destination. In other way a route can be determined when the RREQ reaches either the destination, or an intermediate node with a 'fresh enough' route to the destination. If the intermediate node finds a fresh enough route in its route cache then rather than forwarding RREQ it reply back to the source by RREP. Node receiving the request caches a route back to the originator of the request, so that the RREP are *unicast* from the destination along a path to that originator.

Each node maintains its own sequence number and a broadcast ID. The broadcast ID is incremented for every RREQ the node initiates. RREQ are uniquely identified with the sequence number and node's address. The source node, with its broadcast ID includes the most recent sequence number it has for the destination in the RREQ. Intermediate nodes reply to the RREQ with a RREP packet only if they have a route to the destination whose corresponding destination sequence number is equal or greater than the sequence number contained in RREQ. While forwarding the RREQ, intermediate nodes records address of the neighbors from which the first copy of the broadcast packet is received in its routing table. Node discarded the duplicate copies of the RREQ. Each route entry is associated with a timer, when the timer expire the route entry is discarded from the routing table. As the RREP travels back to the source using the reverse path, all intermediate nodes create a new route entry for destination.

A RERR message is send when a link break is detected in an active route. The RERR message indicates that path to destination is no longer reachable by way of the broken link. A node initiates RERR message under three situations. (i) If it detects a link break for the next hop of an active route in its routing table. (ii) If it gets a data packet destined to a node for which it does not have an active route (iii) If receives a RERR message from its neighbor. Figure 1 shows the message exchange between source and destination in AODV.

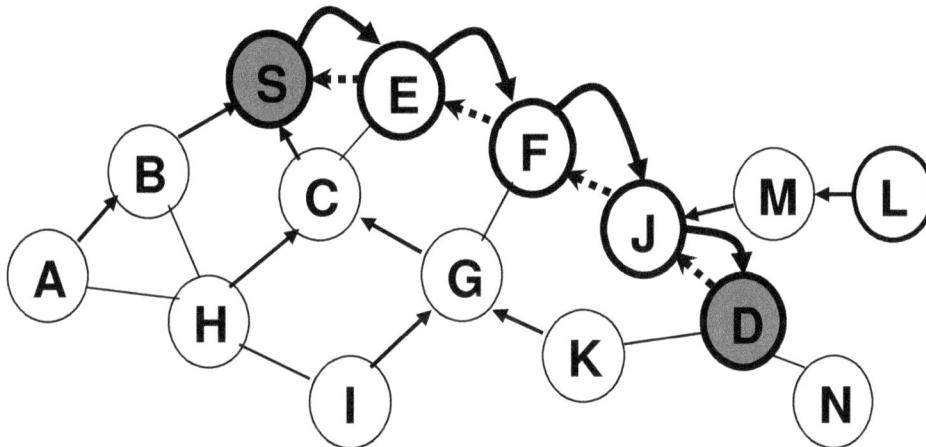

Fig. 1. Message Exchange between Source 'S' to the Destination 'D'

2.2. Dynamic Source Routing (DSR)

Johnson *et al.* proposed DSR which is available as a IETF draft in [8]. DSR is developed for multi-hop wireless networks and is very simple and efficient in term of path findings. The protocol consists of two main phases: (i) Route discovery and (ii) Route maintenance. Each, data packet contains the complete ordered list of nodes along the route in its header. This makes routing trivially loop-free and does not require up-to-date routing information in the intermediate nodes through which the packet is forwarded. This way other nodes along the route can easily cache the routing information for future use. Entries in the route cache are continually updated when new routes are learned. The basic operation of DSR is discussed below.

When a node needs to send a packet to other node it places a complete list of hops to follow in its packet header. It first checks its route cache and if doesn't find the destination in its route cache it start the proceedings to find the path. If it has a route to the destination, it will use this route to send the packets. Otherwise node initiates the route discovery by broadcasting a RREQ packet. RREQ contains sender and destination address. An intermediate node on receiving the packet checks destination address, if node knows the destination address, reply to the source otherwise it appends it's address to the route record and broadcasts it further. A node discards the request it has recently received from another route from the same initiator or the route record already contains its own address. In this way node in DSR prevents looping. Figure 2 and 3 shows the request and reply between source and destination.

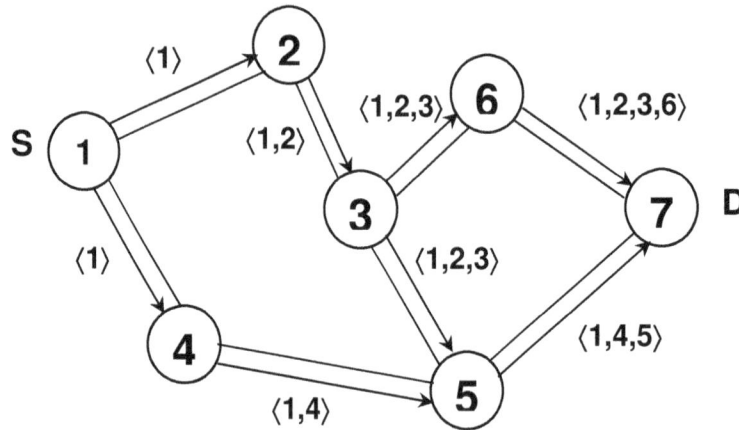

Fig. 2. Request message propagation in DSR, between S to D.

The route maintenance is accomplished through the use of route error packets and acknowledgements. Each node broadcast packets and waits for acknowledgements. If node doesn't receive an acknowledgement after forwarding a packet, it sends an acknowledgement request. If node doesn't receive any reply within a specific time it then generates RERR message and sends it to the sender of the packet.

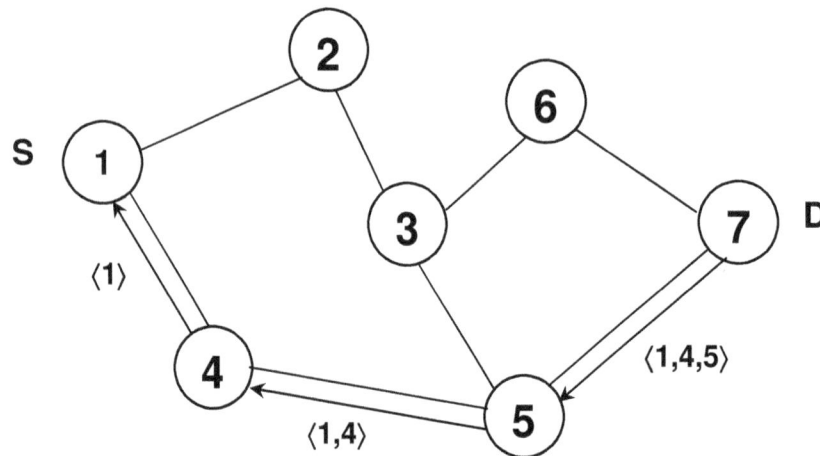

Fig. 3. Reply message propagation between D to S

2.3. Zone Routing Protocol (ZRP)

ZRP [9] is the first hybrid category protocol which effectively combines best features of reactive and proactive routing protocols. It employs concept of proactive routing scheme within limited zone (within the r-hop neighborhood of each node), and uses reactive approach beyond that zone. The two routing schemes used by ZRP are: (i) Intra-zone routing protocol (IARP) and (ii) Inter-zone routing protocol (IERP). The main components of ZRP are discussed below.

The first protocol to be part of ZRP is IARP, which is a pro-active, table-driven protocol. The protocol is used when a node wants to communicate with the interior nodes of it's zone. It allows local route optimization by removal of redundant routes and the shortening the routes if a route with fewer hops has been detected, as well as bypassing link failures through multiple hops, thus leveraging global propagation. IARP is used to maintain routing information and provides route to nodes within zone.

The second protocol which is the part of ZRP is IERP. It uses the *route query* (RREQ)/ *route reply* (RREP) packets to discover a route in a way similar to typical on-demand routing protocols. In ZRP, a routing zone consists of a few nodes within one, two, or a couple of hops away from each other. It works similar to a clustering with the exception that every node acts as a cluster head and a member of other clusters. Each zone has a predefined zone centred at itself in terms of number of hops. Within this zone a table-driven-based routing protocol is used. This implies that route updates are performed for nodes within the zone. Therefore, each node has a route to all other nodes within its zone. If the destination node resides outside the source zone, then an on-demand search-query routing method can be used.

ZRP also uses *Bordercast Resolution Protocol* (BRP). When intended destination is not known, RREQ packet is broadcast via the nodes on the border of the zone. Route queries are only broadcast from one node's border nodes to other border nodes until one node knows the exact path to the destination node. ZRP limits the proactive overhead to only the size of the zone, and the reactive search overhead to only selected border nodes. The IARP in ZRP must be able to determine a node's neighbours itself. This protocol is usually a proactive protocol and is responsible for the routes to the peripheral nodes. Figure 4 shows the components of ZRP.

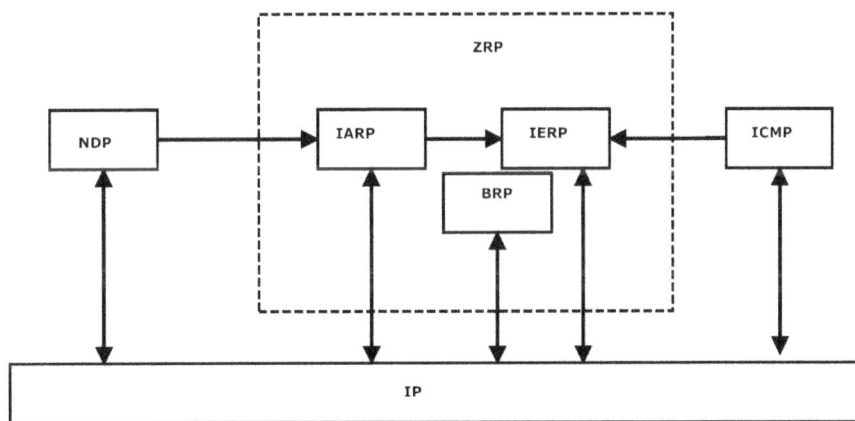

Fig. 4.. Components of ZRP

3. PERFORMANCE EVALUATION

3.1 Network Simulator

Qualnet network simulator [10] is used here for simulating multihop ad hoc wireless network routing protocols. It is a commercial version of GloMoSim [11] and is developed by Scalable Network Technology. It is extremely scalable and can supports high fidelity models of networks of thousands of nodes. In a discrete event simulator like Qualnet, the simulation performed only when an event occurs. It is based on event scheduler, which contains any events that needs to be processed and stepped through. Processing of an event may produce some new events. These new events are inserted into event scheduler. QualNet is modelling software that predicts performance of networking protocols and networks through simulation and emulation. The simulator is written in C++ while it's graphical toolkits are implemented in Java. Every protocols in QualNet starts with an initialization function which takes an inputs and configures the protocol. Event dispatcher activated when an event occurs in a layer. The simulator checks the type of event and calls appropriate event handler to process. At the end of the simulation finalization function is called to print out the collected statistics. Figure 5 shows the event handling process in QualNet.

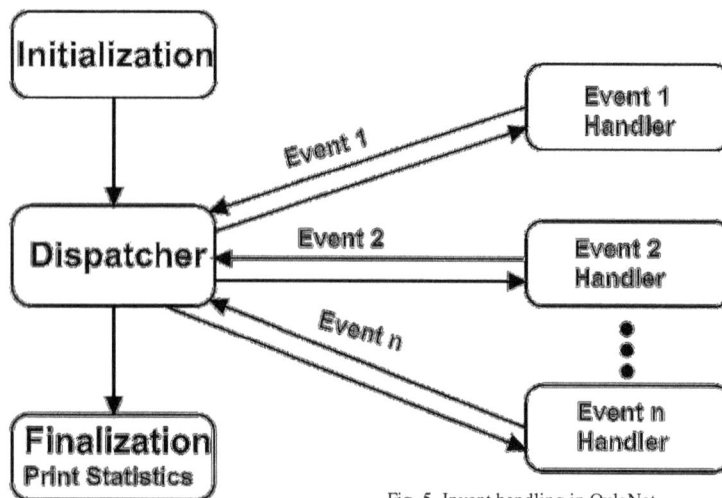

Fig. 5. Invent handling in QulaNet

3.2 Simulation Environment

Our evaluation is based on the simulation of 36 nodes deployed randomly in a square area of (1500m × 1500m) flat surface for a simulation time of 200s. IEEE 802.11 DCF is used as the MAC protocol. The radio model uses bit rate of 2 Mbps and has a radio range of 250 meters. We generated 45 different scenarios with varying movement patterns and communication patterns. Since each method was changed in an identical fashion for which direct comparisons are made from simulator.

Mobility Model

Random way point (RWP) mobility model is used in our simulation which is characterised by a pause time. Node remains stationary for a certain periods of time (known as pause time). At the end of that time node choose for a random destination in (1500m × 1500m) simulation space area. The node moves to the destination at a speed in the range [0, max]. When node reaches the destination it waits for time equal to pause time and starts moving for another destination. It repeats this behaviour for the entire given simulation time. We simulate with 5 different pause times: 0, 25, 50, 100s and 200s. The pause time of 0 seconds represent the continuous motion while pause time of 200 represents no motion. As the movement pattern is very sensitive to performance measurement, we generate 15 different movement patterns, three for each value of pause time. All three routing protocols were run for the same 15 movement pattern. We test with maximum speed movement of 10, 15 and 20 m/s.

Traffic Pattern

For the above comparison of routing protocols constant bit rate (CBR) traffic patterns are used. The network contains four CBR traffic sources and packet size is 512 bytes. Packets are send from source nodes in 1s interval and total 100 items are send from each source. Single communication patterns are taken in conjunction with 15 movement patterns which provides 45 different scenarios.

Energy Model

The energy consumption of radio interface depends upon different type of operation mode. We considered the energy model where energy E consume by a node to transmit k bit can be expressed as:

$$E = P_{active} \times T_{active} + P_{sleep} \times T_{sleep} + P_{idle} \times T_{idle}$$

Where P_{active}, P_{sleep}, P_{idle} are the power consumptions in active, sleep and idle state respectively and T_{active}, T_{sleep} and T_{idle} are the duration of time that a transceiver stays in respective states. In active state a node either transmits or receives packets, while in idle state it waits for traffic to participate. When a node in active state transmits or receives packets the network interface of a node decrements the energy based on certain parameter such as size of the packets, bandwidth of channel, NIC characteristics etc. According to the specification of NIC modelled we consider the energy consumption of 200mA in receiving mode and 230mA in transmitting mode using 3.0 V supply voltage.

3.3 Performance Metrics

Following metrics have been chosen to compare the routing protocols

Throughput (Packet delivery ratio): It is the ratio of packets received at destination to the packet sent by the source node. Protocols with better packet delivery ratio are considered as the efficient protocols; however this parameter depends upon many other factors also.

Average end-to-end delay: It is the average time a packets takes to reach the destination. It considers all types of delay such as queuing delay, route discovery delay, interface delay, etc. It is also known as the average time between sending and successfully receiving a packet. Sometimes it is also known as *path optimality.*

Average jitter: Jitter is referred as measure of the variability over time of the packet latency across a network. When a network has constant latency it has no variation (or jitter). Packet jitter is expressed as an average of the deviation from the network mean latency.

Energy consumption behaviour: Energy is consumed when node transmit and receives packets and processed a packets. Energy also consumed in idle state. Energy consumption of a node mainly depends on the state of a mobile node. The message sending mechanism and carrier sensing methods plays a vital role on energy consumptions along with traffic pattern.

3.4 Simulation Results

We have done simulations using three node movement speeds. A maximum speed of 10, 15, and 20 m/s is used for comparing AODV, DSR and ZRP. Pause time is varied from minimum 0 pause value to maximum value equal to simulation time. Pause time 200 represents that nodes are static and 0 pause time means they are fully mobile.

3.4.1 Throughput at varying pause time

The throughput is calculated as the average number of packets received per amount of time, this is expressed:

*Throughput= (Total bytes received * 8) / (Last packet received - First packet received)*

Figure 6 to 8 shows throughput verses pause time. As expected reactive routing protocols AODV and DSR are giving better throughput as compared to ZRP. Both at high and at low mobility (high mobility when pause time is zero and low mobility when pause time is maximum) and in different movement speed DSR giving better results in comparison to AODV. All the methods deliver a greater percentage of

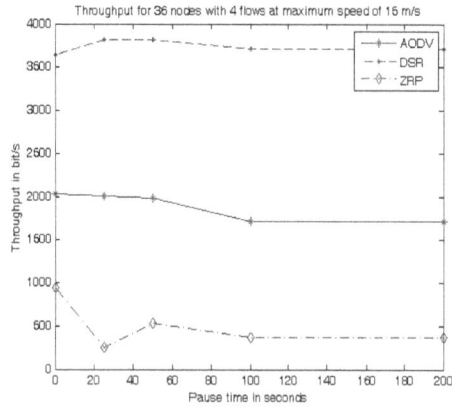

Fig. 6. Throughput at maximum speed of 10 m/s

originated data packets

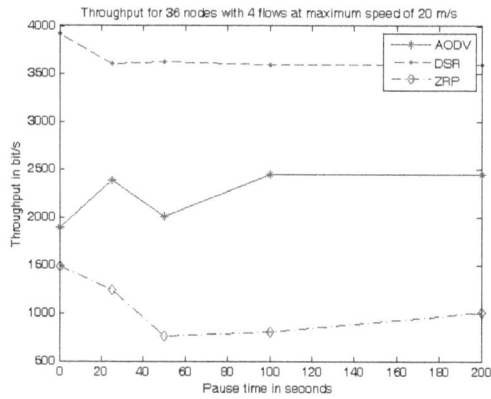

Fig. 7. Throughput at maximum speed of 15 m/s

at low mobility.

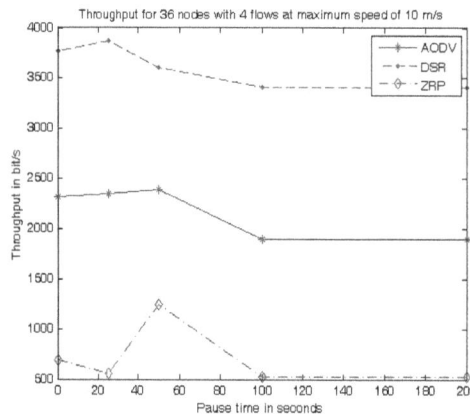

3.4.2 Average end to end delay

The average end to end delay (E2ED) refers the difference between the origination time of the packet and final destination time at which packet reached the destination. This is calculated as:

E2ED= ((Sum of the delays of each CBR packet received)/ number of CBR packets received)

However in our simulator, also consider transmission time for calculating average E2ED in CBR. Figure 9 to 11 shows E2ED at different node speeds. The average E2ED of ZRP is high as compared to DSR and AODV. The delay of DSR and AODV are nearly matched in all mobility and pause time. E2ED of ZRP is more unpredictable due to IARP and IERP protocols working functionality.

Fig. 9. Average end to end delay at maximum speed of 10 m/s   Fig. 10. Average end to end delay at maximum speed of 15 m/s

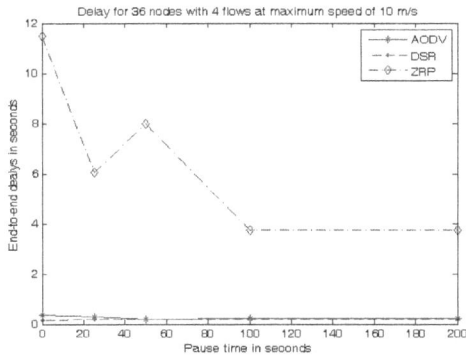

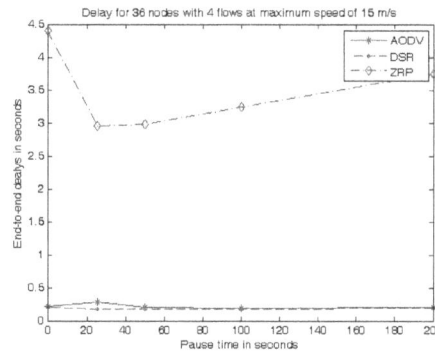

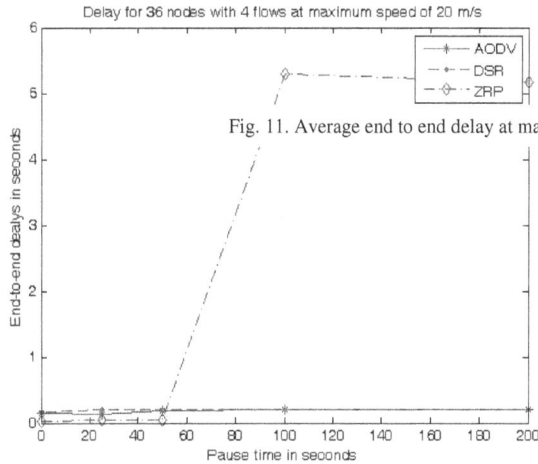

Fig. 11. Average end to end delay at maximum speed of 20 m/s

3.4.3 Average jitter

In order to find the average jitters in the networks following calculations are made. When CBR servers receive N packets from client, i = 1 to N; and if N is greater than 2 then jitter (J) can be calculated as:

$$J (i) = delay (i+1) - delay (i)$$

$$Avg. J = (j (1) + J (2) + ...+J (N-1))/ (N-1)$$

Figure 12, 13 and 14 shows the average jitter at different pause time and node mobility. It is observed that average jitters of AODV at high mobility is less than 0.1s. The average jitters of ZRP are worst in comparison to AODV and DSR.

Fig. 12. Average jitters at maximum speed of 10 m/s   Fig. 13. Average end to end delay at maximum speed of 15 m/s

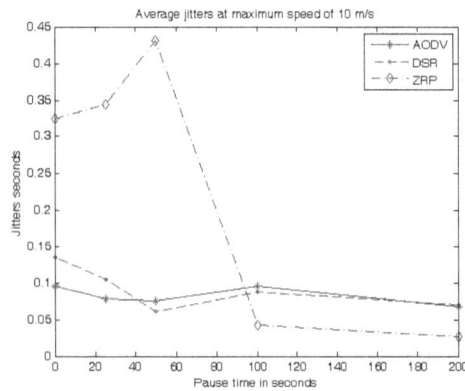

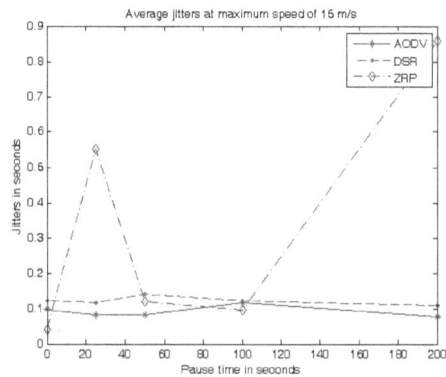

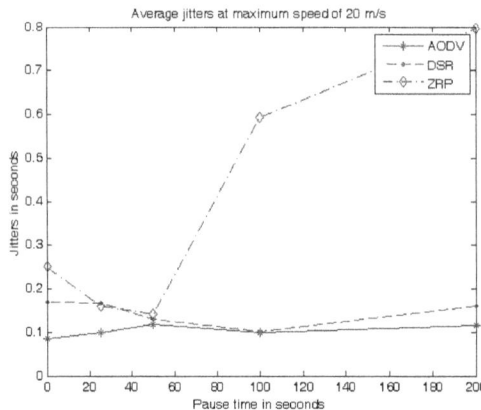

3.4.4   Energy   consumption behaviour

Energy consumptions of three routing protocols at transmit, receive and idle state is depicted through figure 14, to 16. The source nodes of CBR traffic send the packets at the power in which they have assigned. Intermediate nodes between source and destination forward the packets to destination and acts as relay nodes. The nodes those aren't participating in the packet exchange remain in idle for the entire simulation time. In the absence of suitable sleep strategy idle nodes remain active for entire communication periods; hence the idle power consumption in a network is greater in comparison to active state power consumptions. The situation goes to worst condition when most of the nodes are remain in idle state. The energy consumptions in transmit/receive state also vary based on the routing strategy used by the routing protocols.

It is observed that AODV is energy efficient in all cases. AODV uses active route cache path finding for which it consume less energy both in path finding and maintenance process. DSR is energy in-efficient as compared to AODV and ZRP. Energy consumed by DSR is higher both in transmitting and receiving states. One of the main causes of higher energy consumption is it's increase packet header size which increases in each increase of hop counts. Idle state power consumptions is depend on the MAC layer power management strategy for which it is almost independent of routing protocols. All three routing protocol consumes nearly same energy in idle state. We doesn't considered the sleep power consumption in this analysis as sleep state consume very less power [12] and need power management support from MAC layer.

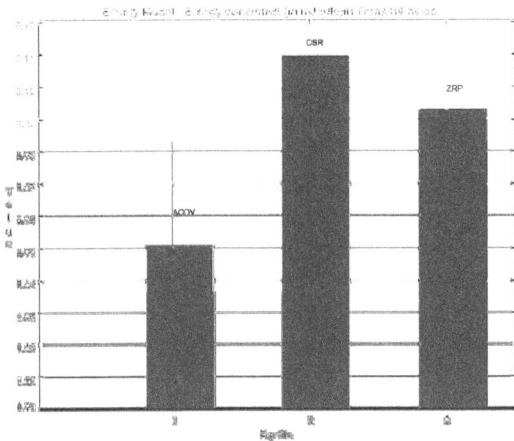

Fig. 15. Energy Consumed (in m Joule) in Transmit mode

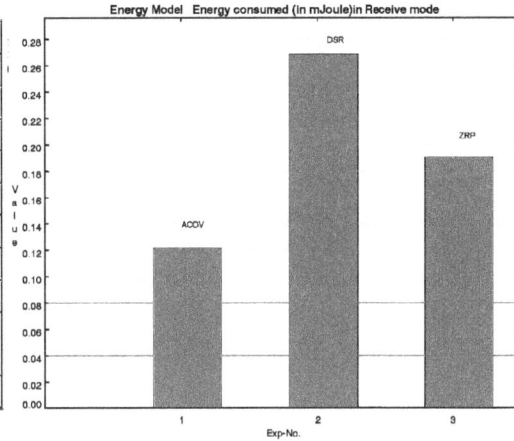

Fig. 16. Energy consumed (in M Joule) in Received mode

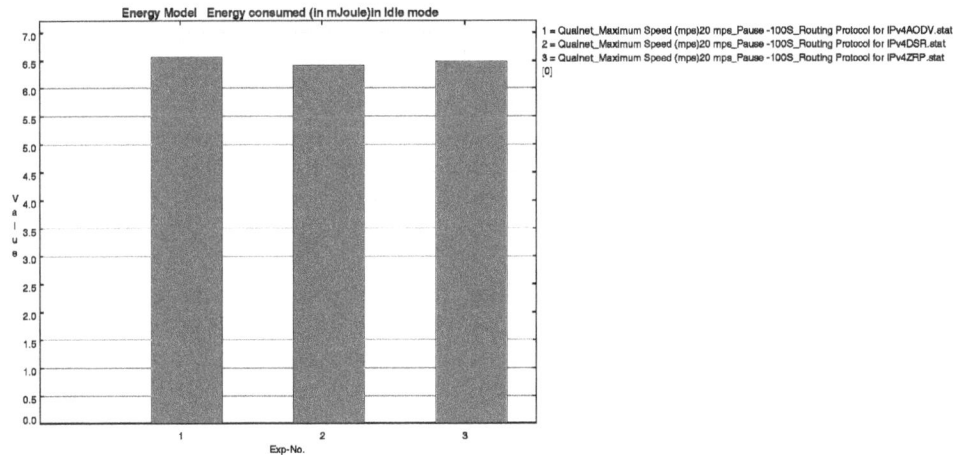

Fig. 17. Energy consumed in idle mode

4. RELATED WORKS

We would like to introduce some study based on comparison of different routing protocols in wireless ad hoc networks. This section focused on literature review pertaining to routing in wireless ad hoc networks.

Broch, *et al.* [13] evaluated four ad hoc routing protocols including AODV and DSR. They used 50 node models with variable mobility and traffic scenarios. Packet delivery fraction, number of routing packets, and distribution of path lengths were used as performance metrics. Simulation results reveal that DSR demonstrated vastly superior routing load performance, and somewhat superior packet delivery and route length performance.

Das *et al.* [14] use MAC and physical layer models to study interlayer interactions and their performance implications. The simulation were done on the *ns-2* [15] and consider the radio model similar to Lucent's WaveLAN radio interface and random waypoint mobility with pause time from 0 to 500 seconds. In two different scenarios, 50 and 100 nodes were utilized with an area size of 1500m/300m respectively 2200m/600m. Even both DSR and AODV share similar on-demand behaviour, their differences in the protocol mechanics can lead to significant performance differentials. The performance differentials are analyzed using varying network load, mobility, and network size. Based on the observations, the authors made some suggestions about how the performance of either protocol can be improved. The simulation results suggest that AODV outperforms DSR in more stressful situations (i.e. larger network, higher mobility). However DSR showed advantage in the general lower routing overhead and in low mobility.

A mix set of reactive, proactive and hybrid protocols are analysed under realistic scenarios by Hsu *et al.* [16]. The paper demonstrates performance comparison of several protocols using the network and traffic configuration from the live exercise. The comparison shows that AODV performed better in term of throughput and lower average delay. The OLSR [17] showed good resilience to link state situation, while packet delivery ratio curve of DSR revels that when the delay in the network grows and network size increase the DSR delivers larger percentage of traffic.

Another work [18] compares performances of AODV, DSDV, TORA [19] based on four performances metrics: average delay, packet delivery fraction, routing load and varying network size. The authors suggest that that AODV exhibits a better behavior in terms of average delay as compared to TORA and DSDV. In less stressful situation, the packet delivery fraction of, the TORA outperforms DSDV and AODV.

In [20] OLSR and LAR1 [21] are compared using RWA mobility. It is found that OLSR has better end to end delay performance in comparison to LAR1. Ref [22] combined the LAR protocol with a probabilistic algorithm and presented a new location based routing protocol called LAR-1P. The modified protocol gives better message transmission at different node density. Authors suggest that their protocol can give better throughput at increasing node density.

Beaubrun and Molo [23] propose a formulation of the routing problem in multi-services MANETs, as well as the implementation of an adaptation of the DSR. Simulation results suggest that DSR enables to provide end-to-end delay less than 0.11 s, as well as packet delivery ratio higher than 99% and normalized routing load less than 13%, for low mobility level and low traffic intensity.

5 CONCLUSIONS

This paper evaluates the performances of three most common routing protocols in wireless ad hoc network. DSR is purely reactive in nature and it's path finding is very effective for which its throughput is high as compared to its main counterpart AODV. However DSR suffers with scalability issue. It's performance at fast moving scenario and at dense network is the major issue of concern. In our case the network density was low for which it gives better packet delivery ratio but we suspect its performance in term of throughput may degrade when the node density will increase. AODV is energy efficient both in transmitting and receiving states. Energy consumption of AODV is better as compared to DSR and ZRP for which it is considered as the most energy efficient protocols. However its performance is not properly compared with the other energy efficient routing protocols such as cone based topology control (CBTC) [24], XTC [25], SPAN [26], etc. ZRP provides very low throughput but its routing overheads is low. In this study we have focused on some metrics but metrics like network life time and battery power utilization is not properly studied at different mobility and node speed. Network life time with different flow is another challenging task can be analysed. In future we will focus in measuring the battery power utilization and network life time analysis at different mobility and traffic load.

REFERENCES

[1]   Tsertou, A.and D.I. Laurenson, " Revisiting the Hidden Terminal Problem in a CSMA/CA Wireless Network". *IEEE Transactions on Mobile Computing*, vol. 7(7), pp.817-831, October 2008.

[2]   Gomez, J. and Campbell, J. "Variable-range transmission power control in wireless ad hoc networks", *IEEE Transaction on mobile computing*, Vol. 6(1), pp. 87–99., 2007.

[3]   Wu, S.L and Tseng, Y.C. "Wireless ad hoc networking, Personal-Area, Local-Area, and the Sensory-Area Networks", *Auerbach publication*, ISBN 10: 10: 0-8493-9254-3, 2007.

[4]   Lou, W. and Wu, J. "Toward Broadcast Reliability in Mobile Ad Hoc Networks with Double Coverage". *IEEE Transaction on mobile computing*, Vol 6(2), pp. 148-163, 2007.

[5]   Perkins, C. E., and Belding-Royer, E. M. "Ad-hoc on demand distance vector routing", *n Proceedings of the IEEE Workshop on Mobile Computing Systems and Applications (WMCSA)* (pp. 90–100). New Orleans, LA, USA, 1999.

[6]   Perkins, C. E., Belding-Royer, E. M., and Das, S., "Ad hoc on demand distance vector (AODV) routing", *IETF RFC 3561.*, 2003.

[7]   Perkins, C. E., and Bhagwat, P. "Highly dynamic destination sequenced distance vector   routing (DSDV) for mobile computers". *In Proceedings of SIGCOMM.* 1994, pp. 234-244, London.

[8]   D. B. Johnson, D. A. Maltz, Y. Hu, and J. G. Jetcheva, "The Dynamic Source Routing Protocol for Mobile Ad Hoc Networks (DSR)." *Routing in Mobile Ad Hoc Networks,*

       http://www.ietf.org/internet-drafts/draft-ietf-manet-dsr-07.txt, Feb 2002., IETF Internet Draft.

[9]   Hass Z.J, Pearlman, M.R and Samar, P, "The Zone Routing Protocol (ZRP) for Ad Hoc Networks" *draft-ietf-manet-zone-zrp-04.txt, 2002.*

[10]   "Qualnet 4.5," http://www.scalable-networks.com,

[11]   UCLA parallel computing laboratory, University of California, "GlomoSim Scalable Mobile Network Simulator", http://pcl.cs.ucla.edu/projects/glomosim

[12]   Ray, N.K and Turuk A.K, Energy Conservation Issues and Challenges in MANETs, *Technological Advancements and Applications in Mobile Ad-Hoc Networks: Research Trends*, IGI Global, 2012, ISBN13: 978-1-4666-0321-9., 2012.

[13]   Broch, J. et al., "A Performance Comparison of Multihop Wireless Ad Hoc Network Routing Protocols," Proc. *IEEE/ACM MOBICOM '98*, Oct. 1998, pp. 85–97.

[14]   Das, S. R., Perkins, C. E., Royer, E. M. and Marina, M. K. "Performance comparison of two on-demand routing protocols for ad hoc networks," *IEEE Personal Communications Magazine, special issue on Mobile Ad Hoc Networks*, Vol. 8(1), pp. 16–29, February 2001.

[15]   Network simulator, http://www.isi.edu/nsnam/ns/

[16]   Hsu, J., Bhatia, S., Takai, M., Bagrodia, R.and Acriche, M.J "Performance of Mobile Ad hoc Networking Routing Protocols in Realistic   Scenarios", *Scalable Network Technologies White Paper, 2004.*

[17]   Clusen, T.H, Hansen, G, Christensen, L.,and Behrmann, G., "The optimized link state routing protocol, evaluation through experiments and simulations", *Proceedings of IEEE Symposium on wireless personal mobile communications 2001,* September 2001.

[18]   Park, V.D., Corson, M.S., "A highly adaptive distributed routing algorithm for mobile wireless networks", *Proceedings of IEEE INFOCOM 1997*, pp.1405-1413, April 1997.

[19]   Vetrivelan, N and Reddy, A.V "Performance Analysis of Three Routing Protocols for

Varying MANET Size" *Proceedings of the International Multi Conference of Engineers and Computer Scientists 2008,* Vol II IMECS 2008, 19-21 March, 2008, Hong Kong.

[20]   *Maurya, A.K, Kumar, A and Singh, D. "RWA mobility model based performance evaluation of OLSR and LAR1 routing protocols" International Journal of Computer Networks & Communications (IJCNC) Vol.3(6), November 2011.*

[21]   Ko, Y.-B. and Vaidya, N. H. "Location-aided routing (LAR) in mobile ad hoc networks," *Proceedings of Mobile Computing and Networking (MOBICOM'98)*, Dallas, TX, USA, 1998, pp. 66–75.

[22]   Al-Bahadili, H and Maqousi, A. "Performance evaluation of the LAR-1P route discovery algorithm" *International Journal of Computer Networks & Communications (IJCNC)* Vol.3(6), November 2011.

[23]   Beaubrun, R. and Molo, B. "Using DSR for Routing Multimedia Traffic in MANETs" *International Journal of Computer Networks & Communications (IJCNC)* Vol.2(1), January 2010.

[24]   Li, Li., Halpern, J.Y., Bahl, P., Yi-Min Wang, Wattenhofer, R. "A Cone-Based Distributed Topology-Control Algorithm for Wireless Multi-Hop Networks", *IEEE Transactions   on Networking*, 13(1), pp. 147-159, 2005.

[25]   Wattenhofer, R. and Zollinger, A. "Xtc: A practical topology control algorithm for ad hoc networks", *Proceedings of Fourth International Workshop on Algorithms for Wireless, Mobile, Ad Hoc and Sensor Networks (WMAN)*. Santa Fe, NM., 2004.

[26]   Chen, B., Jamieson, K., Balakrishnan, H and Morris, R. , Span: An energy efficient coordination algorithm for topology maintenance in ad hoc wireless networks, *ACM Wireless Networks Journal*, 8(5), pp.481-494. 2002

11

Ergodic Capacity Analysis of Cooperative Amplify-and-Forward Relay Networks Over Rice and Nakagami Fading Channels

Bhuvan Modi[1], A. Annamalai[1], O. Olabiyi[1] and R. Chembil Palat[2]

[1]Center of Excellence for Communication Systems Technology Research
Department of Electrical and Computer Engineering,
Prairie View A & M University, TX 77446 United States of America
mrbhuvan2000@yahoo.com, aaannamalai@pvamu.edu, engr3os@gmail.com

[2]Nokia Research Center, Berkeley, CA 94304 United States of America
ramesh.chembil-palat@nokia.com

Abstract

This article investigates the efficacy of a novel moment generating function (MGF) based analytical framework for calculating the ergodic channel capacities of cooperative dual-hop amplify-and-forward (CAF) relay networks under three distinct source-adaptive transmission policies in Rice and Nakagami-m fading environments. The proposed analytical approach relies on a new exponential-type integral representation for the logarithmic function $\ln(\gamma) = \int_0^\infty x^{-1}(e^{-2x} - e^{-2x\gamma})dx$, $\gamma > 0$, that facilitates the task of averaging it over the probability density function (PDF) of end-to-end signal-to-noise ratio (SNR) in a myriad of stochastic fading environments. We also resort to a well-known bounding technique to derive closed-form formulas for the upper and lower bounds for the MGF of dual-hop CAF relayed path SNR in Rice and Nakagami-m channels (since exact evaluation of the MGF of half-harmonic mean SNR in Rice fading is known to be very complicated and numerically unstable). These two attributes dramatically reduces the computational complexity of evaluating the ergodic capacities of CAF multi-relay networks with the optimal joint power-and-rate adaptation (when the channel side information is available at both the transmitter and the receiver) and the optimal rate-adaptation with a constant transmit power (when channel side information is available at the receiver only) policies.

Keywords

cooperative diversity, source-adaptive transmission strategies, ergodic capacity, moment generating function technique, amplify-and-forward relay networks

1. Introduction

The exploding demand for ubiquitous computing and communications in the last decade has exemplified the need for significant improvements in the spectrum utilization efficiency, energy efficiency and the data rates supported by the emerging wireless systems, while ensuring the integrity of data transmission over the noisy and time-varying wireless links. Link adaptation (wherein the source signalling rate, transmitted power level, coding rate, constellation size, packet length, and/or other system parameters are "matched' to the prevailing channel condition) is known to be an effective communication technique for increasing the data rate and spectral efficiency of wireless data centric networks. In recent years, a new communication paradigm known as "cooperative communications" has received a considerable attention owing to its ability to exploit the broadcast nature of wireless transmissions for harnessing a new form of spatial diversity to combat the deleterious effects of multipath fading [1]. Specifically, this form of "user cooperation diversity" facilitates an evolutionary path for reaping the benefits of multiple-input-multiple-output (MIMO) antenna technology with existing small form-factor hand-held devices that are not equipped with an antenna array (e.g., cell-phones, sensor nodes).

The cooperative relaying architecture also offers a modular and flexible solution to meet a prescribed design objective (e.g., data rate, error rate, energy constraint, etc.) and quality of service assurance by enabling the source node to tap into the available resources of local neighbouring nodes to increase its throughput, range, reliability, and covertness. This feature makes it very attractive for a wide range of applications including battlefield communications, first-responder and disaster management networks, cellular communications, wireless sensor networks, vehicular/mobile ad-hoc networks, among many others.

An intermediate node (i.e., relay) in a cooperative relay network may either amplify what it receives (in case of amplify-and-forward relaying protocol) or digitally decodes, and re-encodes the source information (in case of decode-and-forward relaying protocol) before re-transmitting it to the destination node. Other variations of cooperative relaying strategies include compress-and-forward, opportunistic, incremental, variable-gain and fixed-gain (either blind or semi-blind) relaying that are implemented based on the availability of channel side-information (CSI) and the number of active participating nodes for information relaying. In this article, we primarily focus on variable-gain cooperative amplify-and-forward (CAF) relaying protocol because it does not require "sophisticated" transceivers at the relays, although our ergodic capacity analysis may be extended to other categories and variations of cooperative diversity schemes if the moment generating function (MGF) of end-to-end signal-to-noise ratio (SNR) is available. While this protocol can achieve a full diversity by forming a virtual antenna array, there is a loss of spectral efficiency due to its inherent half-duplex operation. This penalty, however, could be "recovered" to some extent by combining the cooperative diversity with a link adaptation mechanism and/or resorting to an opportunistic relaying strategy. The efficacy of integrating link adaptation schemes into the cooperative relay network would be of interest to the system developers of the emerging IEEE 802.16 wireless networks (i.e., the current IEEE 802.16e systems employ adaptive modulation, while the emerging IEEE 802.16j standard specifies the use of cooperative diversity in its multi-hop relay architecture).

But the art of adaptive link layer in cooperative wireless networks is still in its infancy especially when optimized in a cross-layer design paradigm. While there have been extensive prior research on performance analyses of adaptive transmission techniques for non-cooperative wireless networks, majority of the literature on cooperative diversity systems are limited to a constant signalling rate and/or fixed transmit power for all nodes. More recently, the problem of optimal resource allocation in terms of power and bandwidth has been investigated for a three node cooperative wireless network [2]-[3] and for the multi-relay case in [4], although their solutions require the knowledge of CSI of all links (i.e., large overhead especially when the number of nodes in the network is large) and the source rate-adaptation was not considered. As a consequence, it is distinctly different than the adaptive source transmission policies of [5]. Motivated by these observations, [6] and [7] derived bounds for the ergodic capacity of adaptive-link cooperative relay networks with limited CSI (i.e., the destination node only needs to feedback its total received SNR to the source node) in which the rate and/or power level at the source node is adapted according to the channel condition while the relays simply amplify-and-forward or decode-and-forward their received signals to the destination node. In [8], Ikki *et. al.* followed the PDF based analytical approach in [7] to compute the ergodic capacities of an opportunistic relay-selection scheme under different source-adaptive transmission schemes. Nevertheless, all of the aforementioned analyses were limited to Rayleigh fading channels, perhaps for analytical tractability and simplicity. But it is much more realistic to model the channel gains of a practical cooperative relay network as independent and non-identically distributed (i.n.d) Rice or Nakagami-m random variables due to increased likelihood of the presence of strong specular components between the source node and the collaborating relays within a network cluster or in an airborne platform.

Inspired by the works of [9], [10], [11], [12] and [13] for simplifying and unifying the analyses of average symbol error rates and outage probability for a wide variety of digital modulation schemes with diversity receivers over generalized fading channels, we seek to develop a novel and unified analytical framework based on the MGF method for evaluating the ergodic channel capacities of dual-hop CAF relay networks under three distinct adaptive source transmission policies considered in [5] and [7]. Our approach is also motivated by the fact that the MGF of end-to-end SNR is sometimes easier to compute than its PDF and/or it may be readily available (i.e., since it is extensively used for outage probability and error rate analysis!). This development is important and of significant interest because several authors [14]-[16] have recently argued that although the MGF method has been successfully and extensively applied for evaluating the performance of wireless relaying systems in terms of outage probability and error rates, there have been very limited contributions on ergodic capacity of fading relay channels [14, pp. 2286] or explicitly highlighted the complexity of using and generalizing the MGF or the characteristic function (CHF) based approaches for channel capacity computation [15]-[16]. In fact, the lack of significant contributions on ergodic capacity analysis of cooperative relay networks can be attributed to the difficulty of evaluating the exact PDF of end-to-end SNR in closed-form. The authors of [7] and [8] circumvented this difficulty by evaluating the upper and lower bounds for the capacity instead, while [14] relied on the Jensen's inequality to derive an upper bound for the ergodic capacity over Rayleigh fading via the method of moments. More recently, Di Renzo *et. al.* [16] has proposed a general method for channel capacity analysis using a novel integral relation known as E_i –transform. But their analytical framework exhibits the following limitations:

  i.   The integral solutions (see [16, eq. (7) or eq. (8)] for optimal rate-adaptation with constant transmit power (ORA) policy and [16, eqs. (13)-(14) or eq. (28)] for optimal joint power-and-rate adaptation (OPRA) policy) require calculations of the derivatives of the MGF and/or its auxiliary functions, which can be very cumbersome for i.n.d fading statistics especially when the number of relays is large;

  ii.   Numerical results were limited only to independent and identically distributed (i.i.d) Nakagami channels perhaps due to the difficulty in computing a numerically stable MGF for the total received SNR over i.n.d Rice and/or Nakagami-Hoyt fading environments using [17, eq. (5)] along with infinite series expressions summarized in Table 1 of [17]. Even for the Nakagami-m fading environment, the final expressions for the ergodic capacities of two-hop CAF networks with ORA and OPRA policies are in the form of a double-integral and a triple-integral, respectively.

Independently, we have developed yet another unified approach for ergodic capacity analysis of cooperative relay networks over generalized fading channels which we refer to as the "cumulative distribution function (CDF) method" [12], [18]. This approach utilizes the MGF of end-to-end SNR in conjunction with an efficient multi-precision Laplace inversion formula [11], [12] to compute its CDF. Interestingly, its solution for the OPRA transmission strategy appears to be considerably simpler than that of derived in [16]. Moreover, the computational complexity for evaluating the ergodic capacities of ORA and OPRA techniques is in the form of a double-integral if the MGF of total received SNR can be specified in a closed-form.

Motivated by the above arguments and observations, the key contributions of this paper are summarized below:

  i.   We found an exponential-type integral representation for $\ln \gamma$, $\gamma > 0$, in which the conditional fading SNR appears only in the exponent (i.e., facilitates the averaging over fading density functions) similar to the transformation of the conditional error rate expressions as in [9]. Incidentally, our expression (A.5) reduces into [19, eq. (6)] although we show that this integral representation is valid for any $\gamma > -1$ (instead of the $\gamma > 0$ constraint given by [19, Lemma 1]). This distinction is critical since [19, eq. (6)] cannot be used for the evaluating the channel capacity with side information at both the transmitter and the receiver (as argued in [16]) without the correction.

ii. We present a novel MGF-based analytical framework for calculating the ergodic capacities of CAF relay networks with limited CSI under three distinct source-adaptive transmission policies (optimal rate-adaptation with fixed transmit power (ORA), optimal joint power and rate adaptation (OPRA), and truncated channel inversion with fixed signaling rate (TCIFR)) over i.n.d Rice and Nakagami-m fading environments. The corresponding formulas are considerably simpler and more efficient than those reported in [16] and [18].

iii. To facilitate an efficient analysis of ergodic capacity of multi-relay dual-hop CAF networks over i.n.d Rice and Nakagami-m fading channels, we derive both upper and lower bounds for the MGF of half-harmonic mean SNR in closed-form. To the best of our knowledge, ergodic capacity analyses of CAF relay networks over Rice fading environments are not available in the literature. This might be attributed to the difficulty in computing the MGF of harmonic mean SNR for each of relayed paths using the approach presented in [17] (because it involves evaluation of an integral of a product of two infinite series terms with complicated arguments). Even when a closed-form expression for the MGF bound of harmonic mean SNR (see eq. (6)) is available, it cannot be used directly in (14) and (17) or [18, eqs. (11), (15) and (19)] because we need to compute the Marcum's Q-function with complex arguments (which is not supported by the standard built-in functions in MATLAB and MATHEMATICA). We overcome this issue by resorting to a rapidly converging series representation of the Marcum's Q-function (see eq. (10)).

It is also important to highlight that our proposed MGF-based framework is appropriate for the ergodic capacity analysis of CAF relay networks with source adaptive techniques if a bound, an approximation or the exact MGF of total received SNR is available in closed-form. However, the CDF-based framework developed in [18] is more suitable if the CDF of the total received SNR is available in closed-form instead (e.g., opportunistic relay selection at the sender with selection combining at the destination node).

The remainder of this paper is organized as follows. In Section 2, the system and channel models are briefly discussed, including the development of closed-form expressions for the MGF bounds of harmonic mean SNR for dual-hop CAF multi-relay networks in Rice and Nakagami-m fading environments. The ergodic capacity analyses of CAF multi-relay networks under three distinct source adaptive transmission strategies (ORA, TCIFR and OPRA) are presented in Section 3. Selected computational results are presented in Section 4, which is followed by some concluding remarks in Section 5.

2. SYSTEM MODEL AND CHANNEL STATISTICS

Fig. 1 illustrates the link-adaptive cooperative wireless network under consideration. The source node S communicates with the destination node D via a direct-link and through N amplify-and-forward relays, $R_i,\ i \in \{1, 2, ..., N\}$, in two transmission phases. During the initial phase, node S broadcasts signal x to node D as well as to the relays R_i, where the channel fading coefficients between S and D, S and the ith relay node R_i, and R_i and D are denoted by $\alpha_{s,d}$, $\alpha_{s,i}$ and $\alpha_{i,d}$, respectively. During the second phase of cooperation, each of the N relays transmits the received signal after amplification via orthogonal transmissions (e.g., time-division multiple access in a round-robin fashion and/or frequency division multiple access).

If the ith relay amplifier gain is chosen as $G_i = \sqrt{E_s / (E_s |\alpha_{s,i}|^2 + N_0)}$ (where E_s denotes the average symbol energy and N_0 corresponds to the noise variance) and maximal-ratio combiner (MRC) is used to coherently combine all the signals during the two transmission phases, the total received SNR at the output of MRC detector is given by [1]

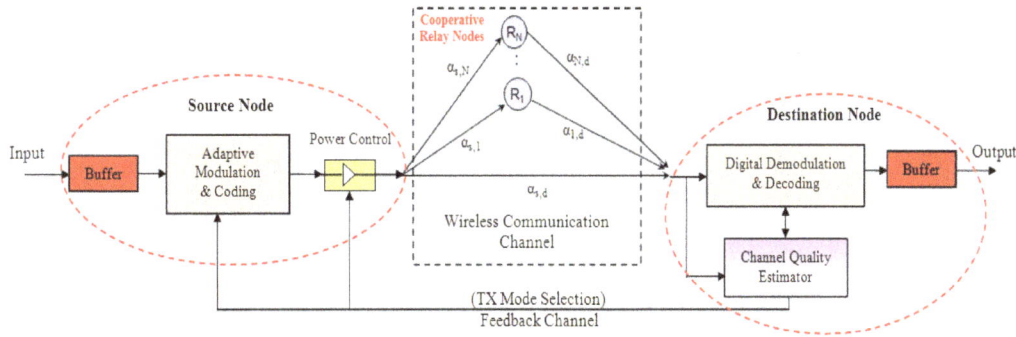

Fig. 1 Link-adaptive cooperative diversity network for ensuring the connectivity and network stability needed to support varying quality-of-service requirements in a wireless network.

$$\gamma = \gamma_{s,d} + \sum_{i=1}^{N} \frac{\gamma_{s,i} \gamma_{i,d}}{1 + \gamma_{s,i} + \gamma_{i,d}} = \gamma_{s,d} + \sum_{i=1}^{N} \gamma_i \approx \gamma_{s,d} + \sum_{i=1}^{N} \gamma_i^{(TB)} \tag{1}$$

where $\gamma_i^{(TB)} = \dfrac{\gamma_{s,i} \gamma_{i,d}}{\gamma_{s,i} + \gamma_{i,d}}$ and $\gamma_{a,b} = |\alpha_{a,b}|^2 E_s/N_o$ corresponds to the instantaneous SNRs of link a-b. The approximation on the right-side of (1) is obtained by recognizing that the instantaneous SNR of a two-hops path can be accurately estimated to be the half harmonic mean of individual link SNRs especially at moderate and high SNR regimes [22]. Suppose $\gamma_{s,d}$, $\gamma_{s,i}$, $\gamma_{i,d}$ are i.n.d random variates, we can immediately show that the MGF of γ in (1) is given by

$$\phi_\gamma(s) = \phi_{\gamma_{s,d}}(s) \prod_{i=1}^{N} \phi_{\gamma_i}(s) \tag{2}$$

where $\phi_{\gamma_{s,d}}(s)$ corresponds to the MGF of SNR of the S-D link while and $\phi_{\gamma_i}(s)$ denotes the exact MGF of end-to-end SNR for a dual-hop relayed path. However, finding $\phi_{\gamma_i}(s)$ in a generalized fading environment can be a very daunting task, with existing results limited only to Rayleigh [23] and Nakagami-m [24] environments. Even in such cases, the final expressions are too complicated for further manipulations. For example, the exact MGF of SNR for a dual-hop CAF with i.n.d Nakagami-m fading statistics derived in [24] involves triple summation terms involving k^{th} derivative of a product of Whittaker functions, which is not easily evaluated using a general computing platform, besides being restrictive to positive integer m values. Other "exact" formulas (i.e., half-harmonic mean bound of the exact end-to-end SNR, $\gamma_i^{(TB)}$) for the MGF of SNR in a relayed path can be found in [21] (for i.n.d Rayleigh channels), [22] (for i.i.d Nakagami-m channels) and [17]. While the MGF-based approach developed in [17] is quite interesting and can be applied to a wide range of fading environments, the resulting integral expressions are often too complicated to compute and/or time-consuming (due to the need to evaluate a nested two-fold integral term with complicated arguments that might include infinite series in some cases such as Rice fading). To circumvent the aforementioned difficulties, both upper and lower bounds have been proposed and developed for $\gamma_i^{(TB)}$, viz.,

$$\gamma_i^{(LB)} = \tfrac{1}{2} \min(\gamma_{s,i}, \gamma_{i,d}) \le \gamma_i^{(TB)} \le \gamma_i^{(UB)} = \min(\gamma_{s,i}, \gamma_{i,d}) \tag{3}$$

Hence performance bounds can be developed by utilizing the bounds for the MGF of total received SNR using the inequality in [7] (i.e., see (3)), viz.,

$$\phi_{\gamma_{s,d}}(s) \prod_{i=1}^{N} \phi_{\gamma_i}^{(UB)}(s) \le \phi_\gamma(s) \le \phi_{\gamma_{s,d}}(s) \prod_{i=1}^{N} \phi_{\gamma_i}^{(LB)}(s) \tag{4}$$

where $\phi_{\gamma_i}^{(UB)}(s)$ and $\phi_{\gamma_i}^{(LB)}(s) = \phi_{\gamma_i}^{(UB)}(s/2)$ correspond to the MGFs of $\gamma_i^{(UB)}$ and $\gamma_i^{(LB)}$, respectively.

Next, we will summarize and/or derive the MGFs of $\gamma_i^{(TB)}$ and $\gamma_i^{(UB)}$ in closed-form for Rayleigh, Rice and Nakagami-m fading environments.

2.1 Rayleigh Fading Channel

The MGF of $\gamma_i^{(TB)}$ in a Rayleigh channel with i.n.d fading statistics is given by [21]

$$\phi_{\gamma_i}^{(TB)}(s) = \left[(1/\Omega_{s,i} - 1/\Omega_{i,d})^2 + (1/\Omega_{s,i} + 1/\Omega_{i,d})s \right]/\Delta^2 + \frac{2s}{\Delta^3 \Omega_{s,i} \Omega_{i,d}} \ln\left(\left(s + \Delta + \frac{1}{\Omega_{s,i}} + \frac{1}{\Omega_{i,d}} \right)^2 \frac{\Omega_{s,i}\Omega_{i,d}}{4} \right) \tag{5}$$

where $\Delta = \sqrt{(1/\Omega_{s,i} - 1/\Omega_{i,d})^2 + 2s(1/\Omega_{s,i} + 1/\Omega_{i,d}) + s^2}$, and $\Omega_{a,b} = E[\gamma_{a,b}]$ denotes the mean link SNR .

2.2 Rice Fading Channel

To the best of our knowledge, analysis and/or computational results for the ergodic capacities of adaptive-link CAF multi-relay networks over Rice fading have not been considered in the literature. This may be attributed to analytical intractability/numerical instability that arises in the computation of the MGF of total received SNR (i.e., (2)) in conjunction with the MGF of half-harmonic mean SNR for each of the relayed paths via [17, eq. (5) and Table 1]. In this case, evaluation of the MGF of $\gamma_i^{(TB)}$ for each relayed paths involves an integration of a product of two infinite series containing modified Bessel functions! Hence in the Appendix D, we derive a tractable and closed-form expression for the MGF of $\gamma_i^{(UB)}$ and $\gamma_i^{(LB)}$ over Rice fading with i.n.d fading statistics, viz.,

$$\phi_{\gamma_i}^{(UB)}(s) = \sum_{\substack{k \in \{(s,i),(i,d)\} \\ j \neq k}} 2A_k e^{-K_k} I\left[\sqrt{2A_j}, \sqrt{2K_j}, \sqrt{2A_k K_k}, 2(s+A_k) \right] \tag{6}$$

where $A_i = 1 + K_i/\Omega_i$, $I[a,b,c,d] = \frac{1}{d} e^{\frac{c^2}{2d}} Q\left(b\sqrt{\frac{d}{d+a^2}}, \frac{ac}{\sqrt{d(d+a^2)}} \right) - \frac{a^2}{d(d+a^2)} e^{\frac{c^2-b^2 d}{2(d+a^2)}} I_0\left(\frac{abc}{d+a^2} \right)$ and the

corresponding MGF for the lower bound may be computed as $\phi_{\gamma_i}^{(LB)}(s) = \phi_{\gamma_i}^{(UB)}(s/2)$.

2.3 Nakagami-m Fading Channel

A simple closed-form formula for the MGF of $\gamma_i^{(TB)}$ over Nakagami-m fading environment with i.i.d fading statistics has been derived in [22], viz.,

$$\phi_{\gamma_i}^{(TB)}(s) = {}_2F_1\left(m, 2m; m + 0.5; -s\Omega/4m \right) \tag{7}$$

where m denotes the Nakagami-m fading severity index. To facilitate performance analyses in a more realistic operating environment with i.n.d fading statistics among the spatially distributed relay nodes, we also derive a closed-form formula for the MGF of $\gamma_i^{(UB)}$ in the Appendix E for a dual-hop CAF relayed path, viz.,

$$\phi_{\gamma_i}^{(UB)}(s) = \sum_{\substack{k \in \{(s,i),(i,d)\} \\ j \neq k}} \frac{\Gamma(m_k + m_j)}{m_k \Gamma(m_k)\Gamma(m_j)} \left(\frac{\Omega_j m_k}{s\Omega_j\Omega_k + \chi_{j,k}} \right)^{m_k} {}_2F_1\left(1 - m_j, m_k; 1 + m_k; \frac{(s\Omega_k + m_k)\Omega_j}{s\Omega_j\Omega_k + \chi_{j,k}} \right) \tag{8}$$

where $\chi_{j,k} = \Omega_j m_k + \Omega_k m_j$. It is also important to highlight that the Gauss hypergeometric function in (8) will reduce into a finite series if $m_{s,i}$ and $m_{i,d}$ are positive integers.

2.4 Outage Probability

Outage probability is defined as the probability that the total received SNR γ (as defined in (1)) falls below a specified threshold value γ_0, i.e., $P_{\text{out}} = F_\gamma(\gamma_0)$. The knowledge of P_{out} or equivalently, the CDF of total received SNR is required in the evaluation of OPRA and TCIFR capacities of CAF relay networks. In general, analytical derivation of the CDF of γ is not necessarily trivial. However, this quantity can be evaluated numerically and efficiently with the aid of Abate's multi-precision Laplace inversion formula [12] once the MGF of γ is found, viz.,

$$F_\gamma(x) \cong \frac{1}{2Z}\phi_\gamma(r)e^{rx} + \frac{r}{Z}\sum_{k=1}^{Z-1}\mathrm{Re}\{\frac{1+j\sigma(\theta_k)}{s(\theta_k)}e^{xs(\theta_k)}\phi_\gamma(s(\theta_k))\}$$ (9)

where $r = 2Z/(5x)$, $\theta_k = k\pi/Z$, $\sigma(\theta_k) = \theta_k + (\theta_k\cot(\theta_k)-1)\cot(\theta_k)$, $s(\theta_k) = r\theta_k(j+\cot(\theta_k))$ and the positive integer Z in (9) can be chosen appropriately to achieve the desired accuracy. It should be emphasized that the above is not the only recommended solution but rather one may also resort to other numerical Laplace inversion techniques such as the Euler summation approach in [11] or via the saddle-point approximation.

Nevertheless, computations of outage probability bounds for CAF multi-relay networks over Rice fading channels using (6) in (9) or [11] pose some difficulties (since they require complex arguments to be passed into the MGF formula (6) that contains the Marcum Q-function). Unfortunately, the built-in function for evaluating the Marcum Q-function on commercial mathematical software packages such as MATLAB does not support complex arguments. To overcome this minor snag, we wrote a new MATLAB routine for computing the generalized Marcum Q-function (that can handle complex arguments) based on its rapidly convergent canonical series representation given by [25, eq. (7)],

$$Q_M(\sqrt{2a},\sqrt{b}) = 1 - \sum_{k=0}^{\infty}\frac{a^k e^{-a}}{k!}\frac{G(M+k,\frac{b}{2})}{\Gamma(M+k)}$$ (10)

where $G(a,x) = \int_0^x t^{a-1}e^{-t}dt$ denotes the lower incomplete Gamma function. Although the saddle point approximation circumvents the requirement for $Q(.,.)$ to handle complex arguments, this method is not very attractive for our application because it requires the first two derivatives of cumulant generating function (CGF) $K_\gamma(s) = \ln(\phi_\gamma(s))$ and also the solution to a root-finding problem that involves first-order derivative of the nonlinear CGF expression.

3. ERGODIC CAPACITY COMPUTATION IN FADING CHANNELS

Ergodic capacity is an important and a basic tool for appraisal, design and optimization of new wireless communication techniques because this metric yields an information-theoretic bound on the achievable average rate for reliable communication over fading channels. It can also be used to gain insights into how and to what degree opportunistic/adaptive transmission strategies (realized by exploiting CSI that may be made available at the transmitter, or the receiver, or both) and diversity schemes can counteract the detrimental effects of fading. Thus in this section, we derive unified analytical expressions (based on a novel MGF method that is different and considerably simpler than that of [16]) for evaluating the ergodic capacities of CAF relay networks in conjunction with three adaptive source transmission techniques over a myriad of fading environments.

3.1 Optimal Rate Adaptation with a Fixed Transmit Power (ORA)

The ergodic capacity of a multi-relay CAF network (with limited CSI feedback) when only the rate at the transmitter is dynamically adapted to the time-varying channel conditions and constant transmit power constraint is given by [7]

$$\frac{\overline{C}_{ORA}}{B} = \frac{1}{N+1}\frac{1}{\ln 2}\int_0^\infty \ln(1+\gamma)f_\gamma(\gamma)d\gamma$$ (11)

where B and N denote the channel bandwidth and number of cooperating relays, respectively. This corresponds to fading channel capacity with side information at the receiver only [5].

Utilizing an "exponential-type" integral representation for $\ln(\gamma+1)$ (please refer to Appendix A for details), we can readily facilitate the averaging problem in (11) given that $\phi_\gamma(\cdot)$ is available in closed-form. Substituting (A.5) or [19, eq. (6)] into (11), we obtain the ergodic capacity in terms of the MGF of γ alone, viz.,

$$\frac{\bar{C}_{ORA}}{B} = \frac{1}{N+1}\frac{1}{\ln 2}\int_0^\infty \frac{e^{-x}}{x}\left[\int_0^\infty (1-e^{-x\gamma})f_\gamma(\gamma)d\gamma\right]dx$$

$$= \frac{1}{N+1}\frac{1}{\ln 2}\int_0^\infty \frac{e^{-x}}{x}\left[1-\phi_\gamma(x)\right]dx \qquad (12)$$

It is also interesting to note that (12) allows us to prove that the ergodic capacity for point-toto-point communication systems increases with the increasing receiver-diversity order regardless of the fading channel model or diversity combining technique employed. This is because the term $1-\phi_\gamma(x)$ in (12) approaches to much closer to unity as receiver-diversity order increases (since $0 \le \phi_\gamma(s) \le 1$ is a monotonically decreasing function with respect to its argument). Gunther [26, pp. 401] suggested that while this is intuitive, it is not easy to prove this trend mathematically for the ORA policy.

3.2 Truncated Channel Inversion with Fixed Rate (TCIFR)

The ORA capacity in Section 3.1 can be achieved using a fixed-power variable-rate coding strategy. However, some real-time applications cannot tolerate the variable delays exhibited by this coding strategy. For these applications, the transmitter may use a fixed-rate coding and adapt its power to keep the total received SNR constant (such that the "channel fading is inverted"). The zero-outage capacity (also known as delay-limited capacity) is given by (13) when the cut-off SNR γ_0 is set to zero. This technique is the least complex to implement given that reliable CSI estimates are available at the transmitter and receiver [5].

However when the channel experiences deep fades, the penalty in transmit power requirement with the channel inversion strategy will be enormous because channel inversion needs to compensate for the deep fades. To address this practical implementation issue, a modified channel inversion policy was also considered in [5]. In the TCIFR technique, channel is only inverted when the received SNR γ is above a certain cut-off fade depth γ_0. In this case, it is easy to show that the outage probability is given by $P_{out} = F_\gamma(\gamma_0)$ (since data transmission is ceased when $\gamma < \gamma_0$) and the corresponding ergodic capacity as

$$\frac{\bar{C}_{TCIFR}}{B} = \frac{1}{N+1}\log_2\left(1+\frac{1}{\int_{\gamma_0}^\infty \gamma^{-1}f_\gamma(\gamma)d\gamma}\right)F_\gamma^c(\gamma_0) \qquad (13)$$

where $F_\gamma^c(x) = 1 - F_\gamma(x)$ denotes the complementary CDF of γ.

Substituting (B.2) (i.e., the PDF of γ is expressed as an inverse Fourier transform integral of its CHF) into (13), we obtain (after some manipulations as discussed in Appendix B)

$$\frac{\bar{C}_{TCIFR}}{B} = \frac{1}{N+1}\log_2\left(1-\frac{1}{\nabla}\right)F_\gamma^c(\gamma_0) \qquad (14)$$

where $\nabla = \frac{1}{\pi}\int_0^\infty \mathrm{Re}\{\phi_\gamma(-j\omega)E_i(-j\omega\gamma_0)\}d\omega$ and the exponential integral with an imaginary argument $E_i(-jc)$ may be evaluated in MATLAB as "$\cos\mathrm{int}(c) - j(-\pi/2 + \sin\mathrm{int}(c))$". Moreover, the cut-off SNR γ_0 in (13) can be chosen to meet a specified P_{out} or to maximize $\bar{C}_{TCIFR}/B$.

3.3 Optimal Joint Power-and-Rate Adaptation (OPRA)

Under an average transmit power constraint, the OPRA strategy seeks to dynamically adapt both the transmission power and rate relative to the channel quality through use of a multiplexed multiple codebook design [5]. This leads to the highest achievable capacity with perfect CSI at the transmitter and the receiver. In [29], the authors' have advocated a simpler approach (from the implementation stand-point) to achieve this capacity using a single codebook (i.e., fixed rate) with variable power transmission because the former is inherently hard to implement. From [5], we have

$$\frac{\overline{C}_{OPRA}}{B} = \frac{1}{N+1}\frac{1}{\ln 2}\int_{\gamma_0}^{\infty}\ln\left(\gamma/\gamma_0\right)f_{\gamma}(\gamma)d\gamma \tag{15}$$

where γ_0 is the optimal cut-off SNR below which the data transmission is suspended. Substituting (A.4) into (15), and changing the order of integration, we get

$$\frac{\overline{C}_{OPRA}}{B} = \frac{1}{N+1}\frac{1}{\ln 2}\int_0^{\infty}\frac{1}{x}\left[e^{-2x}\int_{\gamma_0}^{\infty}f_{\gamma}(\gamma)d\gamma - \int_{\gamma_0}^{\infty}e^{\frac{-2x\gamma}{\gamma_0}}f_{\gamma}(\gamma)d\gamma\right]dx$$

$$= \frac{1}{N+1}\frac{1}{\ln 2}\int_0^{\infty}\frac{1}{x}\left[e^{-2x}[1-F_{\gamma}(\gamma_0)]-\phi_{\gamma}(2x/\gamma_0,\gamma_0)\right]dx \tag{16}$$

where $\int_{\gamma_0}^{\infty}f_{\gamma}(\gamma)d\gamma = 1 - F_{\gamma}(\gamma_0) = F_{\gamma}^c(\gamma_0)$ and $\int_{\gamma_0}^{\infty}e^{-s\gamma}f_{\gamma}(\gamma)d\gamma = \phi_{\gamma}(s,\gamma_0)$ denotes the marginal MGF of random variable γ. Since the desired marginal MGF is typically not available in closed-form, we circumvent this difficulty by computing this term as the "CDF" of an auxiliary function using (9) and (C.2) (additional details are provided in Appendix C), viz.,

$$\frac{\overline{C}_{OPRA}}{B} = \frac{1}{N+1}\frac{1}{\ln 2}\int_0^{\infty}\frac{1}{x}\left[e^{-2x}[1-F_{\gamma}(\gamma_0)]-\phi_{\gamma}\left(2x/\gamma_0\right)+F_{\hat{\gamma}}(\gamma_0)\right]dx \tag{17}$$

where the term $F_{\hat{\gamma}}(\gamma_0)$ can be evaluated using (9) or [11] but with $\phi_{\gamma}(s+2x/\gamma_0)$ instead of $\phi_{\gamma}(s)$. Therefore it is evident that if $\phi_{\gamma}(s)$ is available in closed-form, then the integrand in (17) can be evaluated very efficiently using (9), and computational complexity of (17) is no more complicated that of a double integral.

To achieve the capacity (17), the channel fade level (i.e., CSI) tracked at the receiver must be conveyed to the transmitter on the feedback path for dynamic power and rate adaptation. Since data transmission is suspended when $\gamma < \gamma_0$, this optimal adaptation policy suffers a probability of outage given by $P_{out} = F_{\gamma}(\gamma_0)$, which equals to the probability of no transmission. Moreover, the optimal cut-off SNR γ_0 must satisfy

$$F_{\gamma}^c(\gamma_0) - \gamma_0[1 + \int_{\gamma_0}^{\infty}\gamma^{-1}f_{\gamma}(\gamma)d\gamma] = 0 \tag{18}$$

The integral term in (18) can be evaluated very efficiently by following the development of (14) or (B.3), especially when the MGF of γ is available in closed-form. Hence we can find the optimal cut-off SNR γ_0 by solving the equation $F_{\gamma}^c(\gamma_0) = \gamma_0(1+\nabla)$ for γ_0 numerically. Furthermore, asymptotic analysis of (18) shows that $\gamma_0 \to 0$ when the mean SNR $\Omega \to 0$ because $F_{\gamma}(\gamma) \to 1$ and $f_{\gamma}(\gamma) \to 0$ (i.e., the effect of $\Omega \to 0$ can be predicted from the normalized PDF and CDF curves when its argument $\gamma \to \infty$). Similarly, we observe that $\gamma_0 \to 1$ as $\Omega \to \infty$ because $F_{\gamma}(\gamma) \to 0$ and $\phi_{\gamma}(s) \to 0$. Therefore, the optimal cut-off SNR γ_0 always lies in the interval [0, 1] regardless of the assumptions on fading channel models, diversity techniques used an/or the number of cooperating relay nodes.

Before concluding this section, we would like to highlight that our final ergodic capacity formulas in (12), (14) and (17) for ORA, TCIFR and OPRA schemes, respectively, are all expressed in terms of only the MGF and/or the CDF of the total received SNR. Hence we can efficiently compute the appropriate bounds or tight approximations for the ergodic capacity of CAF relay networks under different source adaptive transmission techniques and fading environments by utilizing the closed-form MGF formulas for the exact half-harmonic mean SNR (i.e., (5) and (7)) or bounds for the half-harmonic mean SNR (i.e., (6) and (8)) in conjunction with (2) and (9). It is evident that our formulas are considerably simpler and more efficient than the MGF-approach based on the E_i-transform proposed in [16].

4. NUMERICAL RESULTS

In this section, selected numerical results are presented for the ergodic capacities of CAF relay networks with different source-adaptive transmission techniques over Rayleigh, Nakagami-m and Rice fading environments. One of the objectives here is to investigate the accuracy, reliability and numerical stability of our proposed analytical framework. The following average link SNRs (arbitrarily chosen) will be used to generate the plots in this paper, unless stated otherwise: $\Omega_{s,1} = \Omega_{2,d} = E_s/N_o$, $\Omega_{s,2} = \Omega_{1,d} = 0.5E_s/N_o$, and $\Omega_{s,d} = 0.2E_s/N_o$.

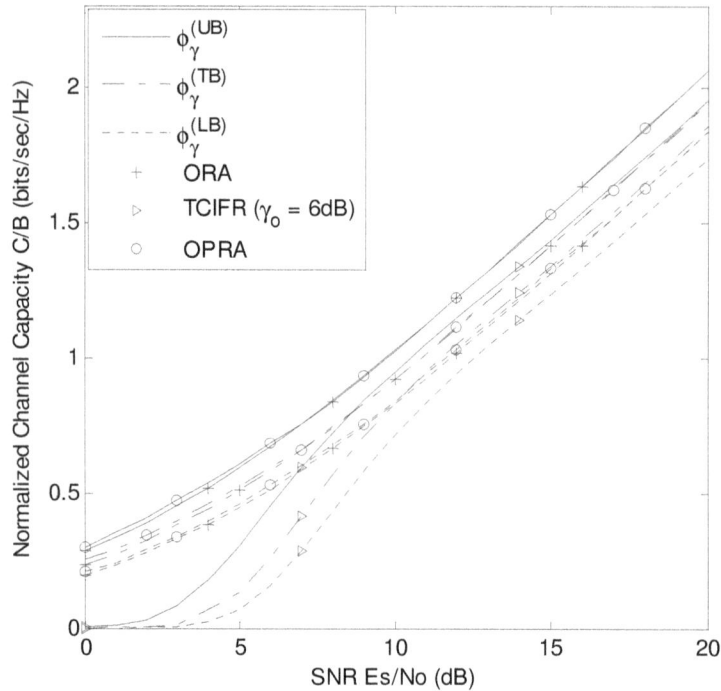

Fig. 2 Ergodic capacities of a cooperative relay network with ORA, OPRA and TCIFR policies in an i.n.d Rayleigh fading channel with two cooperative relays ($N = 2$).

Fig. 2 depicts a comparison between the ergodic channel capacities of three source adaptive transmission policies in an i.n.d Rayleigh fading channels with two cooperative relays. The curves corresponding to the upper and lower bounds were obtained using (4) and (8) by setting $m = 1$ (or using (6) by setting $K = 0$) while the curves corresponding the "tight-bound" case is obtained using (2) in conjunction with (5). It is apparent that the ergodic capacity with both transmitter and receiver CSI is only negligibly larger than the capacity with receiver CSI only (i.e., since there is no significant difference observed in the capacities of OPRA and ORA at high SNR). But the ergodic capacity of TCIFR policy (plotted for the cut-off SNR $\gamma_0 = 6$ dB) is considerably lower than the OPRA and ORA schemes. Although not shown in this figure, we also noticed that the curves corresponding to the "tight bound" are in good agreement with the Monte Carlo simulation results that corresponds to the "exact" ergodic capacity). Moreover, the "exact" ergodic capacity is slightly closer to the lower bound (rather than the upper bound) especially at lower values of E_s/N_o. Although [7] has studied the channel capacities of cooperative relaying system in an i.n.d Rayleigh fading channel, but their framework does not lend itself to the analysis of the "tight bound" case or generalize to any other fading channels, whereas our framework encapsulates all these cases in a unified way (e.g., see Fig. 4 for i.i.d Nakagami-m fading, Fig. 5 for i.n.d Rice fading and Fig. 7 for i.n.d Nakagami-m fading). This in turn demonstrates the generality and utility of our proposed analytical framework, even for the specific instance of ergodic capacity analysis in Rayleigh fading.

In Fig. 3, the ergodic capacity of a CAF relay network with TCIFR policy is plotted as a function of the cut-off SNR at $E_s/N_o = 6$ dB and $E_s/N_o = 15$ dB. It is evident that there exists an optimal choice for the cut-off SNR which maximizes the ergodic channel capacity when E_s/N_0 is fixed. But it should be also emphasized that the selection of γ_0 will directly affect the outage probability performance (i.e., probability of no transmission).

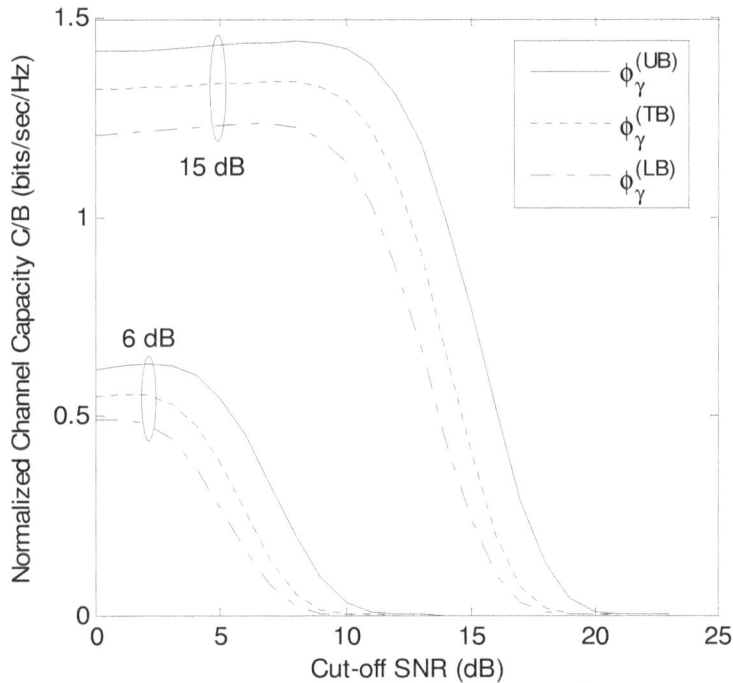

Fig. 3 Ergodic capacity of a CAF relay network with two cooperating relays and TCIFR policy plotted as a function of cut-off SNR γ_0 in an i.n.d Rayleigh fading environment.

Fig. 4 shows the ergodic capacities of a classical 3-node CAF relay network (i.e., $N = 1$) with different source-adaptive transmission schemes over i.i.d Nakagami-m channels (for fading severity indices $m = 0.5, 1, 1.5$ and 2). These performance curves were generated using (2) in conjunction with (7). It should clear by now that our framework does not impose any restrictions of the fading severity index m (i.e., can handle non-integer m values). We observe that the ergodic capacity increases with the increasing value of m for all source adaptive transmission schemes, as anticipated (i.e., ergodic capacity is higher when the channel experiences less severe fading). We also noticed that the gap between the curves corresponding to ORA and OPRA schemes widens as m decreases, although their maximum achievable average transmission rates are quite similar at higher values of E_s/N_o.

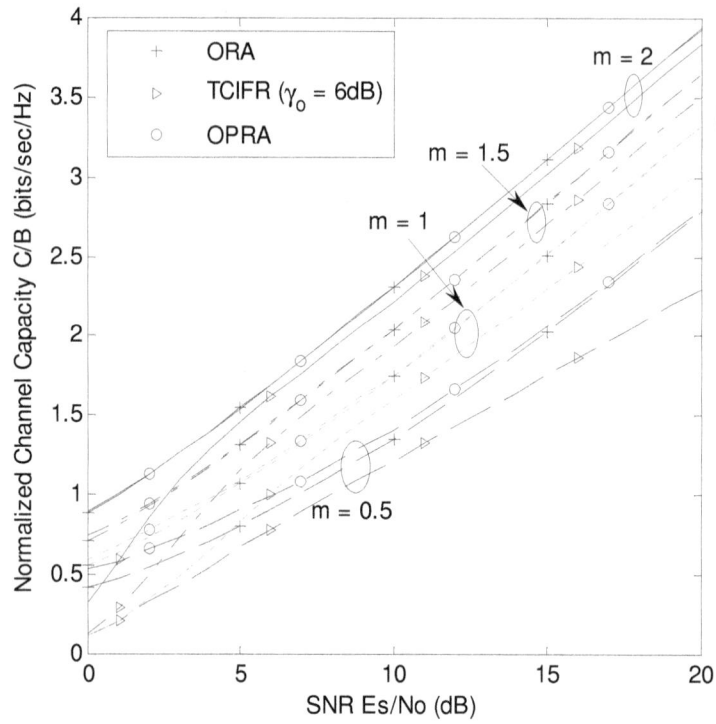

Fig. 4 Ergodic capacities of ORA, OPRA and TCIFR policies over i.i.d Nakagami-m fading environments with a single CAF relay ($N = 1$).

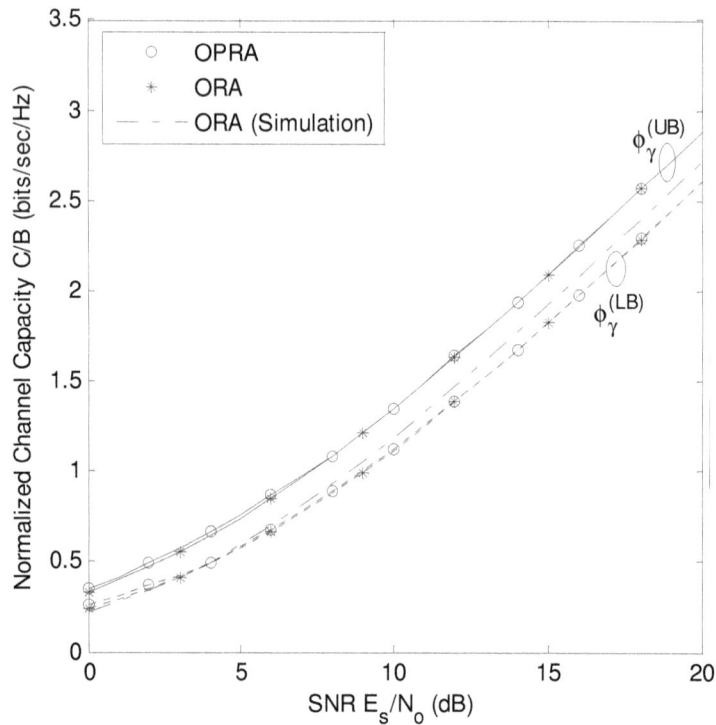

Fig. 5 Ergodic capacities of a CAF relay network ($N = 1$) with ORA and OPRA schemes over an i.n.d Rice fading environment (Rice factor $K = 3$).

Fig. 5 illustrates the ergodic capacity comparison between OPRA and ORA schemes in a 3-node CAF relay network ($N = 1$) over an i.n.d Rice fading environment ($K = 3$). To generate the curves, we have utilized (4) in conjunction with (6). The general trends observed in Fig. 2 are also observed in Fig. 3. The ergodic capacity with OPRA strategy slightly outperforms the ergodic capacity with ORA scheme at the low SNR regime, but their difference diminishes as average link SNR increases. Owing to the difficulty in deriving a tractable analytical expression for the MGF of half-harmonic mean SNR in Rice fading, we validate the tightness of our upper and lower bounds via Monte Carlo simulation. Once again, we observe that the "exact" ergodic capacity (obtained via simulation) is much closer to the lower bound than its upper bound. This example also highlights the utility of our newly derived MGF bound in (6) for the design and appraisal of CAF relay networks over i.n.d Rice fading channels.

In Fig. 6, we investigate the effect of fade distribution on the ergodic capacity of a 4-node CAF relay network in i.n.d Rice fading channels. All the curves were generated by evaluating (12) in conjunction with (2) and (6). The following Rice factors are assumed (arbitrarily chosen) for the spatially-distributed wireless links in a CAF relay network: Case 1: $K_{s,1} = 2$, $K_{1,d} = 4$, $K_{s,2} = 3$, $K_{2,d} = 3$, $K_{s,d} = 1$; Case 2: $K_{s,1} = 1.5$, $K_{1,d} = 3.5$, $K_{s,2} = 2.5$, $K_{2,d} = 2.5$, $K_{s,d} = 1$; Case 3: $K_{s,1} = 2$, $K_{1,d} = 2$, $K_{s,2} = 2$, $K_{2,d} = 2$, $K_{s,d} = 1$; Case 4: $K_{s,1} = 0$, $K_{1,d} = 0$, $K_{s,2} = 0$, $K_{2,d} = 0$, $K_{s,d} = 0$. We have verified that the curve corresponding to the Case 4 matches exactly with the results in [7] (i.e., Rayleigh fading environment). Moreover, we observe that the ergodic capacity can vary considerably (due to the amount of fading experienced on each wireless link) even when the mean link SNRs are kept constant. This observation in turn indicates that the optimal transmit power allocation strategy among nodes in cooperation (i.e., that maximizes the ergodic capacity) should also take into consideration the amount of fading experienced on different wireless links, in addition to compensating for the disparity in their mean link SNRs.

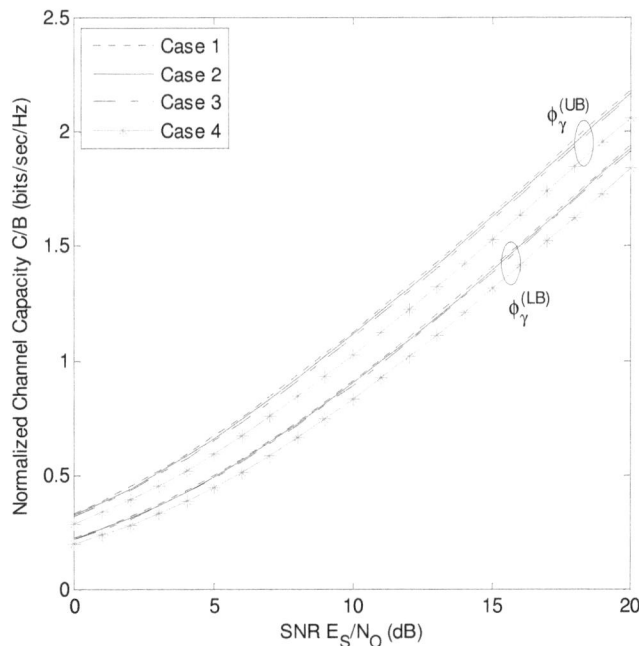

Fig. 6 Effect of the fade distributions on the ergodic capacity of a 4-node CAF relay network ($N = 2$) with ORA policy over i.n.d Rice fading channels.

Fig. 7 investigates the effect of increasing CAF diversity order on the ergodic capacity (with CSI at the receiver only) in an i.n.d Nakagami-m fading environment. The following channel parameters (arbitrarily chosen) have been considered for generating this plot: $m_{s,1} = 2$, $m_{1,d} = 2$,

$m_{s,2} = 5$, $m_{2,d} = 5$, $m_{s,d} = 1$, $\Omega_{s,d} = 0.2 E_s/N_o$, $\Omega_{1,d} = 0.5 E_s/N_o$ and $\Omega_{s,1} = \Omega_{s,2} = \Omega_{2,d} = E_s/N_o$. It is apparent that the ergodic capacity of CAF relay system is considerably higher than the non-cooperative diversity system ($N = 0$) as E_s/N_o gets smaller. However, the opposite trend is observed at larger values of E_s/N_o. This observation can be explained by noting that although CAF relaying protocol can exploit the inherent spatial diversity in wireless broadcast transmissions (which is most beneficial in poor channel conditions such as at the tactical edge or cell boundaries), there is a loss in spectral efficiency due to its inherent half-duplex operation. In fact, there is no incentive for using a cooperative diversity scheme when the source-destination link is good. This observation in turn suggests that we should adapt N based on the channel quality (i.e., increasing N as the channel condition deteriorates to provide additional diversity and maximize the capacity, while decreasing N when better channel conditions prevail).

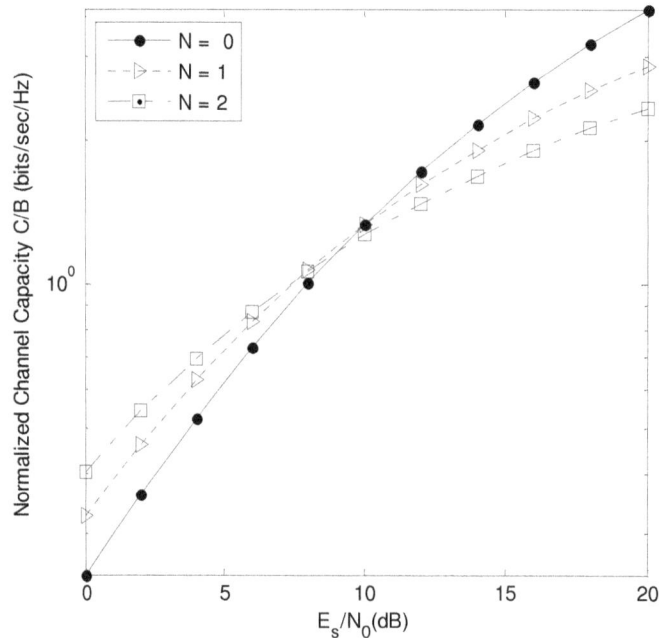

Fig.7 Effect of increasing the number of cooperating relay nodes on the ergodic capacity (upper bound) of a CAF relay network over an i.n.d Nakagami-m fading channel.

5. CONCLUSIONS

Ergodic channel capacity is an important tool for the design and appraisal of new wireless communication techniques devised to improve the spectral efficiency and/or counteract the detrimental effects of multipath fading (e.g., adaptive and opportunistic communication methods). In this article, we develop a novel MGF-based analytical framework that exploits an integral representation of the logarithmic function $\ln \gamma$, $\gamma > 0$, in a "desirable exponential-form" to unify the evaluation of ergodic channel capacities for both cooperative and non-cooperative diversity networks in conjunction with two distinct adaptive source transmission policies (i.e., ORA and OPRA) over a myriad of fading environments. The proposed method (in conjunction with the upper and lower bounds for the MGF of half-harmonic mean SNR in closed-form) also dramatically simplify the ergodic channel capacity calculations for dual-hop CAF multi-relay networks with three distinct adaptive source transmission techniques over Nakagami-m and Rice channels, and facilitates the investigation on the impact of fade distributions and/or dissimilar fading statistics across the spatially distributed communication links on the channel capacity, without imposing any restrictions on the fading parameters. Our analytical framework can be easily extended to characterize the ergodic capacities of other types of cooperative communication techniques including, but not limited to decode-and-forward and opportunistic relay selection strategies, provided the MGF of end-to-end SNR can be found.

APPENDIX A

In this appendix, we outline the derivation of an "exponential-type" integral representation for the logarithmic function $\ln \gamma$ when $\gamma > 0$. Such a representation will facilitate the averaging problem that is typically encountered in the capacity analysis over fading channels, and therefore leads to a unified approach for calculating the ergodic capacity of CAF relay networks in a myriad of fading environments. Utilizing [20, eq. (1.512.2)], we have

$$\ln \gamma = 2 \sum_{\substack{k=1 \\ k \text{ odd}}}^{\infty} \frac{1}{k} \left(\frac{\gamma-1}{\gamma+1} \right)^k = 2 \sum_{\substack{k=1 \\ k \text{ odd}}}^{\infty} \frac{1}{k} (y)^k, \quad \gamma > 0 \tag{A.1}$$

where $y = \dfrac{\gamma-1}{\gamma+1}$. Substituting $y^k = \dfrac{1}{\Gamma(k)} \displaystyle\int_0^{\infty} x^{k-1} e^{-x/y}\, dx$ [20, eq. (3.381.4)] into (A.1), we obtain

$$\ln \gamma = 2 \sum_{\substack{k=1 \\ k \text{ odd}}}^{\infty} \frac{1}{k} \left(\frac{1}{\Gamma(k)} \int_0^{\infty} x^{k-1} e^{-x/y}\, dx \right) dx = 2 \int_0^{\infty} e^{-x/y} \left(\sum_{\substack{k=1 \\ k \text{ odd}}}^{\infty} \frac{1}{k!} x^{k-1} \right) dx \tag{A.2}$$

Recognizing that $\dfrac{1}{x} \, sh\, x = \dfrac{e^x - e^{-x}}{x} = \displaystyle\sum_{\substack{k=1 \\ k \text{ odd}}}^{\infty} \frac{1}{k!} x^{k-1}$ [20, Eq. (1.411.2)], (A.2) can be re-stated as

$$\ln \gamma = 2 \int_0^{\infty} \frac{1}{x} (e^x - e^{-x}) e^{-x \left(\frac{\gamma+1}{\gamma-1} \right)} dx \tag{A.3}$$

Finally using variable substitution $x = z(\gamma-1)$, $dz = \dfrac{dx}{\gamma-1}$, we arrive at (A.4) after some routine algebraic manipulations, viz.,

$$\ln \gamma = \int_0^{\infty} \frac{1}{z} \left[e^{-2z} - e^{-2z\gamma} \right] dz, \quad \gamma > 0 \tag{A.4}$$

It is also obvious from (A.4) that

$$\ln(1 + \gamma) = \int_0^{\infty} \frac{e^{-2z}}{z} \left[1 - e^{-2z\gamma} \right] dz = \int_0^{\infty} \frac{e^{-x}}{x} \left[1 - e^{-x\gamma} \right] dx, \quad \gamma > -1 \tag{A.5}$$

Incidentally, the second term in (A.5) is identical to [19, eq. (6)]. However, it should be emphasized that in Lemma 1 of [19, eq. (6)], the author indicated that his representation is valid only for $\gamma > 0$ instead of $\gamma > -1$ (from our derivation). In fact, if one start the derivation with the power series for $\ln \gamma$ shown in [20, eq. (1.512.3)] (which was used in [19]), then the representation in (A.4) would be valid for any $\gamma \geq 0.5$. In this case, the resulting expression [19, eq. (6)] cannot be used for the ergodic capacity analysis of the OPRA policy. Perhaps for this reason, [16] abandoned the approach in [19], and try to develop yet another MGF method based on the Ei-transform to unify the analysis of ergodic capacity over generalized fading channels.

APPENDIX B

Let $\phi_X(s) = \displaystyle\int_0^{\infty} e^{-sx} f_X(x)\, dx$ and $\Phi_X(j\omega) = \displaystyle\int_0^{\infty} e^{j\omega x} f_X(x)\, dx$ denote the MGF and the characteristic function (CHF) of random variable $X \geq 0$, respectively. In this case, the CHF is related to the MGF as $\Phi_X(j\omega) = \phi_X(-j\omega)$ and the probability density function (PDF) of X (may be expressed as an inverse Fourier transform of its CHF) is given by

$$f_X(x) = \frac{1}{2\pi} \int_{-\infty}^{\infty} \phi_x(-j\omega) e^{-j\omega x} d\omega \tag{B.1}$$

If we express the CHF of random variable X in its polar form $\Phi_X(j\omega) = |\Phi_X(j\omega)| e^{j\theta(\omega)}$, then (B.1) may be re-stated as

$$
\begin{aligned}
f_X(x) &= \frac{1}{2\pi} \int_0^{\infty} \Phi_X(j\omega) e^{-j\omega x} d\omega + \frac{1}{2\pi} \int_{-\infty}^0 \Phi_X(j\omega) e^{-j\omega x} d\omega \\
&= \frac{1}{\pi} \int_0^{\infty} |\Phi_X(j\omega)| \cos(\theta(\omega) - \omega x) d\omega \\
&= \frac{1}{\pi} \int_0^{\infty} \mathrm{Re}\{\Phi_X(j\omega) e^{-j\omega x}\} d\omega
\end{aligned}
\tag{B.2}
$$

Consequently, we can simplify an integral that arises in the ergodic capacity analysis of OPRA and TCIFR techniques as

$$\int_1^{\infty} \gamma^{-1} f_\gamma(\gamma_0 \gamma) d\gamma = \frac{1}{\pi} \int_0^{\infty} \mathrm{Re}\left\{ \Phi_\gamma(j\omega) \left(\int_1^{\infty} \frac{e^{-j\omega \gamma_0 \gamma}}{\gamma} d\gamma \right) \right\} d\omega = -\frac{1}{\pi} \int_0^{\infty} \mathrm{Re}\{\phi_\gamma(-j\omega) E_i(-j\omega\gamma_0)\} d\omega \tag{B.3}$$

with the aid of (B.2) and recognizing that $Ei(-q) = -\int_q^{\infty} \frac{e^{-t}}{t} dt = -\int_1^{\infty} \frac{e^{-qt}}{t} dt$. Although the exponential integral $E_i(x)$ is usually defined for real $x < 0$, it is quite straight-forward to show that this function is also well-defined if its argument is purely imaginary. This is particularly interesting in that our unified expressions for the ergodic capacity with TCIFR policy and the transcendental equation for computing the optimal cut-off SNR γ_0 for the OPRA policy can expressed in terms of $E_i(-jc)$ where $c > 0$ is real. Utilizing the Euler identity, we can express $E_i(-jc)$ in terms of the familiar cosine-integral and sine-integrals, viz.,

$$E_i(\mp jc) = -\int_1^{\infty} \frac{\cos(ct)}{t} \pm j \int_1^{\infty} \frac{\sin(ct)}{t} = \mathrm{ci}(c) \mp j\mathrm{si}(c) \tag{B.4}$$

with the aid of [20, eq. (3.721.2) and eq. (3.721.3)]. Hence, $E_i(-jc)$ can be evaluated in MATLAB using the command line "$\cos\mathrm{int}(c) - j(-\pi/2 + \sin\mathrm{int}(c))$".

APPENDIX C

It is important to note that the knowledge of the marginal MGF of end-to-end SNR may be required while evaluating the ergodic capacity with OPRA policy (e.g., see (16)). However, this quantity is generally not available in closed-form. But if a closed-form expression for the MGF $\phi_\gamma(\cdot)$ is available, we may then use (9) (i.e., multi-precision Laplace inversion formula [12]) or [11] for computing the desired marginal MGF very efficiently as illustrated below. A similar technique was also considered in [16] for computing the truncated MGF of SNR.

Let us define an auxiliary function $f_{\hat{\gamma}}(x) = \exp(-\beta x) f_\gamma(x)$. Hence the marginal MGF of the total received SNR can be evaluated as

$$
\begin{aligned}
\phi_\gamma(\beta, \alpha) &= \int_\alpha^{\infty} e^{-\beta\gamma} f_\gamma(\gamma) d\gamma = \int_\alpha^{\infty} f_{\hat{\gamma}}(\gamma) d\gamma = F_{\hat{\gamma}}(\infty) - F_{\hat{\gamma}}(\alpha) \\
&= \phi_\gamma(\beta) - F_{\hat{\gamma}}(\alpha)
\end{aligned}
\tag{C.1}
$$

where $F_{\hat{\gamma}}(y) = \int_0^y f_{\hat{\gamma}}(x) dx$. It is obvious that $F_{\hat{\gamma}}(\infty) = \int_0^{\infty} e^{-\beta x} f_\gamma(x) dx = \phi_\gamma(\beta)$ and the "MGF" of the auxiliary function can be also expressed in closed-form, viz.,

$$\phi_{\hat{\gamma}}(s) = \int_0^\infty e^{-sx} f_{\hat{\gamma}}(x)dx = \int_0^\infty e^{-(s+\beta)x} f_\gamma(x)dx = \phi_\gamma(s+\beta) \tag{C.2}$$

Therefore we can evaluate the second term $F_{\hat{\gamma}}(\alpha)$ in (C.1) and/or the desired marginal MGF efficiently using Abate's fixed-Talbot method (9) or [11] in conjunction with (C.2).

APPENDIX D

In this appendix, we derive closed-form expressions for the upper and lower MGF bounds for the half harmonic mean SNR (i.e., relayed path of a dual-hop CAF network) over i.n.d Rice fading environments. The PDF of $\gamma_i^{(UB)} = \min(\gamma_{s,i}, \gamma_{i,d})$ for a 2-hop relayed path is given by

$$f_{\gamma_i}^{(UB)}(x) = f_{\gamma_{s,i}}(x)\left[1 - F_{\gamma_{i,d}}(x)\right] + f_{\gamma_{i,d}}(x)\left[1 - F_{\gamma_{s,i}}(x)\right] = \sum_{\substack{k \in \{(s,i),(i,d)\} \\ j \neq k}} f_{\gamma_k}(x)\left[1 - F_{\gamma_j}(x)\right] \tag{D.1}$$

where $f_{\gamma_q}(x)$ and $F_{\gamma_q}(x)$ correspond to the PDF and CDF of fading SNR γ_q respectively, which for the Rice channel are given by [27, pp. 349]

$$f_{\gamma_q}(x) = \left(\frac{1+K_q}{\Omega_q}e^{-K_q}\right)e^{-x(1+K_q)/\Omega_q}I_0\left(2\sqrt{\frac{K_q(1+K_q)x}{\Omega_q}}\right), \ x \geq 0 \tag{D.2}$$

$$F_{\gamma_q}(x) = 1 - Q\left(\sqrt{2K_q}, \sqrt{\frac{2(1+K_q)x}{\Omega_q}}\right) \tag{D.3}$$

where K_q denotes the Rice fading parameter, $\Omega_q = E[\gamma_q]$ corresponds to the mean link SNR, $I_0(.)$ is the zero-order modified Bessel function and $Q(.,.)$ is the first-order Marcum Q-function. Substituting, (D.2) and (D.3) into (D.1), we obtain

$$f_{\gamma_i}^{(UB)}(x) = \sum_{\substack{k \in \{(s,i),(i,d)\} \\ j \neq k}} \left(\frac{1+K_k}{\Omega_k}e^{-K_k}\right)e^{-x\left(\frac{1+K_k}{\Omega_k}\right)}I_0\left(2\sqrt{\frac{K_k(1+K_k)x}{\Omega_k}}\right)Q\left(\sqrt{2K_j}, \sqrt{\frac{2(1+K_j)x}{\Omega_j}}\right) \tag{D.4}$$

Now the MGF of $\gamma_i^{(UB)}$ can be computed as

$$\phi_{\gamma_i}^{(UB)}(s) = \sum_{\substack{k \in \{(s,i),(i,d)\} \\ j \neq k}} \left(\frac{1+K_k}{\Omega_k}e^{-K_k}\right)\int_0^\infty e^{-x\left(s+\frac{1+K_k}{\Omega_k}\right)}I_0\left(2\sqrt{\frac{K_k(1+K_k)x}{\Omega_k}}\right)Q\left(\sqrt{2K_j}, \sqrt{\frac{2(1+K_j)x}{\Omega_j}}\right)dx \tag{D.5}$$

The above integral can be evaluated in closed-form using the identity [28, eq. (46)], which can be simplified into (6) after some standard algebraic manipulations. The MGF of lower bound for half harmonic mean SNR $\gamma_i^{(LB)} = \gamma_i^{(UB)}/2$ is computed as $\phi_{\gamma_i}^{(LB)}(s) = \phi_{\gamma_i}^{(UB)}(s/2)$.

APPENDIX E

In this appendix, we derive the MGF of upper bound for the half harmonic mean SNR of dual-hop CAF relayed path over an i.n.d Nakagami-m fading environment. In this case, the PDF and the CDF of fading SNR γ_q in (D.1) are given by [27, pp. 349], viz.,

$$f_{\gamma_q}(x) = \left(\frac{m_q}{\Omega_q}\right)^{m_q}\frac{1}{\Gamma(m_q)}x^{m_q-1}e^{-\frac{xm_q}{\Omega_q}}, \ x \geq 0 \tag{E.1}$$

$$F_{\gamma_q}(x) = 1 - \frac{\Gamma\left(m_q, \frac{m_q x}{\Omega_q}\right)}{\Gamma(m_q)} \tag{E.2}$$

where m_q is the Nakagami-m fading severity index and $\Gamma(a,x) = \int_x^\infty t^{a-1}e^{-t}dt$ denotes the upper incomplete Gamma function.

Substituting (E.1) and (E.2) in to (D.1), we obtain

$$f_{\eta}^{(UB)}(x) = \sum_{\substack{k \in \{(s,i),(i,d)\} \\ j \neq k}} \frac{1}{\Gamma(m_k)\Gamma(m_j)} \left(\frac{m_k}{\Omega_k}\right)^{m_k} x^{m_k-1} e^{-\frac{x m_k}{\Omega_k}} \Gamma\left(m_j, \frac{m_j x}{\Omega_j}\right) \tag{E.3}$$

Taking the Laplace transform of (E.3), we get the MGF of $\gamma_i^{(UB)}$ as

$$\phi_{\eta}^{(UB)}(s) = \sum_{\substack{k \in \{(s,i),(i,d)\} \\ j \neq k}} \frac{1}{\Gamma(m_k)\Gamma(m_j)} \left(\frac{m_k}{\Omega_k}\right)^{m_k} \int_0^{\infty} x^{m_k-1} e^{-x\left(\frac{s\Omega_k+m_k}{\Omega_k}\right)} \Gamma\left(m_j, \frac{m_j x}{\Omega_j}\right) dx \tag{E.4}$$

The above integral can be evaluated in closed-form using [20, eq. (6.455.1)] and Kummer's transformation identity [20, eq. (9.131.1)], viz.,

$$\int_0^{\infty} x^{\mu-1} e^{-\beta x} \Gamma(v, \alpha x) dx = \frac{\Gamma(\mu+v)}{\mu(\alpha+\beta)^{\mu}} {}_2F_1\left(\mu, 1-v; \mu+1; \frac{\beta}{\alpha+\beta}\right) \tag{E.5}$$

where ${}_2F_1(a, b; c; z)$ denotes the Gauss hypergeometric function.

Next, simplifying (E.4) using identity (E.5), we obtain the MGF of $\gamma_i^{(UB)}$ as shown in (8).

ACKNOWLEDGEMENT

This work was supported in part by funding from the Air Force Research Laboratory/Clarkson Aerospace, and the National Science Foundation (0931679 and 1040207).

REFERENCES

[1] N. Laneman, D. Tse and G. Wornell, (2004) "Cooperative Diversity in Wireless Networks: Efficient Protocols and Outage Behaviour," *IEEE Trans. Info. Theory,* vol. 50, pp. 3062-3080.

[2] A. Madsen and J. Zhang, (2005) "Capacity Bounds and Power Allocation for Wireless Relay Channels," *IEEE Trans. Information Theory*, vol. 51, pp. 2020-2040.

[3] D. Gunduz and E. Erkip, (2007) "Opportunistic Cooperation by Dynamic Resource Allocation," *IEEE Trans. Wireless Communications*, vol. 6, pp. 1446-1454.

[4] Y. Zhao, R. Adve and T. J. Lim, (2007) "Improving Amplify-and-Forward Relay Networks: Optimal Power Allocation versus Selection," *IEEE Trans. Wireless Communications*, vol. 6, pp. 3114-3123.

[5] A. J. Goldsmith and P. Varaiya, (1997) "Capacity of Fading Channels with Channel Side Information," *IEEE Trans. Information Theory*, vol. 43, pp. 1986-1992.

[6] M. Hasna, (2005) "On the Capacity of Cooperative Diversity Systems with Adaptive Modulation," *Proc. Int. Conf. Wireless and Optical Communications Networks*, pp. 432-436.

[7] T. Nechiporenko, K. Phan, C. Tellambura and H. Nguyen, (2009), "Capacity of Rayleigh Fading Cooperative Systems under Adaptive Transmission," *IEEE Trans. Wireless Communications*, vol. 8, pp. 1626-1631.

[8] S. Ikki and M. Ahmed, (2010) "On the Capacity of Relay-Selection Cooperative-Diversity Networks under Adaptive Transmission," *Proc. IEEE Vehicular Technology Conf*, Sept. 2010.

[9] M. Simon and M. Alouini, (1998), "A Unified Approach to the Performance Analysis of Digital Communications over Generalized Fading Channels," *Proc. IEEE*, vol. 86, pp. 1860-1877.

[10] A. Annamalai, C. Tellambura and V. K. Bhargava, (2005) "A General Method for Calculationg Error Probabilities over Fading Channels," *IEEE Trans. Communications*, vol. 53, pp. 841-852.

[11] Y. Ko, M. Alouini and M. Simon, (2000) "Outage Probability of Diversity Systems over Generalized Fading Channels," *IEEE Trans. Communications*, vol. 48, pp. 1783-1787.

[12]   R. Chembil Palat, A. Annamalai and J. Reed, (2008) "An Efficient Method for Evaluating Information Outage Probability and Ergodic Capacity of OSTBC Systems," *IEEE Communications Letters*, vol. 12, pp. 191-193.

[13]   A. Annamalai, C. Tellambura and V. K. Bhargava (2001) "Simple and Accurate Methods for Outage Analysis in Cellular Mobile Radio Systems – A Unified Approach," *IEEE Trans. Communications*, vol. 49, pp. 303-316.

[14]   G. Farhadi and N. Beaulieu, (2009) "On the Ergodic Capacity of Multi-Hop Wireless Relaying Systems," *IEEE Trans. Wireless Communications*, vol. 8, pp. 2286-2291.

[15]   M. S. Alouini, A. Abdi and M. Kaveh, (2001) "Sum of Gamma Variates and Performance of Wireless Communication Systems over Nakagami Fading Channels," *IEEE Trans. Vehicular Technology*, vol. 50, pp. 1471-1480.

[16]   M. Di Renzo, F. Graziosi and F. Santucci, (2010) "Channel Capacity Over Generalized Fading Channels: A Novel MGF-Based Approach for Performance Analysis and Design of Wireless Communication Systems," *IEEE Trans. Vehicular Technology*, vol. 59, pp. 127-149.

[17]   M. Di Renzo, F. Graziosi and F. Santucci, (2009) "A Unified Framework for Performance Analysis of CSI-Assisted Cooperative Communications over Fading Channels," *IEEE Trans. Communications*, vol. 57, pp. 2551-2557.

[18]   A. Annamalai, R. Palat, and J. Matyjas, (2010) "Estimating Ergodic Capacity of Cooperative Analog Relaying under Different Adaptive Source Transmission Techniques," *Proc. IEEE Sarnoff Symposium*, April 2010.

[19]   K. A. Hamdi, (2008) "Capacity of MRC on Correlated Rician Fading Channels," *IEEE Trans. Communications*, vol. 56, pp. 708-711.

[20]   I. Gradshteyn and I. Ryzhik, (1995) *Table of Integrals, Series and Products*, Academic Press.

[21]   W. Su, K. S. Ahmed and K. J. Ray Liu, (2008) "Cooperative Communication Protocols in Wireless Networks: Performance Analysis and Optimum Power Allocation," *Springer Journal Wireless Personal Communication,* vol. 44, pp. 181-217.

[22]   M. Hasna and M. Alouini, (2004) "Harmonic Mean and End-to-End Performance of Transmission System with Relays," *IEEE Trans. Communications*, vol. 52, pp. 130-135.

[23]   R. H. Y. Louie, Y. Li and B. Vucetic, (2008) "Performance Analysis of Beamforming in Two Hop Amplify-and-Forward Relay Networks," *Proc. IEEE ICC'08*, pp. 4311–4315.

[24]   D. Senarante and C. Tellambura, (2010) "Unified Exact Performance Analysis of Two-Hop Amplify-and-Forward Relaying in Nakagami Fading," *IEEE Trans. Vehicular Technology,* vol. 59, pp. 1529-1534.

[25]   A. Annamalai, O. Olabiyi, S. Alam, O. Odejide and D. Vaman, (2011) "Unified Analysis of Energy Detection of Unknown Signals over Generalized Fading Channels," *Proc. IEEE IWCMC'11*, Istanbul, pp. 636-641.

[26]   C. Gunther, (1996) "Comment on 'Estimate of Channel Capacity in Rayleigh Fading Environment'," *IEEE Trans. Vehicular Technology*, vol. 45, pp. 401-403.

[27]   M. K. Simon and M. S. Alouini, (2005) *Digital Communication over Fading Channels*, New York: Wiley, 2nd Edition.

[28]   A. H. Nuttall, (1972) "Some Integrals Involving the Q-Function," *Naval Underwater Systems Center Technical Report 4297*, New London.

[29]   G. Caire and S. Shamai, (1999) "On the Capacity of Some Channels with Channel State Information," *IEEE Trans. Information Theory*, vol. 45, pp. 2007-2019.

12

Empirical performance evaluation of Enhanced throughput schemes of IEEE802.11 technology in Wireless area Networks

Femi-Jemilohun Oladunni .Juliet, Walker Stuart
School Of Computer Science and Electronic Engineering University of Essex Colchester,
Essex, United Kingdom
{ojfemi and stuwal}@essex.ac.uk

ABSTRACT

The success in the growing wireless standards can be measured by the achievement of quality of service (QoS) specifications by the designers. The IEEE802.11 wireless standards are widely accepted as wireless technology for wireless LAN. Efforts have been made over the years by the task group to provide adequate number of QoS enhancement schemes for the increasing numbers of multimedia applications. This paper examines the empirical performances of ad hoc wireless networks deployed on IEEE802.11 standard variants. A survey to some of the QoS schemes incorporated in IEEE802.11 wireless PHY layers were carried out. Then the effects of this enhancement schemes in relation to data throughput and system capacity and reliability in the newest technology deployed on IEEE802.11ac standards was investigated using real time applications and simulation based approaches.

KEYWORDS

Beamforming, IEEE802.11ac, QoS

1. INTRODUCTION

The accessibility to simple, flexible, scalable, cheap and ubiquitous communication provided by IEE802.11 standards had qualified it to be the major wireless technology all over the world. The wave of spectrum licence in the microwaves bands has led to growth explosion in the multimedia applications in the recent times hence; the demand for high throughput, streaming capability in video audio, and web services on the go. The high quality of services required by these applications, such as secured bandwidth, minimal data transfer delay spread, reduced jitter and error rate, have been a huge challenge for IEEE802.11 standards group. The experience in the wireless networks has been that of overcrowding poor quality of service, system degradation, as well as co-channel interference among the co-existing access networks. [1].

In this work, the experiences of the networks deployed on the standards in the real world scenario are being looked into. These standards are operated on the license free microwaves bands of the frequency spectrum. Some works have been carried out to investigate the system performance in IEEE802.11 wireless network systems, but the latest research works on the shielding enhancement techniques for high throughput delivery was not included. Therefore, this work intends to give a broad overview study of the enhancements schemes and their effects as determined by the networks delivery performances based on the coverage areas, received signals, interference between channels, and throughput capability in the IEEE802.11standards variants.

The rest of this work is organised as follows: Section two gives the overview of the related published work on IEEE802.11 WLAN performances. The background study of IEEE802.11 and the various shielding techniques for quality of service enhancement are addressed in section three. The empirical investigations of co-located networks performances of IEEE802.11 standards deployed on microwave bands and the newest technology: IEEE802.11ac, and development and implementation of beamforming algorithm were reported in section four. The empirical models of the measurement results of the signal strength distribution of the networks on microwave bands, the upcoming technology: IEEE802.11ac and discussions were presented in section five. The conclusion was given in section six.

.

2. RELATED WORKS.

In literature, there are some works on System performance evaluation and interference mitigation techniques in WLAN deployed on IEEE802.11 standards. However, very few of these authors have engaged physical devices in their works for real time applications assessment. Some of these are highlighted below.

2.1. Related works of WLAN indoor coverage.

Different techniques have been adopted in the previous works to achieve the target. Some of these techniques are: continuous wave transmitter with power meter receivers. [2] used a theoretical hybrid model that combined a two-dimensional ray tracing model with a direct transmitted ray (DTR) model to predict the radio coverage on single floors in multifloored buildings. Another technique is Broadband Pulse channel sounding. [3], measured the pulse response of network at 2.4GHz inside a building, he modulated the carrier with a repetitive 5ns pulses and transmitted it through a wideband biconical antenna. Also signal strength percentage reported by wireless LAN card, [4] used measured data and empirical models to predict the propagation path loss in and around residential area for the UNII band at 5.85GHz. While [5] developed throughput prediction model using the available measurement software   products called SiteSpy and LanFielder to measure wireless LAN throughput and other network performances criteria and recorded the results in a precise site-specific manner. His measurement was limited to IEEE802.11b/n wireless system. [6] examined the impact of structural wall shielding on the system performance of an interference limited CDMA system using ray-tracing propagation models based on Geometric Optics and Uniform Theory of Diffraction to estimate the received signal. They observed a logarithmic relationship between the average output probability and the extent of a shielding in a typical single-floor office environment. They concluded that efficient deployment of shielding will enhance both system capacity and signal strength of wireless system. [7], in their work employed analytical approach using the Okumura's model to determine the propagation range of IEEE802.11 Radio LAN's for outdoor applications. This work used a lot of assumptions and does not reveal the performance range in the real application scenario. To evaluate the performance of 802.11b, [8] used two laptops equipped with Orinoco WLAN cards, and two PCs with D-Link WLAN cards. The combination of equipment from different chipset manufacturer may not favour adequate reading required. In [9] Discrete-event simulation of MAC portion of the IEEE802.11 protocol was used to evaluate the performance of the standard. [10-12] also used empirical methods to show the coverage area and the signal strength inside the coverage area to reveal the technology that gives better WLAN. [13] using computer simulation of a typical 802.11b/e access to an IP core network through an access point in an infrastructure WLAN.in the same manner, [14] conducted field performance comparison between IEEE802.11b and power line based communication network. They concluded that power line network outperforms IEEE802 .11b. [15-18] is similar in some aspects to what is done in this work. They used the experiment to validate the analytical performance models for IEEE802.11b .It was noted in their results that analytical result estimate experimental data with a mean error of 3-5%. In [19] Primary-Secondary transmission scheme was used to mitigate interference by beam forming and increase system capacity in WLAN systems where matlab simulation was engaged. [20] Investigated the efficiency of multi-cast beamforming optimization over IEEE802.11n WLAN. [21] Proposed the combination of advanced distributed beamforming techniques at physical layer for the improvement of the overall network capacity. Analytical modelling to evaluate the performance of ad hoc networks with M-element array was conducted by [22]. The upper and lower bounds on the transmission capacity in the multiple antenna ad hoc networks with multi-stream transmission and interference cancellation at the receiver was derived by [23] to increase system capacity through beamforming technique. Similar to our work, [24] gave the overview of the newest technologies promised to deliver multi-gigabits throughput: Ieee802.11ac/ ad, they described the channelization PHY design, and MAC modifications and beamforming in the standards.[25] proposed a joint adaptive beamforming algorithm for interference suppression in GNSS(Global Navigation Satellite System) receivers: first the orthogonal projection is used to suppress strong interference by producing nulling in the array pattern, while the maximum post-correlated C/No constrained is used to process the interference-free signal to further enhance the signal quality. [26] designed an advance interference resilient scheme for asynchronous slow frequency hopping wireless personal area networking and time division multiple access cellular system for interference reduction in wireless system. Different approaches have been engaged by different authors in their works to analyse the performances of wireless networks deployed on the IEEE802.11 technology but none has employed real time application approach, especially in the newest technology deployed in IEEE802.11ac for the empirical performances assessment. Hence this work differs from all in this aspect.

3. IEEE802.11

This is a set of standards for implementing WLAN (wireless local area networking) communications in the 2.4, 3.6, and 5 GHz frequency bands. The first in this group is IEEE80.11-2007 which has been subsequently amended. Other members are 802.11b and 802.11g protocols. These standards provide the basis for wireless network products using the Wi-Fi brand. IEEE 802.11b and g use the 2.4 GHz ISM band and are operated on WiFi channels. WiFi channel are grouped into 14 overlapping channels. Each channel has a spectral bandwidth of 22MHz, (though the nominal figures of the bandwidth of 20MHz are often given). This channel bandwidth exists for all standards even though different speed is available for each standard: 802.11g standard has 1, 2, 5, or 11 Mbps while 802.11g standard has 54Mbps. The difference in speed depends on the RF modulation scheme used. The adjacent channels are separated by 5MHz with the exception of 14 with the centre frequency separated from channel 13 by 12MHZ. From figure 1, it is obvious that a transmitting station can only use the fourth or the fifth channel to avoid overlapping. Most often, Wifi routers are set to channel 6 as the default, hence channels 1, 6, and11 have been adopted generally and particularly in Europe being non-overlapping channels for wireless transmission in the ISM band. This band can provide up to 11Mbps [8-10]

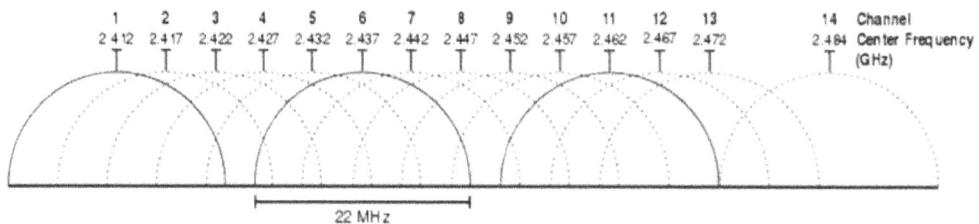

Figure 1. Structure of Wi-Fi Channels in the 2.4 GHz band [11, 12]

The UNII (unlicensed national information infrastructure) band is allocated for two radio devices: the unlicensed part is classified into A (5150-5350) MHz and B (5470-5725) MHz; the higher band which is licensed is C (5725-5859) MHz used in the installation of Fixed wireless Access (FWA) services, while A is for indoor mobile and B for indoor and outdoor WLANs. 5GHz U-NII bands which offers 23 non-overlapping channels, can provide up to 54Mbps in the WLAN and it is deployed in the IEEE802.11a/n standards.[11, 12]

Table 1. Wi-Fi channels in the 5GHz[27]

Channel	Frequency /GHz	Band(A,B)	Maximum EIRP
36	5.18	A	200mW
40	5.20	A	200mW
44	5.22	A	200mW
48	5.24	A	200mW
52	5.26	A	200mW
56	5.28	A	200mW
60	5.30	A	200mW
64	5.32	A	200mW
100	5.50	B	1W
104	5.52	B	1W
108	5.54	B	1W
112	5.56	B	1W
116	5.58	B	1W
120	5.60	B	1W
124	5.62	B	1W
128	5.64	B	1W
132	5.66	B	1W
136	5.68	B	1W
140	5.70	B	1W
149	5.745	C	4W
153	5.765	C	4W
157	5.785	C	4W
161	5.805	C	4W

Among these channels, there are eight widely used channels. These are divided into two called lower band UNII-1(36, 40, 44, and 48) and upper band UNII-3(149, 153, 157, and161). The remaining 15 channels are divided into lower UNII-2 having four channels, and upper UNII-2, which has 11 channels. The reason for the four-unit gaps is that the channels are 20MHz wide, while there is 5MHz spacing between one channel and the other. The upper bands uses 20 times as much power as the lower bands and the more power used to send signal, the likelihood of interference between networks [13].

3.2. IEEE808.11ac

In order to address the higher data rate throughput capacity require by the new technological growth in wireless applications, the IEE task group in 2008,developed an amendment to the IEEE802.11 PHY and MAC layer . IEEE802.11ac has an improving delivery capacity beyond its counterpart, IEEE802.11n.It is the next evolution of the Wi-Fi standard with the capacity to deliver multiple (High Definition) HD video streams simultaneously and provide improvements over 802.11n.It can cover 33m propagation distance by using 80MHz on the 5GHz band[28-30] This new technology proposed by IEEE802.11- VHT ,aims at higher throughput of 1Gbps and with the frequency band of 6GHz or below excluding 2.4GHz frequency bandwidth of 80MHz and 160MHz option building on the 40MHz available in 802.11n.For spectrum efficiency, MU-MIMO technology is employed in this standard to achieve the aimed target throughput of 1Gbps.An obvious technical problem is envisaged in this technology. Due to its huge frequency bandwidth which will only allow four frequency channels , will lead to severe inter cell interference due to frequency channel shortage among multiple basic service sets.TGac and TGad were established in Sep 2008 and Dec 2008 respectively, targeting completion of standardization late in 2012[31-33] The final approval of the IEEE802.11ac standard amendment is expected in December 2013, though as the moment, initial products with basic feature of 802.11ac has started to emerge on the market[30].

Table 2. Feature Enhancement Schemes Comparison: 802.11n/802.11ac [34]

	IEEE802.11n	IEEE802.11ac
Frequency Band	2.4 and 5 GHz	5GHz only
Channel Widths	20, 40 MHz	20,40,80MHz, 106MHz optional
Spatial Streams	1 to 4	1 to 8 total up to 4 per client
Multi- User MIMO	No	Yes
Single Stream (1x1) Maximum Client Data Rate	150Mbps	450Mbps
Three Stream (3x3) Maximum Client Data Rate	450Mbps	1.3Gbps

3.3. Amendments to PHY and MAC layers for improved QoS

3.3.1. Higher Order Modulation Scheme

The maximum modulation in .11n is 64QAM constellation with six bits of coded information. This has been increased to 256QAM constellation with eight coded bits.256QAM seems to be the current practical limit of digital modulation scheme, though efforts are on to increase to 512 and 1024 QAM. There are 256 possible states in this scheme, with each symbol representing eight bits. This makes this scheme very spectral efficient, the symbol rate is 1/8 of the bit rate. This is revealed in the fig.3.1, 33% increase is realized when 256QaM is used instead of the 64QAM. On the other hand, the constellation of this scheme are closely packed together, thus, more susceptible to noise and distortion, leading to higher bit error rate in higher modulation scheme [35, 36]

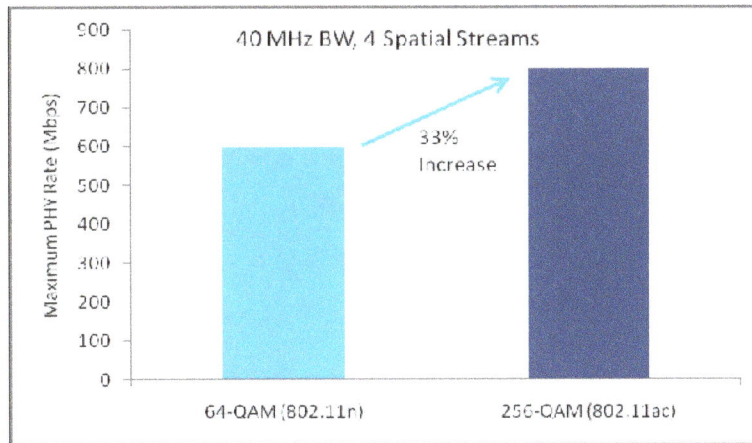

Figure 2. Performance enhancement in 802.11ac over 802.11n [37]

3.3.2. MIMO

MIMO (Multiple in Multiple out) techniques improves communication reliability as well as increase data throughput through spatial division multiplexing (SDM), without necessarily expanding the frequency band by using multiple antenna at the transmitter and receivers ends This results in signal interference. SVD-MIMO (singular value decomposition –multiple in multiple out) system can help in suppressing the signal interference as a result of lots of antenna by beamforming. This is effective in large scale MIMO systems [38, 39] while orthogonal division frequency multiplexing is a powerful tool to equalize received signals under multipath fading environments. The combination of these techniques is currently undergoing a great deal of attention in wireless communications[40, 41] Multiple-input multiple-output orthogonal frequency multiplexing (MIMO-OFDM) is a powerful technology that enhances communication capacity and reliance. This is being used in 802.11n WLAN system, defined as four spatial streams in spatial division multiplexing (SDM), it is expected that the upcoming standardization of 802.11ac will be defined in eight spatial streams in single-user MIMO (SU-MIMO[42]

3.3.3 Adaptive Beamforming Technology

This is one of the latest technologies in wireless communication system. It uses Smart Antenna techniques to produce multiple beams concentration to enhance signal of interest, and at the same time places nullity on the direction of interfering signal. The Beamformer varies a weight vector for this adaptive process with the ultimate goal to improve communication channel. The figure1b is the setup model for adaptive beamforming. The mathematical algorithms are shown in the following equations [25]

$0 < \mu < 2 \frac{E[|u(n)|^2]}{E[|e(n)|^2]} D(n)$, Where

$E[|e(n)|^2]$ is the power of the difference between the output and reference signals μ –step size $E[|u(n)|^2]$ is power of incoming signal and $D(n)$ is the deviation

The error signal is given by;

$$e(n) = d(n) - w^{\wedge H}(n)u(n) \tag{1}$$

$$\hat{w}(n+1) = \hat{w}(n) + \frac{\mu}{\|u(n)\|^2} u(n)e^*(n) \tag{2}$$

$$y(n) = w^H(n)x(n), \tag{3}$$

$$e(n) = d(n) - y(n) \tag{4}$$

Estimated of weight vector at time n+1,

$$w(n+1) = w(n) + \mu x(n)e^*(n)$$

$$= w(n) + \frac{\mu}{\|x(n)\|^2} x(n)e^*(n)$$

$$= w(n) + \frac{\mu}{\|x(n)\|^2} x(n)\left\| d(n) - \frac{H}{w}x(n)\right\|^* \tag{5}$$

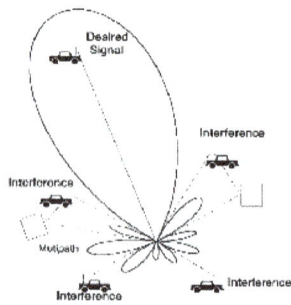

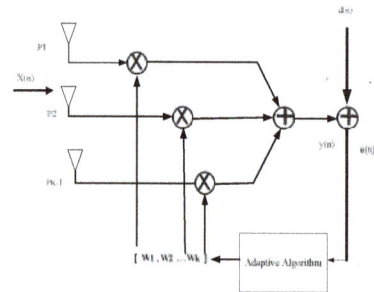

(6)

(a)   (b)

Figure 3. (a) Antenna Array Beamforming, (b) Block Diagram of Adaptive Beamforming [28]

4. METHODOLOGY

4.1. Empirical Measurements of wireless transmission on 2.4GHz channels

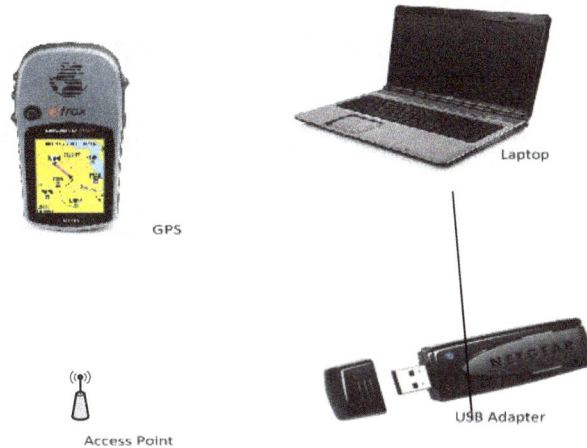

Figure 4. Microwave bands transmission experimental setup

The field measurements of radio signals at different locations from the various access points were taken with a net sniffer called Inssider, a 5GHz dual band dongle, GPS, and a laptop. Measurements were repeated about 25 times in each location for the different networks. While conducting the experiments, a huge amount of data was recorded in files and the data records include the time information (t), MAC address of AP, Signal Strength (SS) information, transmitter channel of AP, service set identifier (SSID).

The measured values were saved in kml file format from the personal computer and later extracted to Excel .Xml. Figs. 5-7 are the empirical models from the acquired data .Out of 23 non-overlapping channels in UNII band, only two were employed in the wireless network service of the University. They are channels 36 and 48, which also engaged in frequency reuse pattern. This band has higher data rate of 54Mbps hence higher throughput delivery compared to its unlicensed counterpart in the ISM band. The transmission distribution measurement of this band was also conducted and the simulations carried out. The following models are the outcomes.

4.2. Empirical Measurements of wireless transmission on IEEE802.11ac

To assess performance of 802.11ac, an ad-hoc system with the latest technology in wireless standards was used. The AirStation 1750 (Buffalo) is amongst the first routers to implement the newest WiFi standard: IEEE802.11ac. It promises throughput of 1300Mbps and also additional 450Mbps in the 802.11n radio.[30] The setup for the experiments are shown in figs 1&2 using WLI-4H-D1300 and WZR-D 1800H for the transmission using 5GHz channels, For comparison, in fig two, the client was changed to a WNDA 3100 N600 (Net gear) dual-band dongle using the 5GHz channels of 802.11n The measurement of throughputs over a range of distances up to 30m was observed on the PC connected to the wireless transmission link. The PCs were equipped with LANSpeedTest software for the monitoring of the throughputs.

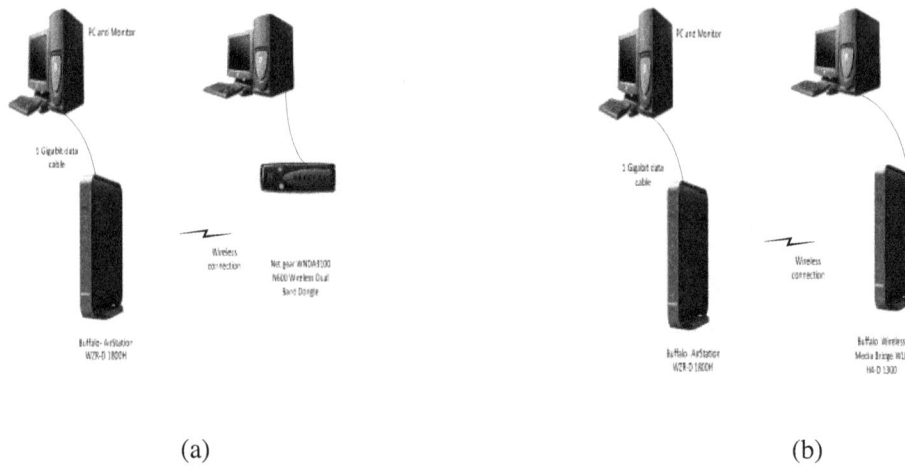

(a)   (b)

Figure 5. Buffalo AirStation AC1300/N900 Gigabit Dual Band Wireless Router and Client. (a) With IEEE802.11ac Standard, (b) with Net gear WNDA3100N600 Wireless Dongle Receiver on IEEE802.11n

4.3. Matlab Simulation-Based Evaluation of Beamforming in IEEE802.11ac

Table 4. Parameters for Simulation

Carrier frequency	5GHz
Sampling frequency	2FC
Desires transmitted angle	Pi/4
Interferer transmitted angle	Pi/4
Step size	0.32
Number of element antenna array	8

5. EMPIRICAL MODELS OF THE WIRELESS TRANSMISSION ON 2.4, 5GHz, IEEE802.11A, AND MATLAB SIMULATIONS

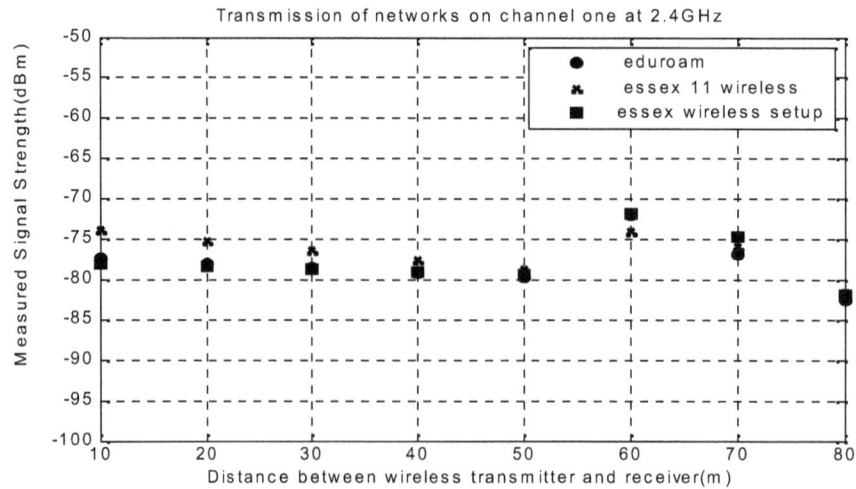

Figure 6. Channel One Transmission at 2.4GHz Coverage Distribution

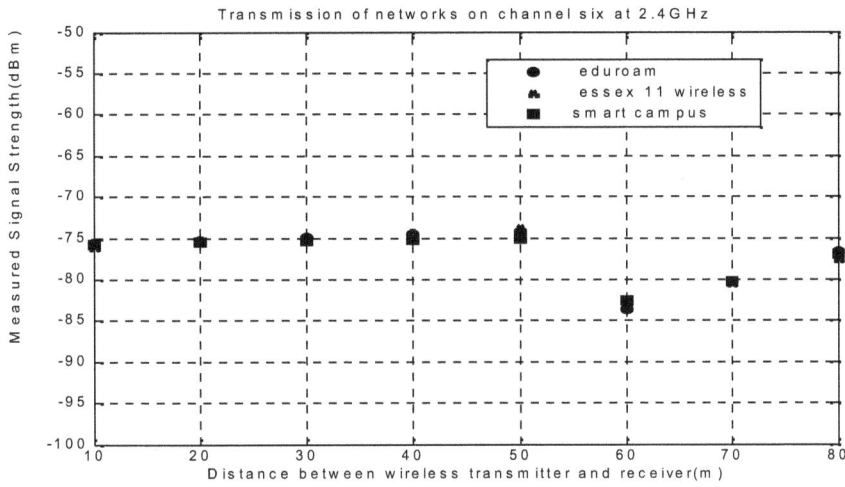

Figure 7. Channel Six Transmissions at 2.4GHz Coverage Distribution

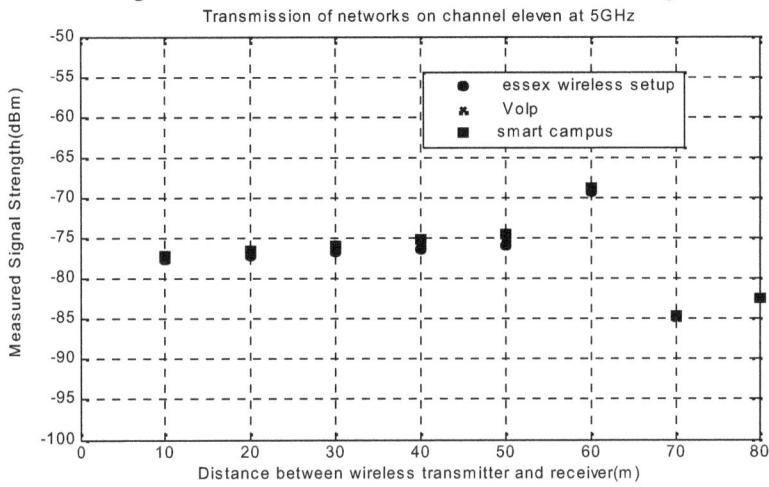

Figure 8. Channel Eleven Transmissions at 2.4GHz Coverage Distribution

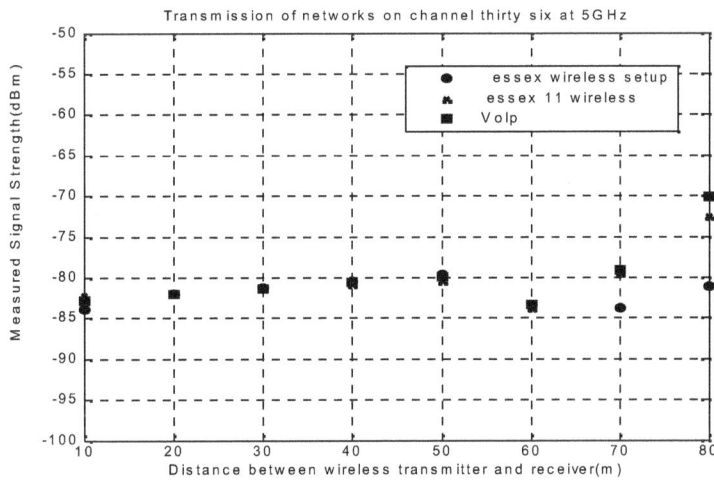

Figure 9. Channel 36 Transmissions at 5GHz Coverage Distribution

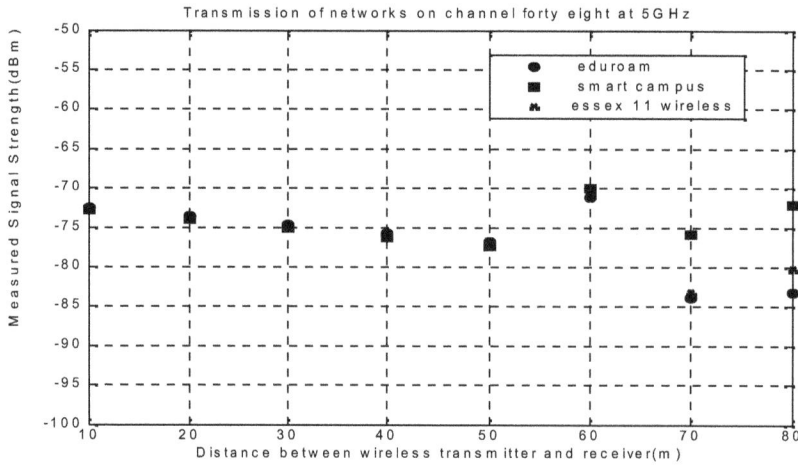

Figure 10. Channel 48 Transmissions at 5GHz Coverage Distribution

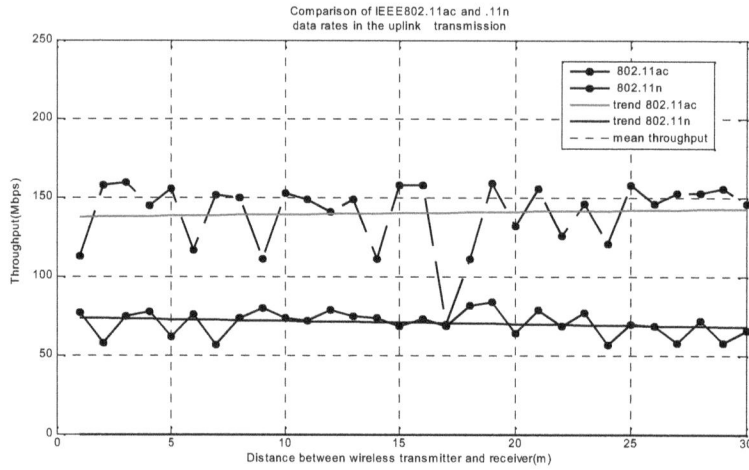

Figure 11.   Uplink Transmission Throughputs with AirStation Receiver and Dongle Receiver

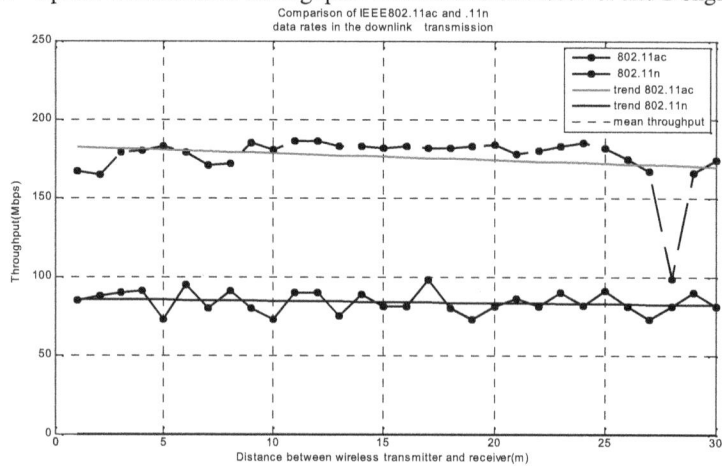

Figure 12. Downlink Transmission Throughputs with AirStation Receiver and Dongle Receiver

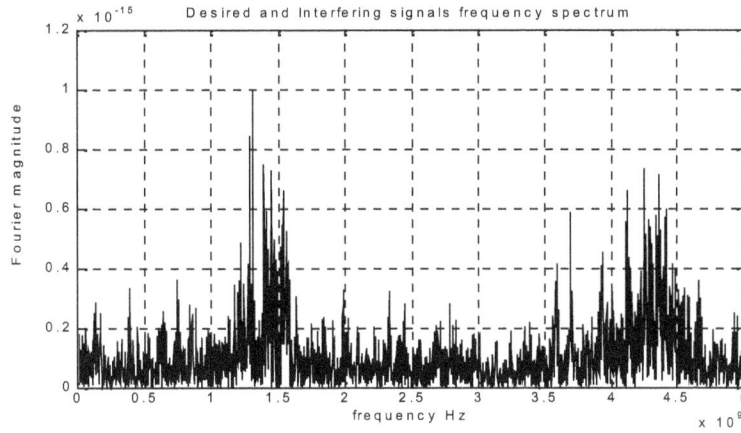

Figure 13. Frequency Spectrum of Interfering and Desired Signals

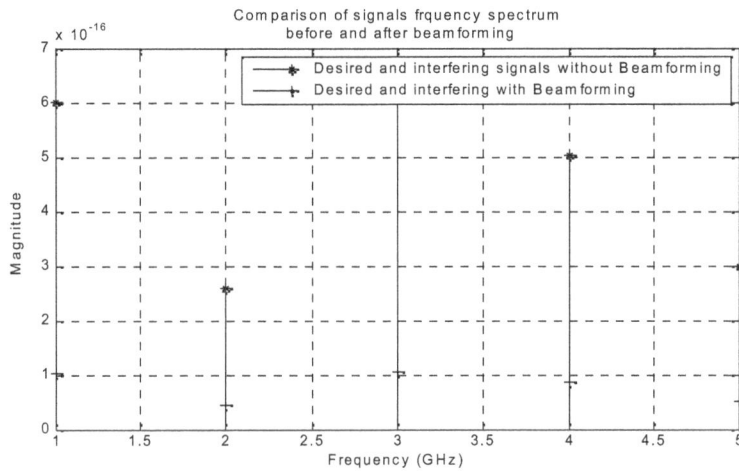

Figure 14. Frequency Spectrum before and after Beamforming Technique

5.2. Discussion

Measurements of transmission on microwaves bands were taken at twelve different locations in the University of Essex campus using the set up in figure 4.At each location, 25 signal strengths sampling from the access points (Aps) were recorded. Simulations of the measured values were carried out using the Excel.XML program, Haversine formula, and Frii's propagation law. The models are presented in figures 6-10. Some distance apart, show a measure of overlap. As shown in the figures, none of the access points was protected against interference. Each of the access points is degraded in throughput by certain percentage due to interference from other neighbouring access points. It is noted that eduroam and essex wireless setup access points were degraded by almost 100% in throughput.

The setup shown in figure 5 was used for wireless transmission on standard IEEE802.11ac deployed on frequency of 5GHz. The router was fixed in a position while the client/mobile terminal was moving at an equidistant points separated by 1m until a total coverage of 30m was reached in a laboratory hall at the University of Essex, where the measurement was conducted. The throughput levels received at each measure point is collected by a PC running with application software called LANSpeed Test software, as we moved along various points of measurement to monitor the throughput from the client. Figures 11-12 show the level of uplink and downlink throughputs of the Buffalo AirStation AC1300/N900 Gigabit Dual Band Wireless router and client when it is configured to work only in 802.11ac, and WNDA3100 N600 wireless dual-band dongle receiver when configured to work in 802.11n. As the figures show, the propagation square law was minimal and more revealed in IEEE802.11n configuration than IEEE802.11ac. There is appreciable measure of improved throughput in IEEE802.11ac over its counterpart IEEE802.11n. This established the benefits of beamforming technology in preserving the data to the targeted consumers. The matlab based simulation to investigate the effect of beamforming as a QoS enhancement scheme is revealed in figures13 and 14. The magnitude of the desired and interfering signals is shown in red before beamforming algorithms were implemented, while the blue colour signifies the magnitude of the desired signal after beamforming in figure 14. The drastic reduction in the magnitude is as a result of interference suppression through the shielding technique: beamforming technique, applied while the reception of desired signal is enhanced, hence improved quality of service.

6. CONCLUSION

The success in the growing wireless standards can be measured by the achievement of quality of service specifications by the designers. It becomes imperative to ascertain the performances of wireless devices. [31] Though. the theoretical peak data rates of 600Mbps and 1.3Gbps promised in IEEE802.11ac and IEEE802.11n respectively.[32] may not be realizable due to propagation impairment contending with wireless transmission in the real world scenario , nevertheless there is a considerable data throughput improvement on the counterpart 802.11n. The experiments have shown that IEEE802.11ac outperforms its lower counterpart IEEE802.11n in throughput delivery. This is as a result of the beamforming technology incorporated into this standard as a shielding technique for enhancement of system performance. The beamforming technology could be conceivably be modified to have better performance. The simulation based evaluation depicted in figures 13 and 14 revealed the effectiveness of beamforming technique.

ACKNOWLEDGEMENTS

This work is sponsored by the Federal Government of Nigeria through the Tertiary Education Trust Fund (TETFUND) scheme.

REFERENCES

[1]   Q. Ni, L. Romdhani, and T. Turletti, "A survey of QoS enhancements for IEEE 802.11 wireless LAN," Wireless Communications and Mobile Computing, vol. 4, pp. 547-566, 2004.

[2]   S. Chen, N. Ahmad, and L. Hanzo, "Smart beamforming for wireless communications: a novel minimum bit error rate approach," 2002.

[3]   H. Liu, H. Darabi, P. Banerjee, and J. Liu, "Survey of wireless indoor positioning techniques and systems," Systems, Man, and Cybernetics, Part C: Applications and Reviews, IEEE Transactions on, vol. 37, pp. 1067-1080, 2007.

[4]   K. K. Leung and B. J. Kim, "Frequency assignment for IEEE 802.11 wireless networks," in Vehicular Technology Conference, 2003. VTC 2003-Fall. 2003 IEEE 58th, 2003, pp. 1422-1426.

[5]   D. C. K. Lee, M. J. Neve, and K. W. Sowerby, "The impact of structural shielding on the performance of wireless systems in a single-floor office building," Wireless Communications, IEEE Transactions on, vol. 6, pp. 1787-1695, 2007.

[6]   A. Baid, "Multi-radio interference diagnosis in unlicensed bandsusing passive monitoring," Rutgers University-Graduate School-New Brunswick, 2011.

[7]   K. Kim, "Interference mitigation in wireless communications," 2005.

[8]   G. Ofori-Dwumfuo and S. Salakpi, "WiFi and WiMAX Deployment at the Ghana Ministry of Food and Agriculture," Research Journal of Applied Sciences, vol. 3, 2011.

[9]   A. Sandeep, Y. Shreyas, S. Seth, R. Agarwal, and G. Sadashivappa, "Wireless Network Visualization and Indoor Empirical Propagation Model for a Campus WI-FI Network," World Academy of Science, Engineering and Technology, vol. 42, 2008.

[10]   J. J. van Rensburg and B. Irwin, "Wireless Network Visualization Using Radio Propagation Modelling," 2005.

[11]   Y. Lee, K. Kim, and Y. Choi, "Optimization of AP placement and channel assignment in wireless LANs," 2002, pp. 831-836.

[12]   A. Mishra, E. Rozner, S. Banerjee, and W. Arbaugh, "Exploiting partially overlapping channels in wireless networks: Turning a peril into an advantage," 2005, pp. 29-29.

[13]   S. Choi, K. Park, and C. Kim, "On the performance characteristics of WLANs: revisited," 2005, pp. 97-108.

[14]   D. Porcino and W. Hirt, "Ultra-wideband radio technology: potential and challenges ahead," Communications Magazine, IEEE, vol. 41, pp. 66-74, 2003.

[15]   X. Guo, S. Roy, and W. S. Conner, "Spatial reuse in wireless ad-hoc networks," 2003, pp. 1437-1442 Vol. 3.

[16]   S. Mare, D. Kotz, and A. Kumar, "Experimental validation of analytical performance models for IEEE 802.11 networks," in Communication Systems and Networks (COMSNETS), 2010 Second International Conference on, 2010, pp. 1-8.

[17]   A. Grilo and M. Nunes, "Performance evaluation of IEEE 802.11 e," in Personal, Indoor and Mobile Radio Communications, 2002. The 13th IEEE International Symposium on, 2002, pp. 511-517.

[18]   M. J. Ho, M. S. Rawles, M. Vrijkorte, and L. Fei, "RF challenges for 2.4 and 5 GHz WLAN deployment and design," 2002, pp. 783-788 vol. 2.

[19]   T. Murakami, R. Kudo, Y. Asai, T. Kumagai, and M. Mizoguchi, "Performance evaluation of distributed multi-cell beamforming for MU-MIMO systems," 2011, pp. 547-551.

[20]   C. Papathanasiou and L. Tassiulas, "Multicast Transmission over IEEE 802.11 n WLAN," in Communications, 2008. ICC'08. IEEE International Conference on, 2008, pp. 4943-4947.

[21]   Y. Lebrun, K. Zhao, S. Pollin, A. Bourdoux, F. Horlin, S. Du, and R. Lauwereins, "Beamforming techniques for enabling spatial-reuse in MCCA 802.11 s networks," EURASIP Journal on Wireless Communications and Networking, vol. 2011, pp. 1-13, 2011.

[22]   K. Fakih, J. F. Diouris, and G. Andrieux, "Analytical evaluation on the performance of ad hoc networks when using beamforming techniques," in Communications, 2008. ICC'08. IEEE International Conference on, 2008, pp. 2337-2342.

[23]   R. Vaze and R. W. Heath, "Transmission capacity of ad-hoc networks with multiple antennas using transmit stream adaptation and interference cancellation," Information Theory, IEEE Transactions on, vol. 58, pp. 780-792, 2012.

[24]   P. Xia, X. Qin, H. Niu, H. Singh, H. Shao, J. Oh, C. Y. Kweon, S. S. Kim, S. K. Yong, and C. Ngo, "Short range gigabit wireless communications systems: potentials, challenges and techniques," 2007, pp. 123-128.

[25]   K. M. alias Jeyanthi and A. Kabilan, "A Simple Adaptive Beamforming Algorithm with interference suppression," International Journal of Engineering and Technology, vol. 1, pp. 1793-8236, 2009.

[26]   S. M. Alamouti, "A simple transmit diversity technique for wireless communications," Selected Areas in Communications, IEEE Journal on, vol. 16, pp. 1451-1458, 1998.

[27]   N. V. Kajale, "Uwb and wlan coexistence: A comparison of interference reduction techniques," University of South Florida, 2005.

[28]   M. Matsuo, R. Ito, M. Kurosaki, B. Sai, Y. Kuroki, A. Miyazaki, and H. Ochi, "Wireless transmission of JPEG 2000 compressed video," 2011, pp. 1020-1024.

[29]   M. Matsuo, M. Kurosaki, Y. Nagao, B. Sai, Y. Kuroki, A. Miyazaki, and H. Ochi, "HDTV over MIMO wireless transmission system," 2011, pp. 701-702.

[30]   E. Perahia and M. X. Gong, "Gigabit wireless LANs: an overview of IEEE 802.11 ac and 802.11 ad," ACM SIGMOBILE Mobile Computing and Communications Review, vol. 15, pp. 23-33, 2011.

[31]   K. Nishikawa, "Ultra High-speed Radio Communication Systems and Their Applications-Current Status and Challenges."

[32]   R. Liao, B. Bellalta, and M. Oliver, "DCF/USDMA: Enhanced DCF for Uplink SDMA Transmissions in WLANs."

[33]   M. Park, "IEEE 802.11 ac: Dynamic Bandwidth Channel Access," 2011, pp. 1-5.

[34]   R. Watson. (May 2012, Understanding the IEEE802.11ac Wi-Fi Standard. White Paper, 10.

[35]   A. Technology. (2001, Digital Modulation in Communications systems An introduction, 48.

[36]   Q. Incorporated. (2012, IEEE802.11ac: The Next Evolution of Wi-Fi Standards. 15.

[37]   J. Montavont and T. Noel, "IEEE 802.11 handovers assisted by GPS information," 2006, pp. 166-172.

[38]   K. Nishimori, N. Honma, T. Seki, and K. Hiraga, "On the Transmission Method for Short Range MIMO Communication," Vehicular Technology, IEEE Transactions on, pp. 1-1, 2011.

[39]   T. Kaji, S. Yoshizawa, and Y. Miyanaga, "Development of an ASIP-based singular value decomposition processor in SVD-MIMO systems," 2011, pp. 1-5.

[40]   S. Yoshizawa, S. Odagiri, Y. Asai, T. Gunji, T. Saito, and Y. Miyanaga, "Development and outdoor evaluation of an experimental platform in an 80-MHz bandwidth 2× 2 MIMO-OFDM system at 5.2-GHz band," 2010, pp. 1049-1054.

[41]   H. Kano, S. Yoshizawa, T. Gunji, T. Saito, and Y. Miyanaga, "Development of 600 Mbps 2× 2 MIMO-OFDM baseband and RF transceiver at 5 GHz band," 2010, pp. 891-894.

[42]   S. Yoshizawa, D. Nakagawa, N. Miyazaki, T. Kaji, and Y. Miyanaga, "LSI development of 8× 8 single-user MIMO-OFDM for IEEE 802.11 ac WLANs," 2011, pp. 585-588.

13

Smart handover based on Fuzzy Logic Trend in IEEE802.11 Mobile ipv6 Networks

Joanne Mun-Yee Lim[1] Chee-Onn Chow[2]

[1]Department of Engineering, UCTI (APIIT), Malaysia
`jlmy555@gmail.com`

[2]Department of Electrical Engineering, University Malaya, Malaysia
`cochow@um.edu.my`

ABSTRACT

A properly designed handoff algorithm is essential in reducing the connection quality deterioration when a mobile node moves across the cell boundaries. Therefore, to improve communication quality, we identified three goals in our paper. The first goal is to minimize unnecessary handovers and increase communication quality by reducing misrepresentations of RSSI readings due to multipath and shadow effect with the use of additional parameters. The second goal is to control the handover decisions depending on the users' mobility by utilizing location factors as one of the input parameters in a fuzzy logic handover algorithm. The third goal is to minimize false handover alarms caused by sudden fluctuations of parameters by monitoring the trend of fuzzy logic outputs for a period of time before making handover decision. In this paper, we use RSSI, speed and distance as the input decision criteria of a handover trigger algorithm by means of fuzzy logic. The fuzzy logic output trend is monitored for a period of time before handover is triggered. Finally, through simulations, we show the effectiveness of the proposed handover algorithm in achieving better communication quality.

KEYWORDS

Handover, Mobile IPv6, Fuzzy Logic, RSSI, Speed, Distance

1. INTRODUCTION

The basic reason to handover is to prevent communication quality deterioration or disconnection of services by connecting to the Internet at all times. Communication quality degradation can be minimized by choosing the right moment to initiate handover. One of the major challenges in triggering handover at the right moment is choosing the reliable parameters for decision making. There are three main reasons that contribute to the communication quality degradation during handover due to increase false handover trigger alarms. First, due to multipath and shadow effects, misrepresentations of RSSI readings are more common on the communication network quality. Second, the random movement of a mobile node contributes to the misrepresentations of parameters' readings. Third, sudden change of parameters' readings might cause frequent false handover alarms. Therefore, the parameters chosen and the handover algorithm used are the key aspects in the development of solutions supporting mobility scenarios. The parameters chosen should be able to predict communication degradation precisely and trigger timely and reliable handovers by monitoring signal strength and location factors in order to ensure communication quality is either maintained or improved.

Three objectives are identified in this paper in order to achieve improved communication quality during handover. The first objective is to reduce misrepresentations of RSSI readings due to multipath and shadow effect by using additional input parameters. The second objective is to control the handover decisions depending on the users' mobility by employing location factors in handover decision making process. The third objective is to minimize false handover alarms cause by sudden fluctuations of parameters' readings by monitoring the output trend of fuzzy logic for a period of time before making handover decision. With RSSI, speed and distance as the input parameters, a fuzzy logic based handover decision algorithm is developed. This is accomplished by applying fuzzy logic with to evaluate the criteria simultaneously and to initiate handover processes effectively. In our work, by means of simulation, we compare the proposed handover algorithm with some of the existing methods. There are two main contributions in this paper. First, we developed a fuzzy logic based handover algorithm which incorporates RSSI, speed and distance as the input parameters to trigger handovers efficiently and to achieve improved communication quality performance. Second, the outputs of fuzzy logic are monitored for a period of time to observe its pattern of changes so that false handover trigger alarms can be avoided. This paper is organized as follows. Section 2 presents the related work. Section 3 describes the proposed handover scheme. Section 4 provides the details of simulation studies. Section 5 analyzes the results and presents a comparative study with the existing triggering schemes. Section 6 concludes this paper.

2. RELATED WORK

Many studies have investigated various numbers of ways to improve handover performance. In this section, we describe the existing handover trigger schemes.

The algorithm developed in [1] was based on received signal strength indicator (RSSI) measurements to predict the next state of mobile node. A new prediction technique called RSSI Gradient Predictor was used to detect the change of state in the RSSI values. The predictor was proved to be efficient for applications of real time video in a wireless network that combined wireless fidelity (WiFi) and worldwide interoperability for microwave access (WiMAX) technologies.

In [2], a handoff ordering method based on packet success rate (PSR) for multimedia communications in wireless networks was proposed. A prediction was done on the remaining time for each session to reach its minimum PSR requirement. Results showed that PSR could effectively improved the handoff call dropping probability with little increased of the new call blocking probability.

In [3], the different aspects of handoffs designs and performance related issues were discussed. A vertical handoff decision function (VHDF) that provided handoff decisions while roaming across heterogeneous wireless network was implemented. VHDF utilized cost of service, security, power consumption, network conditions and network performance as the parameters in making handover decisions. VHDF managed to increase the throughput especially in situations where the background traffic varied.

A handoff algorithm which triggered handovers based on both, distance from a mobile station to the neighbouring base stations and relative signal strength (RSS) measurement was proposed [4]. The algorithm performed handoff when the measured distance exceeded a given threshold and when the RSS exceeded a given hysteresis level. However, the performance of the proposed algorithm was less efficient under worst case conditions in comparison to other algorithms. This could be resolved by employing a high accuracy location method such as differential global positioning system (GPS) or real time kinematic global positioning system (GPS).

A fuzzy normalization concept in handover decisions within a heterogeneous wireless network was introduced [5]. The fuzzy inputs, relative signal strength (RSS), velocity and system loading were used as the input parameters for the proposed fuzzy normalization (FUN) – handover decision strategy (HODS), FUN-HODS. Finally, FUN-HODS proved to be able to balance the system loading while reducing handover failures which were usually caused by mobile node's velocity and weak RSS.

In [6], the authors stated that the performance degradation in mobile nodes (MNs) were usually due to reduction of signal strength caused by mobile nodes' movement, intervening objects and radio interference with other wireless local area networks (WLANs). By employing file transfer protocol (FTP) and voice over internet protocol (VoIP) applications, the usage of signal strength and number of frame retransmission as the handover triggers were investigated through experiments in a real environment. The results showed that signal strength could not promptly and reliably detect the degradation of communication quality in both FTP and VoIP communications when the signal strength was affected by MN's movement or intervening objects. However, the number of frame retransmissions was capable of detecting the degradation of communication quality of a wireless link due to MN's movement and intervening objects. Therefore, it was suggested that the number of frame retransmissions, unlike signal strength, was able to detect the communication quality degradation caused by radio interference and reduction of signal strength.

3. FUZZY LOGIC TREND (RSSI, SPEED AND DISTANCE), FL TREND (RSD)

In this section, we describe the proposed handover trigger scheme, Fuzzy Logic Trend (RSSI, Speed and Distance), FL Trend (RSD). The list of precise steps and the order of computations in the FL Trend (RSD) is shown in figure 1. The FL Trend (RSD) starts at iteration of i=0 every time the mobile node attaches itself to a new access point. Subsequently, FL Trend (RSD) starts obtaining RSSI, speed and distance readings at every interval of B. The RSSI, speed and distance collected are used as the input parameters in the FL Trend (RSD) as the representation of the current communication quality. Next, FL Trend (RSD) used these input readings to collect respected input scores based on fuzzy logic input membership functions as shown in figure 2(a) – (c). The range of distance, speed and RSSI shown in figure 2 (a) – (c) are based on a human walking or running scenario in a campus environment. Based on fuzzy logic rule matrix shown in table 1, the respected output scores are obtained for each rule matrix. These output scores are used to obtain the final output score which is used to decide whether to handover or not to handover. The final output score is calculated based on equation (1):

$$Score = \frac{(-100)*Not\ handover\ score + (100)*Handover\ score}{Not\ handover\ score + Handover\ score} \qquad (1)$$

where not handover score is calculated from equation (2):
$$Not\ handover\ score = \sqrt{NH1^2 + NH2^2 + \cdots \ldots \ldots \ldots . + NH13^2} \qquad (2)$$

and handover score is calculated from equation (3):
$$Handover\ score = \sqrt{H1^2 + H2^2 + \cdots \ldots \ldots \ldots . + H14^2} \qquad (3)$$

A sliding window is set to monitor the FL Trend (RSD)'s output scores trend as shown in figure 3. If within N intervals, a threshold of $TH_{HO\ has}$ not been reached, the algorithm adds to the iteration i and continues to monitor the input parameters, RSSI, speed and distance. Otherwise, the handover process is initiated. After the handover trigger has been initiated, next, a delay of W, calculated in equation (4) where speed is the speed of the moving MN and distance, D is the distance that the MN has travelled is introduced. This delay prevents mobile node from triggering multiple handovers after the node attaches to a new access point with the intention that ping pong effect could be prevented.

$$Delay, W = \frac{Distance, D}{Speed} \qquad (4)$$

If the speed is low, the delay is longer as it foresees that the mobile node takes longer time to move to the edge of the coverage. However, if the speed is high, the delay is shorter because the probability of the mobile node moving to the edge in a short period of time is higher. Thus, a shorter delay ensures that the parameters are closely monitored. After this delay period, the value is set back to zero as the mobile node has now been connected to a new access point. The speed and distance in the equation can be obtained using methods proposed by [7] that uses the difference between two consecutive signal updates and free space path loss (FSPL) equations to estimate the speed of a node. Using FSPL equation, the covered distance can easily be calculated if the attenuation of signal is known. Using the time difference between two mobile nodes at two different positions, speed is calculated. In another method, information about the angle of arrival of the signal is used to calculate the estimated speed using cosine theorem. The estimation of speed can be calculated at both nodes; the mobile node or the network node. The travelling period can be obtained using the time difference between two different nodes at two different positions. With this information, the speed and distance of the node can be calculated.

Table 1
FL Trend (RSD) Rule Matrix

Rule	Input of Fuzzy Logic			Output
	RSSI	Speed	Distance	
1	High	High	High	Not handover, NH1
2	High	Medium	High	Not handover, NH2
3	High	Low	High	Not handover, NH3
4	Medium	High	High	Handover, H1
5	Medium	Medium	High	Handover, H2
6	Medium	Low	High	Handover, H3
7	Low	High	High	Handover, H4
8	Low	Medium	High	Handover, H5
9	Low	Low	High	Handover, H6
10	High	High	Medium	Not handover, NH4
11	High	Medium	Medium	Not handover, NH5
12	High	Low	Medium	Not handover, NH6
13	Medium	High	Medium	Handover, H7
14	Medium	Medium	Medium	Handover, H8
15	Medium	Low	Medium	Not handover, NH7
16	Low	High	Medium	Handover, H9
17	Low	Medium	Medium	Handover, H10
18	Low	Low	Medium	Handover, H11
19	High	High	Low	Not Handover, NH8
20	High	Medium	Low	Not handover, NH9
21	High	Low	Low	Not handover, NH10
22	Medium	High	Low	Not Handover, NH11
23	Medium	Medium	Low	Not Handover, NH12
24	Medium	Low	Low	Not handover, NH13
25	Low	High	Low	Handover, H12
26	Low	Medium	Low	Handover, H13
27	Low	Low	Low	Handover, H14

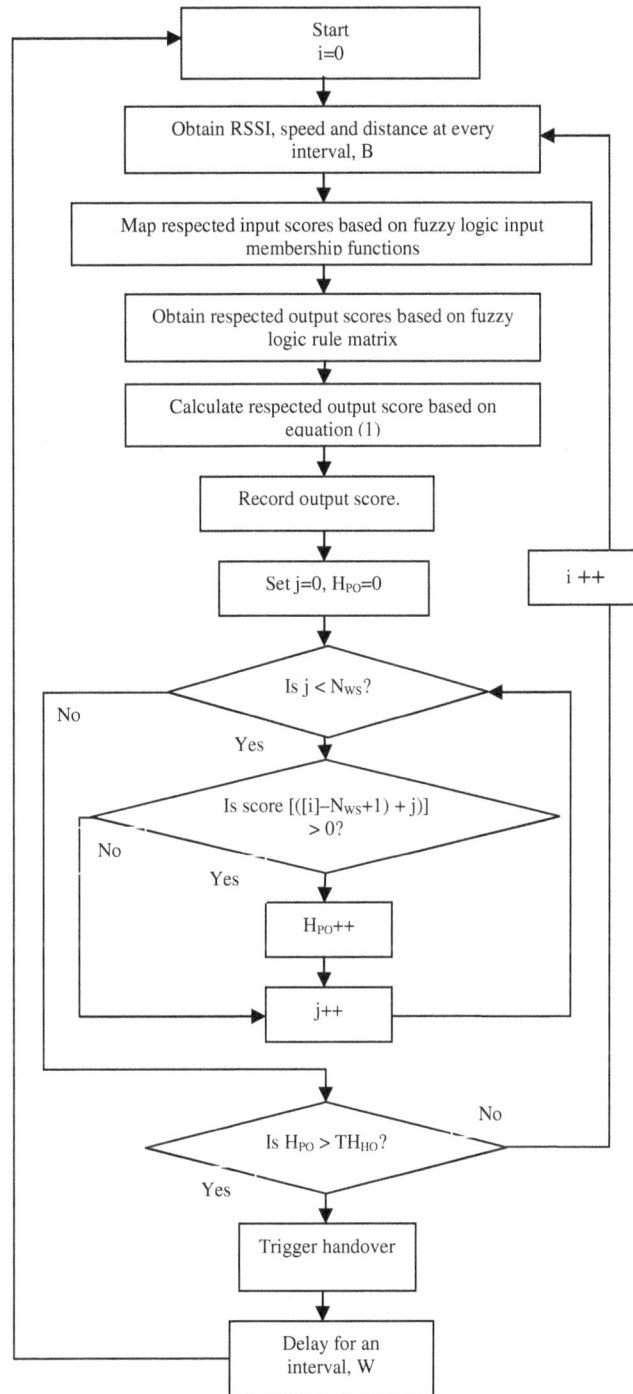

Figure 1: FL Trend (RSD) Flow Chart

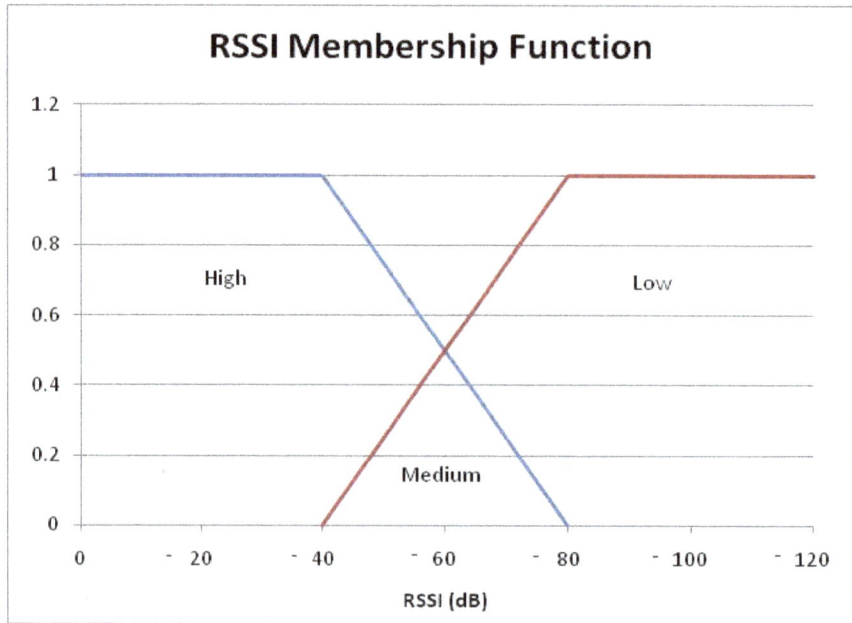

Figure 2(a): RSSI Membership Function

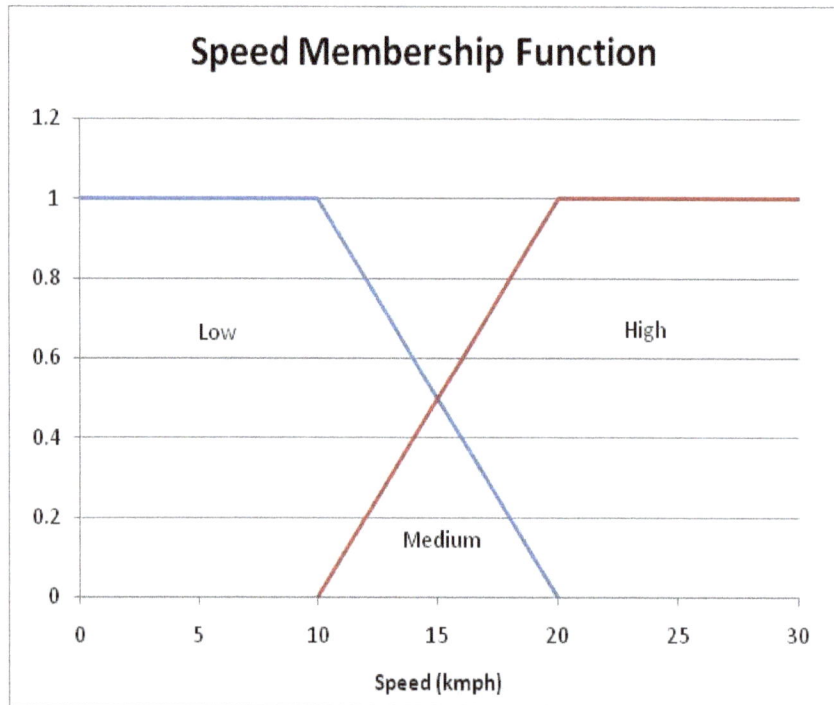

Figure 2(b): Speed Membership Function

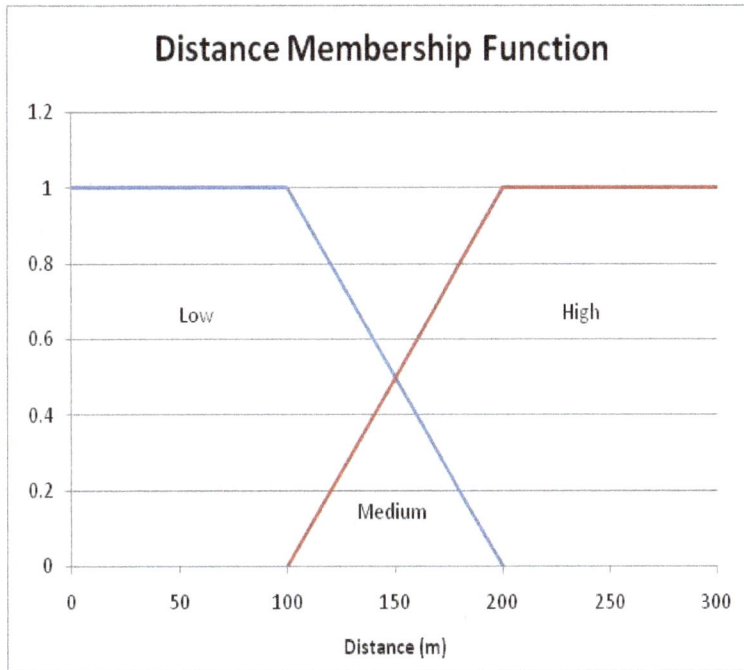

Figure 2(c): Distance Membership Function

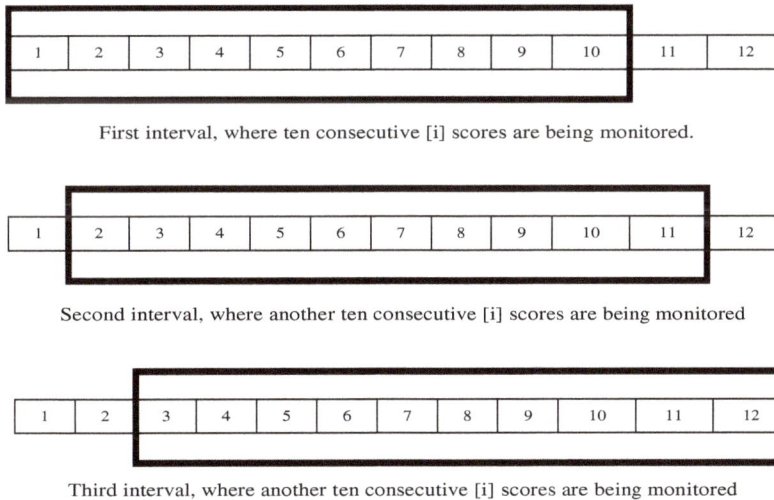

First interval, where ten consecutive [i] scores are being monitored.

Second interval, where another ten consecutive [i] scores are being monitored

Third interval, where another ten consecutive [i] scores are being monitored

Figure 3: Timing Chart of a Sliding Window where $N_{WS}=10$.

4. SIMULATION STUDIES

In order to evaluate the performance of FL Trend (RSD), we designed and implemented the following scenario as shown in figure 4. Mobile node (MN) traversed randomly in the assigned space within a time limit. The simulation was run at different speeds ranging from

5kmph to 30kmph. The low speed range (5kmph to 15kmph) defined in this simulation is to replicate the human walking scenario whereas the high speed range (16kmph to 30kmph) defined in this simulation is to replicate a human running or a vehicle driving in an urban area scenario. The access points were intentionally placed overlapping one another to generate interference. The reason behind this implementation was to create a realistic simulation scenario and also to test the ability of different parameter to detect communication degradation. This simulation was run on Omnet 4.0 simulator [8]. Mobile node started from the home agent and traversed randomly across the network. Once a handover triggering signal was received, mobile node performed handshake process with the selected access point. CN received updates from MN. MN then started sending traffic flow to the new interface where MN was connected. Subsequently, MN began sending data transmission to the corresponding node (CN) at a constant bit rate with the packet size of 1000 bytes at the interval of F, 0.5s whereas CN began sending data transmission at a constant bit rate with the packet size of 1000bytes at the interval of C, 0.08s. When MN detected the necessity to handoff once more, the handover algorithm was triggered and MN continued the handover process to ensure the continuity of communication session. The simulation was run on ten different scenarios and the averaged results were recorded. The results obtained were tabulated and averaged. The parameters employed in this simulation were shown in table 2. In this comparative study, we evaluated four handover trigger algorithms; RSSI Threshold, Change of RSSI, FR Threshold and FL Trend (RSD). RSSI Threshold scheme's threshold was set at 75dB [9]. An analytical model derived in [10] was used to estimate the threshold for RSSI Threshold scheme link going trigger. Based on Fritz path loss model in equation (5):

$$\left[\frac{Prx(d)}{Prx(do)}\right] dB = -10\beta \log\left[\frac{d}{do}\right] \qquad (5)$$

where Prx(d) was the received signal power level in Watts, Prx(do) was the received power at the close-in reference distance, d was the distance between the transmitter and receiver, do was determined using the free space path loss model and β was the path loss exponent, the threshold was calculated. With this, a threshold of 86dB was obtained. Hence, according to [9, 10] , a threshold of 75dB was used as the threshold limit for RSSI Threshold algorithm.

Table 2: Simulation Parameters

Parameter	Value
Transmitter Power	50mW
Wavelength (λ)	0.125m
Path Loss Exponent	2
Radio Carrier Frequency	2.4GHz
Minimum Channel Time	1s
Maximum Channel Time	3s
CN Data Interval, C	0.08s
MN Data Interval, F	0.5s
Distance, D	100m
Update Interval, B	0.1s
Sliding Window Size, N_{ws}	10
Threshold, TH_{HO}	7

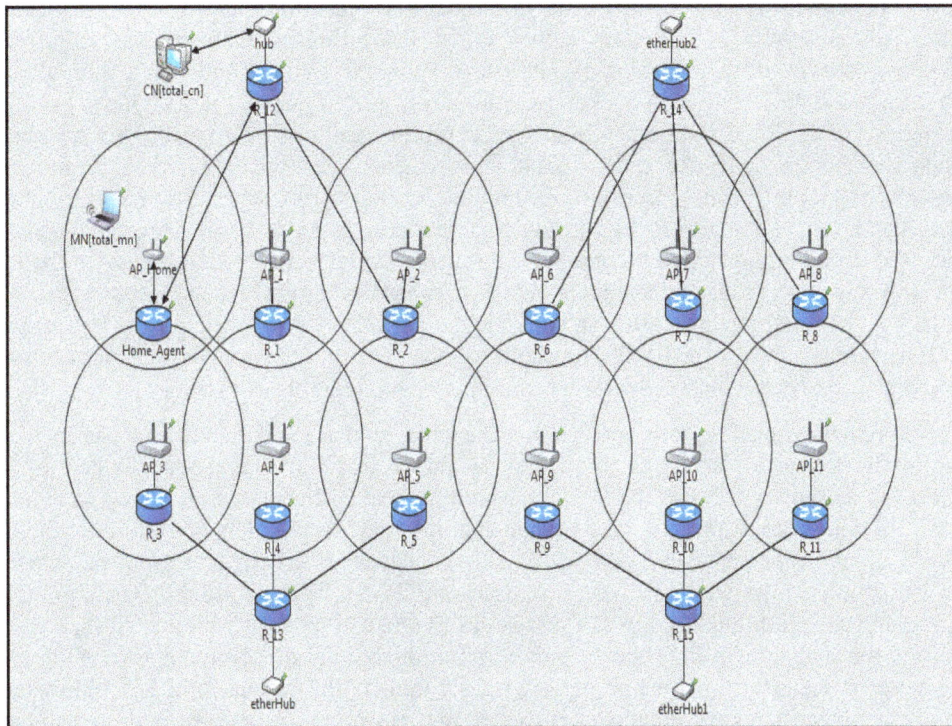

Figure 4: Simulation Scenario

5. RESULTS AND DISCUSSION

In this section, we evaluate four handover trigger algorithms. Through performance analysis we analyze the existing handover trigger algorithms and compare them with our proposed handover trigger algorithm. To achieve seamlessness in wireless networks, the algorithm has to satisfy multiple objectives. We advocate that efficient handover trigger algorithms should have the following characteristics to support seamless handovers in a wireless network. These characteristics include being able to guarantee low packet loss, high throughput and low packet delay during handover. All these should be achieved with minimal number of handovers in order to attain improved communication quality. We study the performance of the algorithms with data traffic. The evaluations of different schemes cover the following dimensions: (i) Number of handovers. (ii) Average packet loss. (iii) Average throughput. (iv) Minimum packet delay. (v) Maximum packet delay. (vi) Mean packet delay. Number of handovers depicts the process of transferring an ongoing data session; therefore it is crucial to be evaluated as increased of number of handovers affects the communication quality experienced by the users. Packet loss causes the transmitted packets to fail to arrive at their designated destination which in turn result in communication errors. Average throughput defines the average success rate of packets delivery over a period of time, thus high throughput is desirable to achieve high data rates. Minimum packet delay is the minimum latency of each link where low minimum packet delay is preferred as it indicates shorter time required by a packet to reach its designated destination. Maximum packet delay represents the maximum latency of each link where high maximum packet delay is not favored as it indicates that the communication link is probably experiencing congestion or abrupt change of communication quality. Mean packet delay defines the average latency of packets where high mean packet delay indicates bad communication quality which requires MN to perform a handover in order to maintain the ongoing communication session.

It is pointed out that RSSI threshold is unable to guarantee improved communication quality. This is shown in our comparative study. RSSI threshold attains high number of handovers as shown in figure 5, high packet loss as shown in figure 6 and 7, low throughput as shown in figure 8 and 9 and high packet delay as shown from figure 10 to 15. Due to multipath and shadow fading, RSSI sometimes fluctuates abruptly even when the established connection between the mobile node and access point is still adequate. Thus, this forces unnecessary handovers to take place which decreases the communication performance. RSSI Threshold only triggers handovers when the RSSI threshold is reached. Instead, RSSI Threshold does not monitor the RSSI readings throughout a period of time. Therefore, a sudden change in the RSSI readings would have caused handovers to be triggered when in actual scenario, it is not necessary as the RSSI readings are just experiencing a sudden drastic drop for a short period of time. Hence, solely based on RSSI Threshold to trigger handovers has the potential to cause unnecessary handovers which degrades the communication quality performance.

Through evaluations, we observe that the Change of RSSI achieves low packet loss as shown in figure 6 and 7 and high throughput as shown in figure 8 and 9 at the cost of high packet delay as shown in figure 10 to 15 and high number of handovers as shown in figure 5. This is especially true in the low speed range (5kmph to 15kmph). The conventional methods use RSSI as the handover trigger parameter. This is because RSSI triggers handover when the signal strength becomes weak. However, at times, the RSSI gives wrong representation of the current communication link quality as RSSI is easily affected by multipath and shadow fading. Therefore, the change of RSSI is being monitored through a period of time before the decision to handover is made in the Change of RSSI algorithm. If the change of RSSI indicates the necessity to trigger handover, handover is carried out. However, the drawback of using change of RSSI is that unnecessary handover might be triggered due to the incapability of RSSI to detect radio interference which causes degradation of communication quality. Therefore, we observe high number of handovers in Change of RSSI at all range of speed which results in the increment of overall packet delay. However, more handovers results in higher probability of mobile node being handover to close proximity access points. Thus, this improves the throughput and packet loss slightly within the low speed range as less packet delay is incurred due to nearer access point connectivity. At high speed range (16kmph to 30kmph), mobile node moves around at higher speeds which results in abrupt change of RSSI in a short period of time. This results in increase number of handovers. Subsequently, this in turns incurs higher packet delay as the distance between the connected access point and mobile node changes abruptly due to the increase of speed. The increase probability of unnecessary handovers causes handovers to take place to the less efficient access points. Due to the abrupt change in the mobility of mobile node caused by the increase of speed, the probability of wrongly chosen access point as a result of false handover trigger alarms is higher. Therefore, the false handover trigger alarms causes handovers to take place to the inefficient and distant access points. This in turn causes high packet loss and low throughput at the high speed region.

It was shown in [6] that FR Threshold has the ability to detect radio interference much better as compared to RSSI readings. As pointed out in the low speed range, FR Threshold shows 5% to 20% lower number of handovers as shown in figure 5, lower packet loss as shown in figure 6 and 7 and higher throughput as shown in figure 8 and 9 in comparison with FL Trend (RSD) at the cost of higher packet delay as shown from figure 10 to 15. However, at the high speed region, FL Trend (RSD) outperforms FR Threshold. This is because FR Threshold does not monitor the frame retransmission readings throughout a period of time. Therefore, sudden change of frame retransmission would have caused unnecessary handovers to take place which causes degradation of communication quality. At high speed region, mobile node moves drastically in a high speed manner which causes constant abrupt change of frame retransmission readings. This causes frame retransmission threshold to be reached constantly, thus causing unnecessary handovers to be triggered. This in turn causes high packet loss and low throughput which is exceptionally obvious in the high speed region. FL Trend (RSD) outperforms FR Threshold in the high speed region in eliminating unnecessary handovers by taking location and speed of mobile node into consideration as the input parameters. Besides that, FL Trend (RSD) monitors the change of fuzzy logic outputs in a known period of time which eliminates inefficient handover decisions by avoiding unnecessary handovers that are triggered due to sudden change of parameters readings.

According to figure 5, FL Trend (RSD) is able to perform better in guaranteeing lower number of handovers as shown in figure 5; lower packet loss as shown in figure 6 and 7 and higher throughput as shown in figure 8 and 9. This is especially true in the high speed range (16kmph to 30kmph). Our comparative study shows that FL Trend (RSD) is able to attain lower packet delay as seen from figure 10 to 15. FL Trend (RSD) considers RSSI, speed and distance as the inputs to obtain accurate output decisions of whether handover is necessary. RSSI fluctuates abruptly due to multipath and shadow fading. However, RSSI readings are less affected by radio interference. As such, RSSI readings might not be dependable in situations where radio interference is strong. However, in such situation where radio interference and multipath fading exist, with distance and speed as the input parameters, the location of mobile node can be identified. With this information, any misinterpretations of RSSI can be minimized. Therefore, with the use of these three parameters, RSSI, speed and distance, unnecessary handovers are minimized. Hence, FL Trend (RSD) manages to achieve lower number of handovers, lower packet loss, lower packet delay and higher throughput which is especially obvious in the high speed region.

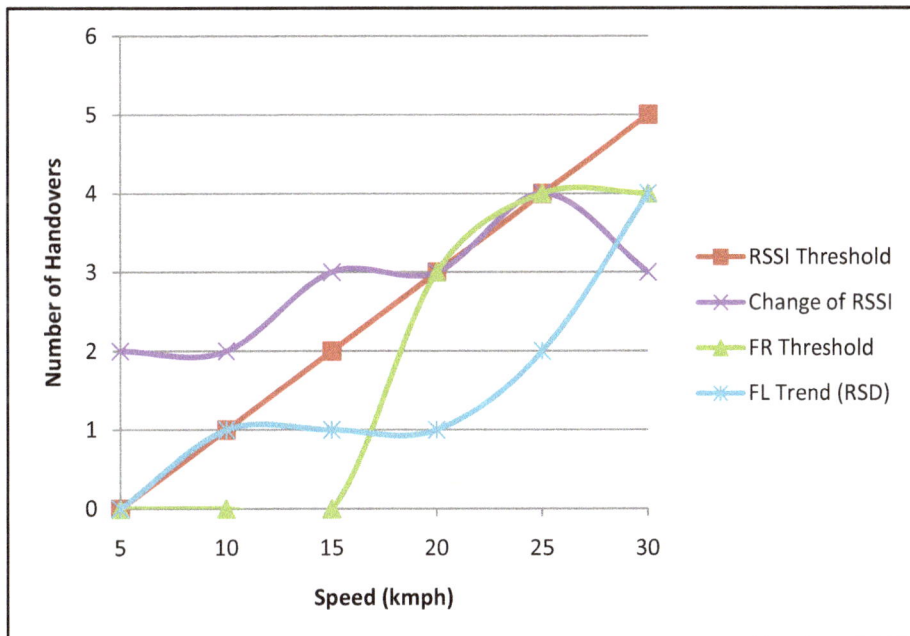

Figure 5: Number of Handovers versus Speed (kmph) at Mobile Node, MN

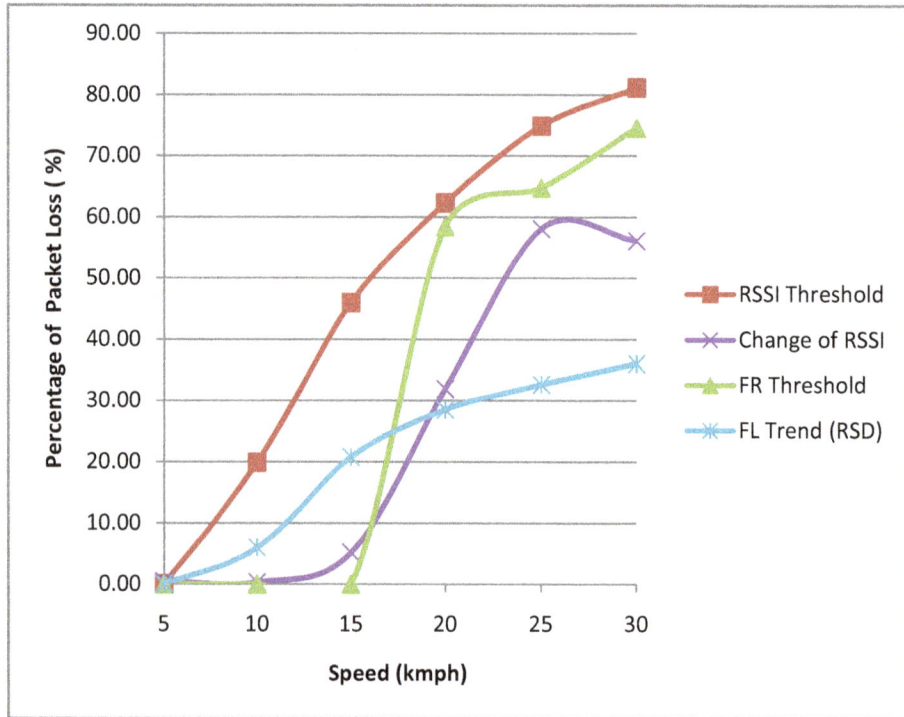

Figure 6: Percentage of Packet Loss (%) versus Speed (kmph) at Mobile Node, MN

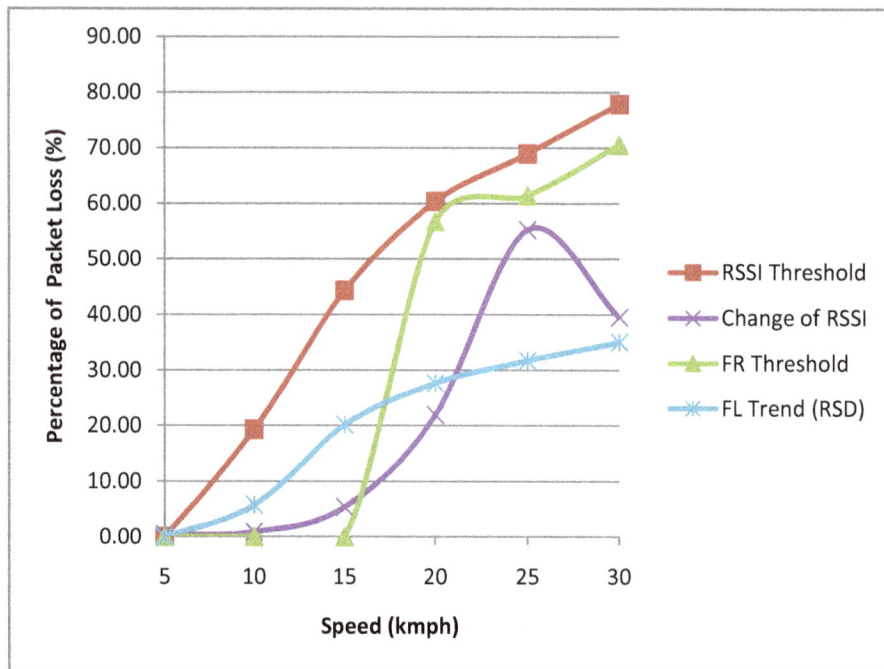

Figure 7: Percentage of Packet Loss (%) versus Speed (kmph) at Corresponding Node, CN

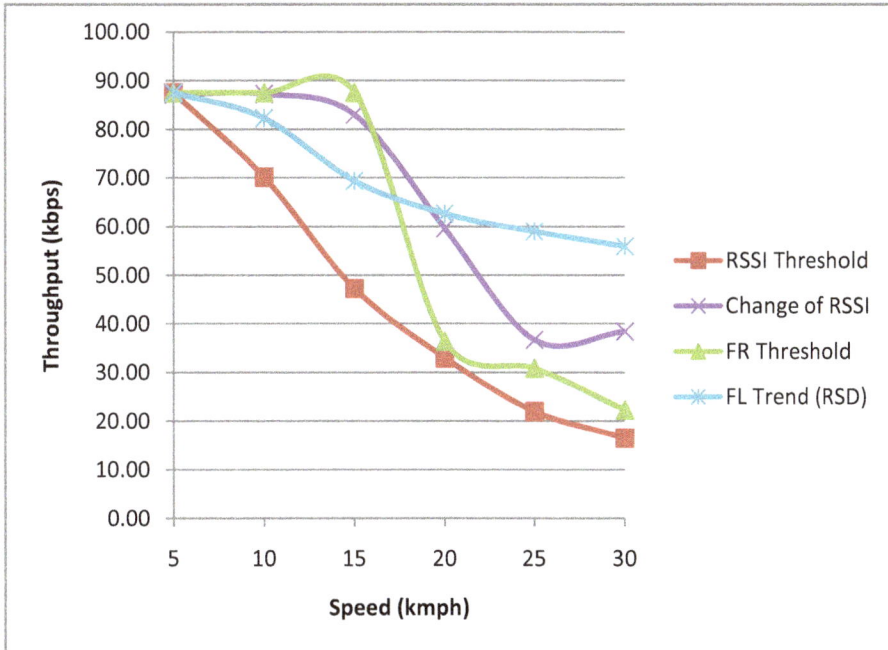

Figure 8: Throughput (kbps) versus Speed (kmph) at Mobile Node, MN

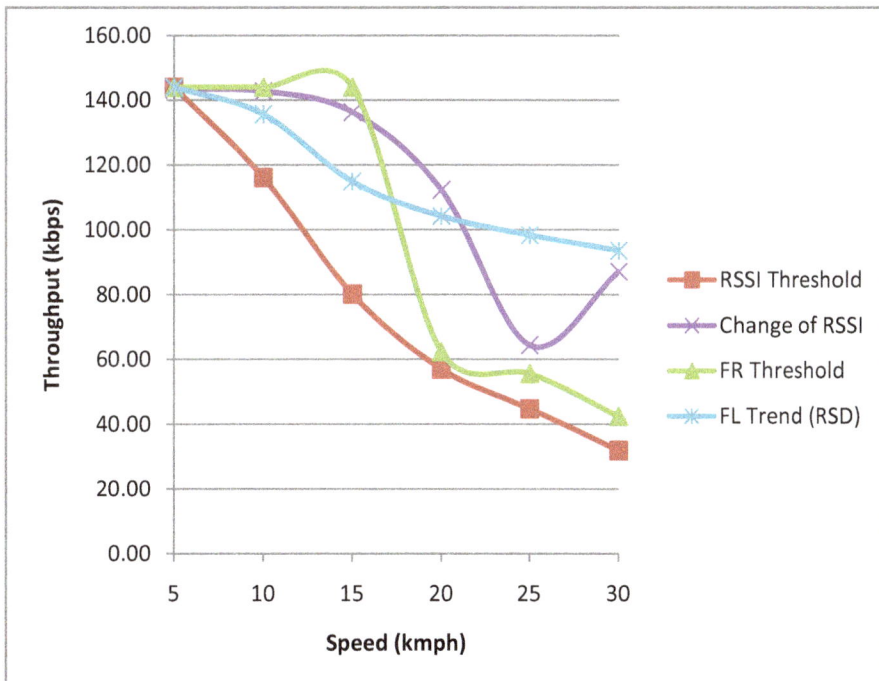

Figure 9: Throughput (kbps) versus Speed (kmph) at Corresponding Node, CN

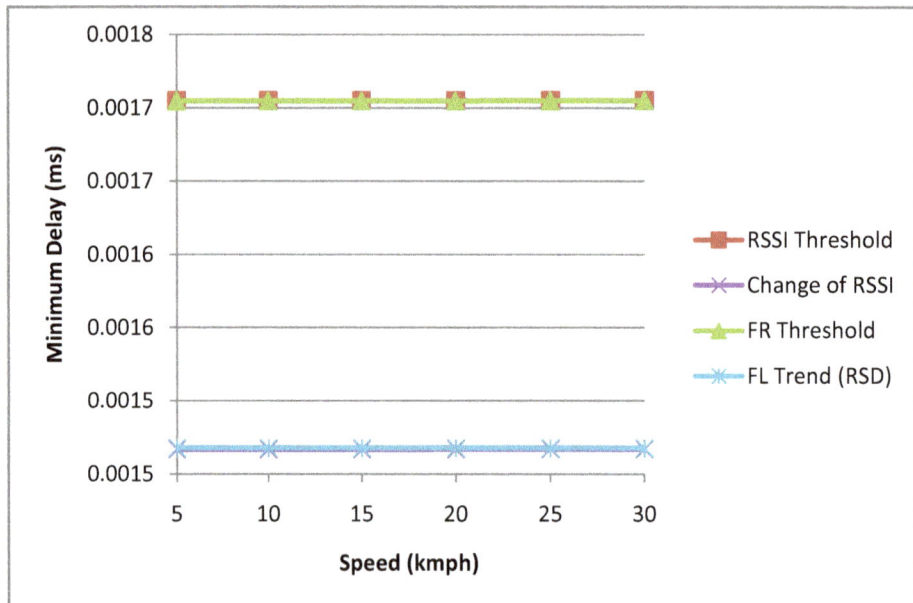

Figure 10: Minimum Delay (ms) versus Speed (kmph) at Mobile Node, MN

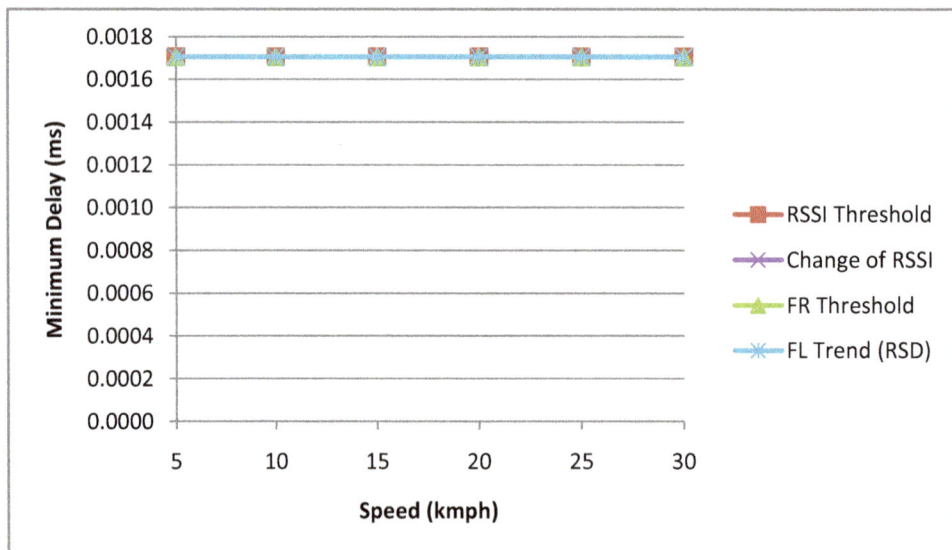

Figure 11: Minimum Delay (ms) versus Speed (kmph) at Corresponding Node, CN

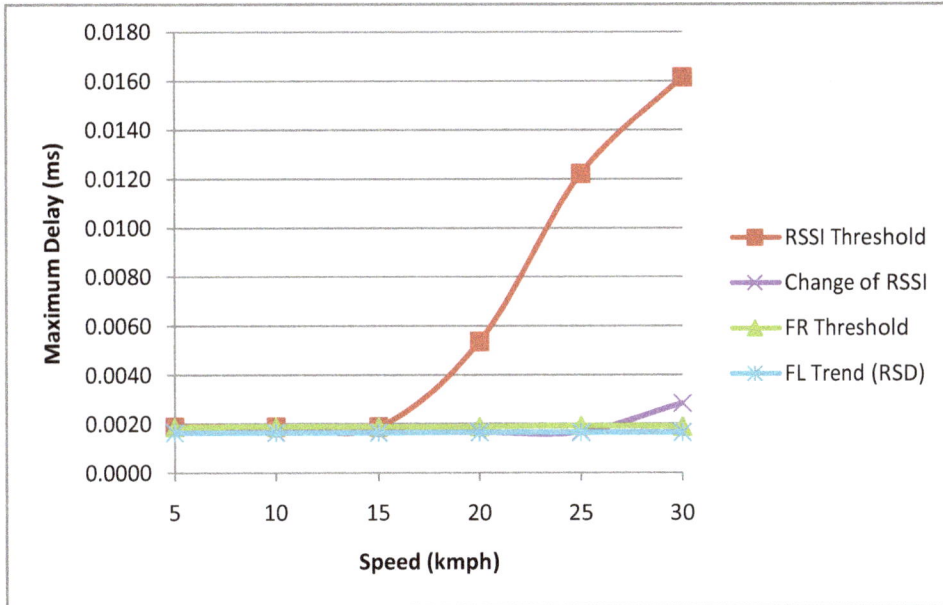

Figure 12: Maximum Delay (ms) versus Speed (kmph) at Mobile Node (MN)

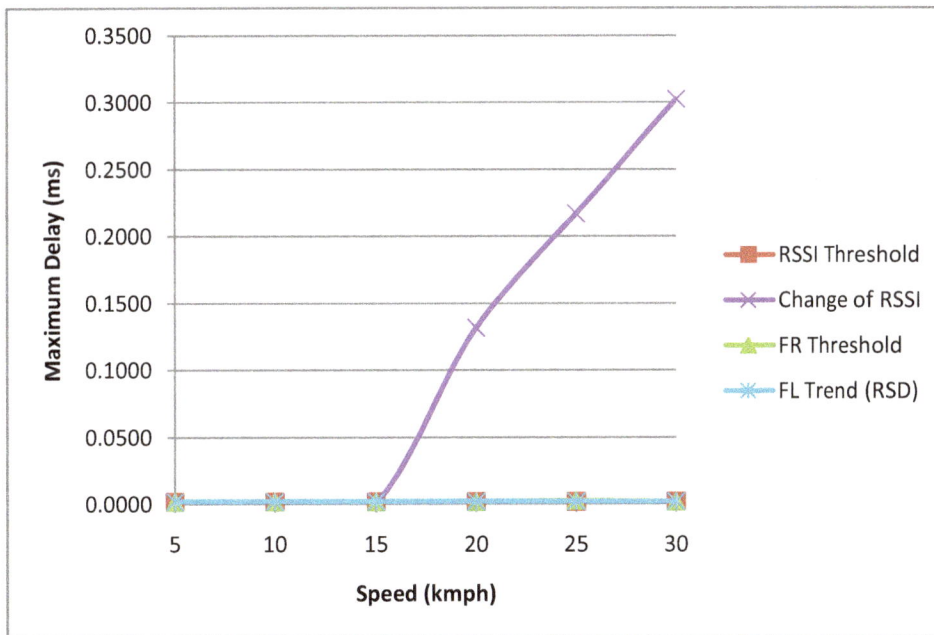

Figure 13: Maximum Delay (ms) versus Speed (kmph) at Corresponding Node, CN

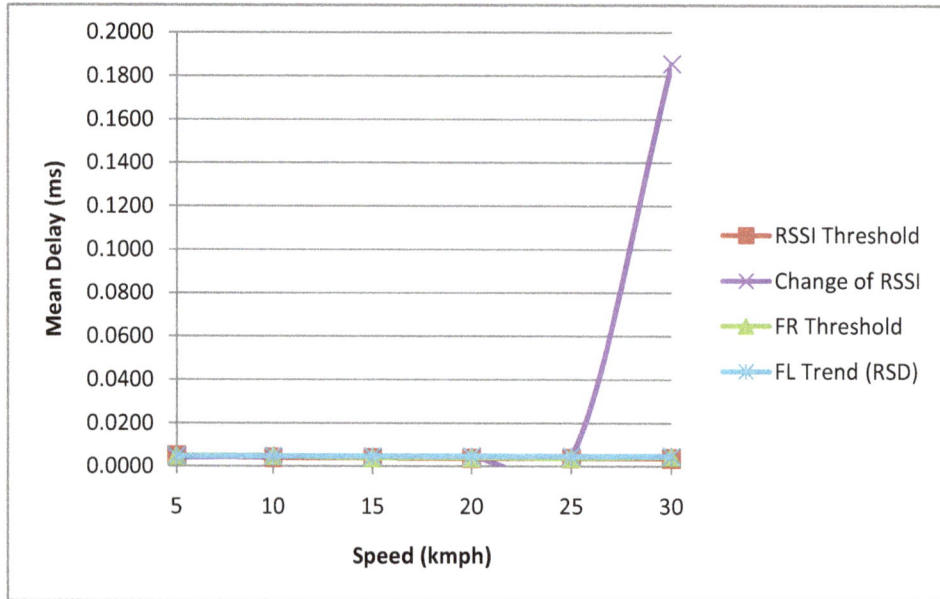

Figure 14: Mean Delay (ms) versus Speed (kmph) at Mobile Node, MN

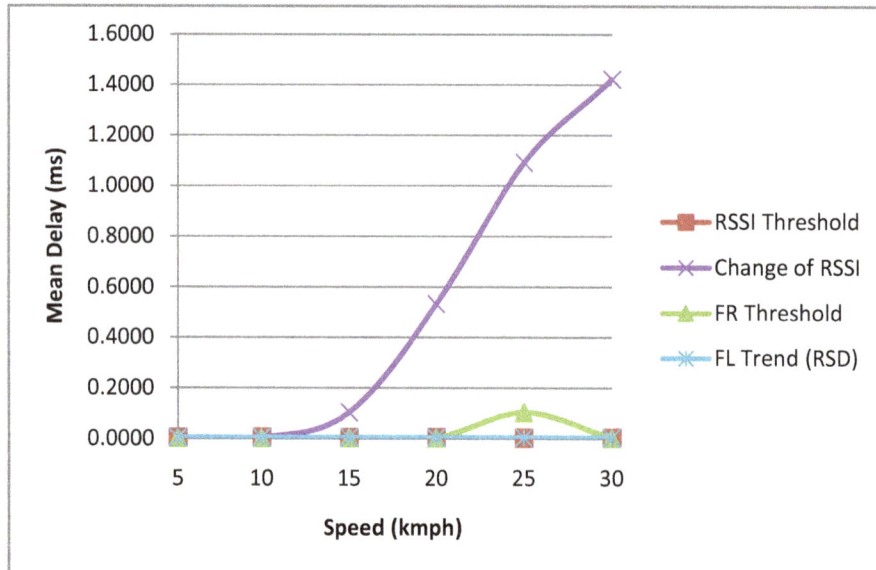

Figure 15: Mean Delay (ms) versus Speed (kmph) at Corresponding Node, CN

6. SUMMARY

In this section, we present a comparative performance of four handover trigger algorithms; RSSI Threshold, Change of RSSI, FR Threshold and FL Trend (RSD). Handovers are required to maintain the communication quality of a session. Traditionally it was pointed out that RSSI has the capability of detecting communication quality degradation and triggers handovers. In a wireless network, RSSI changes abruptly due to many factors such as multipath and shadow fading. Therefore, a more challenging problem is to incorporate several parameters to assist RSSI to minimize misrepresentations of the communication quality. In this context, we propose to use RSSI, speed and distance as the decision factors in a fuzzy logic system to optimize performance quality. In addition, we also propose to monitor the output trend of fuzzy logic in FL Trend (RSD) before making handover decisions in order to minimize unnecessary handovers due to sudden fluctuation in the parameters' readings. In comparison with RSSI Threshold, FL Trend (RSD) achieves lower number of handovers, lower packet loss, higher throughput and lower packet delay at all speed ranges. In comparison with Change of RSSI, at low speed (5kmph to 15kmph) FL Trend (RSD) achieves lower number of handovers and lower handover delay at the cost of 5% to 20% higher packet loss and 5% to 20% lower throughput. However, at the high speed range (16kmph to 30kmph), FL Trend (RSD) outperforms Change of RSSI in terms of number of handovers, packet loss, throughput and packet delay. In comparison with FR Threshold, in the low speed range, FL Trend (RSD) achieves lower packet delay at the cost of 5% to 20% lower number of handovers, higher packet loss and lower throughput. However, at high speed, FL Trend (RSD) outperforms FR Thresholds with lower number of handovers, lower packet loss, higher throughput and lower packet delay. The slight decrease of FL Trend (RSD) performance in the low speed region is generally due to the inefficiency of RSSI to detect radio interference. However, this is not the case in the high speed region because in the high speed region, distance readings contribute more towards decision making in FL Trend (RSD) which helps FL Trend (RSD) to make better and more effective handover triggers to achieve improved communication quality.

7. CONCLUSION

In wireless networks, handovers are necessary to maintain the communication quality. Traditionally, handovers occur when the signal strength of the serving access point drops below a certain threshold value. However, in mobile IPv6 environment, users are predicted to move constantly. Therefore, location factors are necessary to be considered when making handover decisions besides signal strength because location factors give an insight of the location of mobile node from the access points. Addressing the drawbacks of sudden change in parameters in a radio propagation environment, the outputs fuzzy logic in FL Trend (RSD) is being monitored for a fixed period of time before handover decisions are made. In this paper, we showed that FL Trend (RSD) outperforms the existing handover trigger algorithms in terms of number of handovers, packet loss, throughput, minimum packet delay, maximum packet delay and mean packet delay. This is especially obvious in the high speed topologies (16kmph to 30kmph). Additionally, in the low speed topologies (5kmph to 15kmph), FL Trend (RSD) outperforms RSSI Threshold in all aspects, achieves lower number of handovers and packet delay at the cost of 5% to 20% increase in packet loss and decrease in throughput as compared to Change of RSSI and achieves lower packet delay at the cost of 5% to 20% increase in the number of handovers, increase in packet loss and decrease in throughput as compared to FR Threshold. In general, FL Trend (RSD) achieves improved communication quality by attaining lower number of handovers, lower packet loss, lower packet delay and higher throughput which is especially obvious in the high speed region. Analysis and simulations show that FL Trend (RSD) is able to create timely and reliable handover triggers to achieve improved communication quality performance which is certainly apparent in the high speed region.

REFERENCES

[1]   Kholoud Atalah, Elsa Macias and Alvaro Suarez, "A proactive horizontal handover algorithm based on RSSI supported by a new gradient predictor," Ubiquitous Computing and Communication Journal, vol. 3, 2008.

[2]   Tsung-Nan Lin and Po-Chiang Lin, "Handoff ordering using link quality estimator for multimedia communications in wireless networks," Proceedings of Global Telecommunications Conference, Taipei, Taiwan, 2005.

[3]   Nidal Nasser, Ahmed Hasswa and Hossam Hassenein, "Handoff in fourth generation heterogeneous networks," IEEE Communications Magazine, vol. 44, pp. 96-103, 2006.

[4]   Ken-Ichi Itoh, Soichi Watanabe, Jen-Shew Shih and Takuro Sato, "Performance of handoff algorithm based on distance and RSSI measurement," IEEE Transactions on Vehicular Technology, vol. 51, pp. 1460-1468, 2002.

[5]   Shih-Jung Wu, "Fuzzy-based handover decision scheme for next-generation heterogeneous wireless networks," Journal of Convergence Information Technology, vol. 6, pp. 285-297, 2011.

[6]   Kazuya Tsukamoto, Takeshi Yamaguchi, Shigeru Kashihara and Yuji Oie, "Experimental evaluation of decision criteria for WLAN handover: signal strength and frame retransmission," The Institute of Electronics, Information and Communication Engineers Transactions, vol. 12, pp. 3579 – 3590, 2007.

[7]   Valentin Rakovic, Vladimir Atanasovski and Liljana Gavrilovska, "Velocity aware vertical handovers," Proceedings of Applied Sciences in Biomedical and Communication Technologies, Bratislava, Slovakia, pp. 1-6, 2009.

[8]   Faqir Zarrar Yousaf, Christian Bauer and Christian Wietfeld, "An accurate and extensible mobile IPv6 (xMIPv6) simulation model for Omnet++," in Proceedings of 1st International Conference on Simulation Tools and Techniques Communication, Networks and Systems & Workshops, Belgium, 2008.

[9]   Vivek Mhatre and Konstantine Propagiannaki, "Using smart triggers for improved user performance in 802.11 wireless networks," Proceedings of the 4th International Conference on Mobile Systems, Applications and Services, New York, United States, pp. 246-259, 2006

[10]   S. Woon, N. Golmie and Y.A. Sekercioglu, "Effective link triggers to improve handover performance," Proceedings of Personal, Indoor and Mobile Radio Communications, Helsinki, Finland, pp. 1-5, 2006.

14

Effect of Multipath Fading and Propagation Environment on the Performance of a Fermat Point Based Energy Efficient Geocast Routing Protocol

Kaushik Ghosh[1], Partha Pratim Bhattacharya[2] and Pradip K Das[3]

Faculty of Engineering & Technology, Mody Institute of Technology & Science (Deemed University), Lakshmangarh, Dist. Sikar, Rajasthan – 332311, India

[1] kghosh.et@mitsuniversity.ac.in,
[2] ppbhattacharya.et@mitsuniversity.ac.in,
[3] pkdas@ieee.org

ABSTRACT

Energy efficiency is a much talked about thing in the domain of geocast routing protocols for Wireless Ad Hoc and Sensor Networks (WASNs). Fermat point based protocols are capable of reducing the energy consumption of a WASN by reducing the total transmission distance in a multi hop-multi sink scenario. Presently, there are quite a handful of them but many of them have not considered the effect of changing propagation environment around the considered network while measuring the performance of the protocol. Congested environment around a WASN increases the chance of multipath propagation and it in turn introduces multipath fading. In this paper, the effects of both of these factors are considered on the performance of I-Min routing protocol designed for WASNs.

KEYWORDS

WASN, Geocasting, Energy efficiency, Multi path fading, Propagation environment.

1. INTRODUCTION

Geocast routing protocols have become an integral part in the domain of routing for Wireless Ad Hoc and Sensor Networks (WASNs). It is well known that these kind of networks demand special attention towards energy consumption due to their inherent nature and places of their deployment. Energy is in fact a function of the inter nodal distances (d^n), where n is called path loss exponent. Quite a few protocols of said type have been proposed over the years by numerous authors but the radio models used in most of them have undermined the effect of changing propagation environment around the network while determining its energy expenditure. A hullabaloo around a WASN is sure to introduce multipath fading, which again demands some modification in the radio model proposed for a free space condition. In this paper we have considered both these parameters to find out the degree of variation on the performance of a protocol that doesn't consider either of these two parameters. Here, the geocast routing protocol under consideration is the I-MIN protocol [1]. In [1] the radio model used considered data size and distance between the nodes as the functions of energy consumed by a node while transmitting. Energy consumed by a node while receiving was however a function of the received data volume only. Not only this, the network was assumed to be operating in a free space only (n=2), as was considered by many other protocols [2], [3], [4],

[5].In this paper we have modified the radio model used in [1] by incorporating both of the parameters. From Frii's free space equation we get

$$P_r = P_t * (G_t * G_r * \lambda^2)/(16 * \pi^2 * d^2 * L)$$

Here, P_r and P_t are the power required for receiving and transmitting respectively over a distance of d. So it is clear that the power consumed by a node while receiving data is dependent upon the received signal strength from the sender along with the distance between them. Now, in order to include that factor in P_r, it is not sufficient to know P_t alone. Because multipath fading [6] causes the signal strength at the receiving ends to decrease by some order which cannot be predicted using the above equation alone. The nature of multipath fading is not as simple as an exponential decay shown in most of the elementary text books. Findings in [6] shows the signal strength to decrease as a result of multipath fading with distance but the nature of the graph is bit complicated. In this paper we have thus tried to modify the radio model in [1] with the findings of [6] and changing the values of n to all integer values between 2 and 6. Coming to the distance part, it becomes evident that we need to find a way or the other to reduce the total transmitting distance where the nodes seldom changes their deployed position. A Fermat point based scheme helps the most in this regard. There are however techniques other than Fermat point based ones for reducing d and thereby minimizing the energy expenditure of a WASN.

The remaining sections include related works (2), a brief discussion of the I-MIN protocol (3), effect of changing propagation environment around the network and multipath fading on that protocol (4), results(5) and finally, conclusion and future works(6).

2. RELATED WORKS

The introduction of Fermat point based routing in geographic routing protocols had been made for the first time in [7]. In [7] we get a detailed discussion on locating the Fermat point for a triangular region in a geometric way. [3] have shown to do the same by finding the Global Minima which ensures a more generalized approach of finding the Fermat point for any number of geocast regions. However, the forwarding technique in [3] couldn't ensure a loop free solution. This problem was eliminated in [1] and the results were thus better when it came to energy consumption and total number of hops encountered by a data packet before it reached its final destination. Exploiting the concept of Fermat point for energy reduction is done in [8] as well. Data dissemination for multi sink scenario through Fermat point is again shown in [9] using a graph based approach. In fact Fermat point based approach has been used by the authors since it is capable of reducing the transmission distance in a multi sink-multi hop environment. However, many other techniques other than the using the Fermat point for distance reduction has also been explored. Of all the different techniques, **Location Aided Routing** (LAR) [10] can be termed as the ancestor of many of the present day geocast routing protocols. LAR professed for reducing the number of transmissions by determining the geographic region of the destination. However, action on encountering routing holes was unknown till authors in [11] proposed the idea of perimeter routing for the same. Reducing the distance to reduce the energy consumed through a non Fermat point based approach has been shown in [12] as well. Although outdated, [13] gives a detailed survey on geocast routing protocols. [14] on the other hand most probably have introduced the concept of geocasting itself in the field of MANETs. The aforementioned works have mainly focused on reducing the transmission distance between the nodes without disturbing their initial deployment to attain an energy friendly nature.

Techniques other than distance reduction for energy consumption have also been used in WASNs [15-18]. But of all the protocols discussed, none have discussed the effect of changing propagation environment or multipath fading on the amount of energy consumed by the network.

In this paper we have discussed how change in propagation environment around a WASN can affect the performance of energy efficient routing protocols, even when all other factors remain unchanged.

3. I-MIN SCHEME

The I-Min scheme has been described in [1] in details, yet for sake of continuity let us re describe it in brief. The said scheme uses a Fermat Point based data forwarding technique [1-3], [7]. It is a protocol designed for static WASNs with a number of geocast regions to transmit a message to. The Global Minma algorithm [3] is used for finding out the Fermat point of the polygonal/triangular region under consideration (figure 1a). The results in [1] have shown it to outperform some of its predecessors like [2] and [3] when energy and delay are the measuring parameters (Fig 1). The reason for a stark difference in performance between I-MIN and its predecessors are due to the fact that in neither of them the authors have talked about a loop free solution which in fact, I-MIN did.

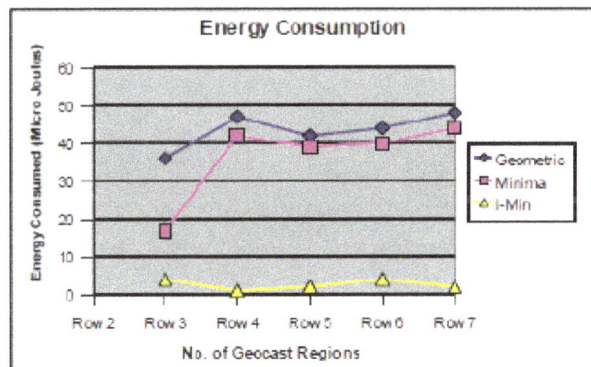

Fig 1a. Energy consumption in the three different schemes.

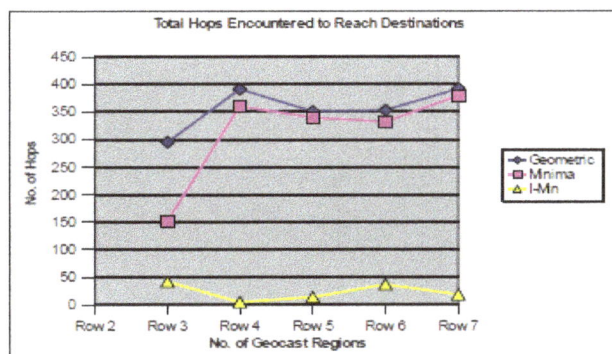

Fig 1b. Comparison between three schemes with respect to number of hops encountered by a packet to travel from a given source to destination.

The protocol under consideration uses a localized greedy approach while forwarding data (Fig 2b) i.e. a node forwards a packet to such a node amongst its neighbors which is closest to the destination. Although, one of the major problems of this kind of an algorithm is local maximization [11], yet we expect it not to happen in our case as the network under consideration has been taken dense enough to have a **hole** in it. Moreover, another improvement over the forwarding method of I-Min scheme has been made here.

The next forwarding node here is selected not merely on the basis of its distance from the destination but its residual energy has also been taken into consideration. In figure 1b we see that the next node is selected from a list of neighbors by comparing their distance from the destination. In this paper we would like to modify the forwarding technique through the following equation

$$max\_id = res\_energy*u_1 - dist*u_2.$$

The parameter **max_id** is calculated for all the neighbors of a node and the one with highest value of **max_id** becomes the next forwarding node.

Minima Algorithm

Input: Coordinates of the sender node and that of different geocast regions.
Output: (fx, fy); Coordinates of the Fermat_Point.
Tdist : Total distance traveled by the packet.
Tpow: The sum of power consumed by all the intermediate nodes to forward the packet to m geocast regions

```
1.   max_x=MAX_x (Sx, GRX(N));
2.   max_y=MAX_y (Sy, GRY(N));
3.   min_x=MIN_x (Sx, GRX(N));
4.   min_y=MIN_y (Sx, GRY(N));
5.   dx=0;  /*         Initialize dx */
6.   dy=0   /* Initialize dy */
7.   flag=0; /* To check the Fermat Point*/
8.   for ( i=min_x; i<max_x; i++ )
9.   { if (flag==1) break; /* Fermat point found */
10.  for ( j=min_y; j=max_y; j++ )
11.  { x=i;
12.  y=j;
13.  for ( k=0;k<n;k++ )
14.  { dx+= termdx ( x,GRX(k),y,GRY(k) );
15.  dy+= termdy ( y,GRY(k),x,GRX(k) );
16.  } /* end of for loop (line-11)*/
17.  if ( dx==0 && dy==0 )
18.  { flag=1; /* Fermat point found */
19.  break;
20.  }
21.  dX=0; dY=0;
22.  } /* end of for loop (line-8) */
23.  }/* end of for loop (line - 6) */
24.  if( flag==1 )
25.  { fx=x; fy=y; } /* Fermat point */
26.  Tdist= Total_Dist ( Sx, Sy, GRX(N), GRY(N), fx, fy );
27.  Tpow= Total_Pow ( Tdist )
```

Figure 2a. Global Minima Algorithm.

I-Min Algorithm

```
Input: Node_id of the destination.
Output: Boolean.
D_ID: Node_id of the destination.
NN: Number of neighbors of the forwarding
node.
Neigh_ID= Node_id of a neighbor.

  1. if (flag==0)
  2. next_hop=MAX_DIST(D_ID, NN, neigh[]);
  3. else
  4. next_hop=D_ID;
  5. for(i=0; i<NN; i++)
  6. {if (D_ID==Neigh_ID)
  7. flag=1;
  8. else
  9. flag=0; }
```

Figure 2b. I-MIN forwarding scheme.

Here, **res_energy** is the residual energy of a node and **dist** is its distance of the node from the destination or Fermat point. This is because, for a node lesser the distance from destination and higher the residual energy, greater is its potential to be the next forwarding node amongst a group of nodes. Both these parameters are known to a node when a neighbor piggybacks them with data packets in case there is any change in those parameters. u_1 and u_2 are two weight age functions where $u_1 > u_2$. This way we make the scheme energy efficient as it increases the probability that a node with higher residual energy is selected even if its distance from destination is somewhat more as compared to that for another node with a lesser value for residual energy.

4. EFFECT OF PROPAGATION ENVIRONMENT AND FADING

In this section we will consider separately the effect of propagation environment and multipath fading on the radio model used in [1] and then would combine them to present a new radio model to be used in WASNs.

4.1. Effect of Propagation Environment

The radio model in [1] has considered the space between the constituent nodes of the network to be free space only. In fact it is the model used in [5] as well. The model is given below for clarity

$$E_{TX}(m, d) = m * E + m * \epsilon * d^2$$

$$E_{RX}(m) = m * E$$

Where,

$E = 50nJ/bit$ and $\epsilon = 10\,pJ/bit\,/m^2$.
E_{TX}=Energy consumed for transmission.
E_{RX}=Energy consumed for reception.
d=distance between the transmitting and receiving node.
ϵ=Permittivity of free space.
m=Number of bits.

However, the scenario in real life applications may not be that simple. Value of path loss exponent n may range from 2 for free space to 6 for congested city. We have thus taken a whole range of values considering n=2,3,4,5 and 6. At every step we find the energy to increase in the order of 100. Thus it becomes evident how much more the distance between the nodes is going to affect the performance of a protocol when the same experiment is repeated in a city rather than in rural area or free space.

We are thus generalizing the above model as

$E_{TX}(m,d) = m*E + m*\acute{\varepsilon}*d^n$

$E_{RX}(m) = m*E.$

With all other things retaining their meanings, $\acute{\varepsilon}$ is permittivity of air as medium.

4.2. Effect of Fading

In this section we modify the radio model of [1] yet further. Other than effect of propagation environment, what is going to have an impact on the performance of energy aware protocols, is the fading due to multipath propagation of the signals. In practical systems, hardly there is any LOS link and we have to depend on multipath signals. The signal strengths from two transmitting nodes in a multipath fading environment are given in Fig 3[6].

From the results in [6] we get the received signal strength at a node to be -50 dBm when the transmitting node is 100 meters away. Since the typical transmitting range of a sensor node is 100 meters or so, we include this -50 dBm of signal strength in the receiver ends to modify our radio model. To precise, the distance of a forwarding node from its immediate predecessor is always taken as 100 meters for the sake of simplicity. Because, otherwise things would get complicated since at every hop, the said distance varies arbitrarily between any values within 100 meters.

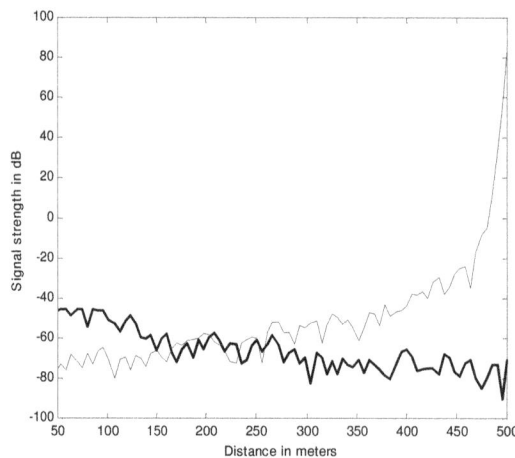

Node A   **Node B**

Fig 3. Received signal strength.

So, if we try to make that consideration, it won't be possible to include the effect of multipath fading in the radio model. The new radio model thus turns out to be

$P_{TX}(m',d) = m'*E + m*\acute{\varepsilon}*d^n$
$P_{RX}(m') = m'*E + 10^{-5}.$

Where,

P_{TX} = Power required for transmission
P_{RX} = Power received at the receiver
m = Data rate
$\dot{\varepsilon}$ = Permittivity
d= Distance between nodes
n = path loss exponent whose value lies between 2 to 6.

The factor 10^{-5} accounts for the amount of received power at the receiver's end (-50 dB), when the distance from transmitter is 100 meters (see figure 2). Since the typical transmitting range of a sensor node is 100 meters or so, we include this -50 dB of signal strength in the receiver ends to modify our radio model. To be precised, the distance of a forwarding node from its immediate predecessor is always taken as 100 meters for the sake of simplicity.

5. RESULTS

The results here show that after modifying the radio model of [1] with considerations for changed propagation environmental effects and multipath fading, the consumption of energy in a geocast routing protocol will vary considerably (Fig 4). Figure 4 shows the condition when the numbers of geocast regions are only 3. Higher the number of geocast regions, larger is the total distance that a data packet has to travel and thereby greater is the effect of propagation environment combined with the effect of multipath fading on the performance of an energy aware algorithm. This condition is depicted in figure 4.

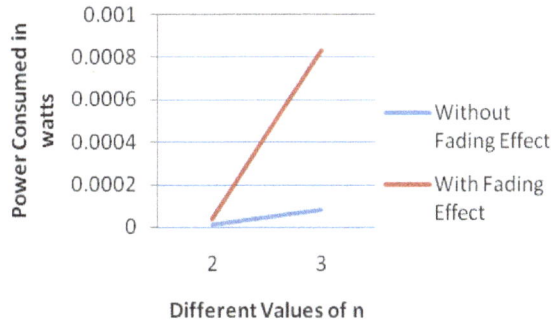

Fig 4a. Effect of propagation environment and multipath fading on energy consumption (n=2 and 3)

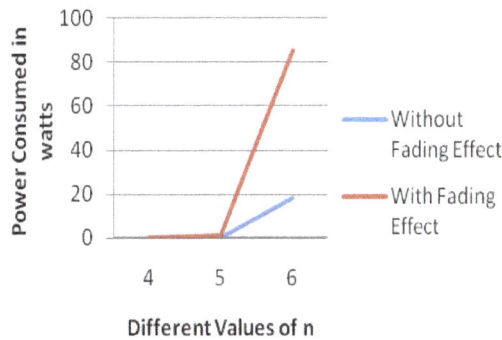

Fig 4 b). Effect of propagation environment and multipath fading on energy consumption (n=4,5 and 6)

The figure above depicts that the energy consumption of a network increase considerably when the value of n reaches 6. Although that is an extreme case, yet the results have shown that even a moderate value of n (say 4, which is most common) the energy consumption of a network is to increase by an order of 1000 when compared to the same network under free space condition (n=2).

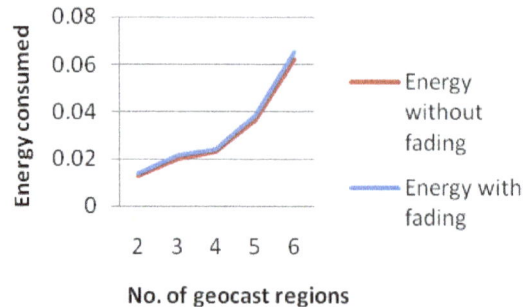

Fig 5. Energy consumption vs. number of geocast region (n=4)

From Figure 5 we find out that though small, but consideration of fading effect is increasing the energy consumption of a WASN when all other factors including the total number of hops encountered, remain the same.

6. CONCLUSIONS

Results in this paper are good enough to show that the effect of propagation environment and multipath fading are something which can never be ignored while forming a radio model for energy aware geocast routing protocols. The same protocol would consume larger amount of energy while operating in a congested environment than in a free space. As future work, one can think of a scheme where the effect of fading can be incorporated in the radio model for exact distance between the nodes than an approximated one as in the present paper. This in fact can be added on with the effect of pulsating data stream. The same amount of data received by a receiver over an intermittent period of time may result in better battery utilization than received in a single iteration. Modifying the protocol in that direction may put some further light on its behaviour.

ACKNOWLEDGEMENTS

The authors would like to acknowledge Mody Institute of Technology & Science, Rajasthan, for providing immense support towards the present work.

REFERENCES

[1]   Kaushik Ghosh, Sarbani Roy and Pradip K. Das "*I*-Min: An Intelligent Fermat Point Based Energy   Efficient Geographic Packet Forwarding Technique for Wireless Sensor and Ad Hoc Networks", *International Journal on Applications of Graph Theory in Wireless Adhoc Networks and Sensor Networks(GRAPH-HOC)*, Vol.2, No.2, June 2010 pp 34-44.

[2]   S.H. Lee and Y.B. Ko. "Geometry-driven Scheme for Geocast Routing in Mobile Adhoc Networks", *IEEE Transactions for Wireless Communications*, VTC Spring 2006: 638-642.

[3]   Kaushik Ghosh, Sarbani Roy and Pradip K. Das, "An Alternative Approach to find the Fermat Point of a Polygonal Geographic Region for Energy Efficient Geocast Routing Protocols:Global Minima Scheme", *AIRCC/IEEE NetCoM* 2009.

[4]   W.B. Heinzelman, A.P. Chandrakasan and H. Balakrishnan, "An Application-specific Protocol Architecture for Wireless Micro Sensor Networks", *IEEE Transactions on Wireless Communications*, Vol.1, Issue 4, October 2002.

[5]   I-Shyan Hwang and Wen-Hsin Pang, "Energy Efficient Clustering Technique for Multicast Routing Protocol in Wireless Adhoc Networks", *IJCSNS*, Vol.7, No.8, August 2007.

[6]   P. P. Bhattacharya, P. K. Banerjee, "User Velocity Dependent Call Handover Management", *International Journal HIT Transaction on ECCN*, Vol 1, No. 3, pp 150-155, July 2006.

[7]   Young-Mi Song, Sung-Hee Lee and Young-Bae Ko "FERMA: An Efficient Geocasting Protocol for Wireless Sensor Networks with Multiple Target Regions" T. Enokido et al. (Eds.): *EUC Workshops* 2005, LNCS 3823, pp. 1138 – 1147, 2005. © IFIP International Federation for Information Processing 2005.

[8]   K-F Ssu, C-H Yang, C-H Chou and A-K Yang "Improving RoutingDdistance for Geographic Multicast with Fermat Points in Mobile Adhoc Networks", *Computer Networks* 53 (2009) pp 2663-2673.

[9]   P-J Chuang and B-Y Li "Fermat Point Based Data Dissemination in Sensor Networks", *Journal of the Chinese Institute of Engineers*, Vol. 32, No. 7 pp. 959-966 (2009).

[10]   Y.B. Ko and N. H. Vaidya, "Location-aided routing (LAR) in Mobile Ad Hoc Networks", *ACM/Baltzer Wireless Networks (WINET) journal*, vol. 6, no. 4, 2000, pp. 307-321.

[11]   B.Karp and H.Kung, "GPSR: Greedy perimeter stateless routing for wireless networks", *ACM/IEEE MobiCom*, August 2000.

[12]   Y.Dong, W-K Hon, D.K.Y. Yau, J-C Chin, "Distance Reduction in Mobile Wireless Communication;Lower Bound Analysis & practical Attainment",*IEEE transactions on Mobile Computing*, vol. 8 no. 2, February 2009.

[13]   C.Maihofer, "A survey of geocast routing protocols", *IEEE Communications Survey & Tutorials*,vol.6, no.2, Second Quarter 2004.

[14]   Young-Bae Ko and Nitin H. Vaidya "Geocasting in Mobile Ad Hoc Networks: Location-Based Multicast Algorithms ", *Proceedings of the Second IEEE Workshop on Mobile Computer Systems and   Applications* (WMCSA), February 1999.

[15]   Lynn Choi, Jae Kyun Jung, Byong-Ha Cho and Hyohyun Choi, "*M-Geocast: Robust and Energy- Efficient Geometric Routing for Mobile Sensor Networks*", SEUS 2008, LNCS 5287,pp. 304–316, IFIP (International Federation for Information Processing) 2008.

[16]   A.K.Sadek, W.Yu and K.J.Ray Liu, "*On the Energy Efficiency of Cooperative Communications in Wireless Sensor Networks*", ACM transactions on Sensor Networks, vol. 6, N0.1: December 2009.

[17]   M. Zorzi and R R Rao, "*Geographic Random Forwarding(GeRaF) for Adhoc & Sensor Networks:Energy & Latency Performance*",IEEE transactions on Mobile Computing, vol.2 no.4, October-December 2003.

[18]   A.G. Ruzzelli, G.'Hare and R. Higgs, "*Directed Broadcast eith Overhearing for Sensor Networks*",   ACM Transactions on Sensor Networks, Vol.6, No.1, December 2009.

15

LOCALIZATION IN SENSOR NETWORK WITH NYSTROM APPROXIMATION

Shailaja Patil[1] and Mukesh Zaveri[2]

Department of Computer Engineering,
Sardar Vallabhbhai National Institute of Technology,
Surat, India,
p.shailaja@coed.svnit.ac.in[1]; mazaveri@coed.svnit.ac.in[2]

ABSTRACT

The recent innovations in wireless technology and digital electronics have opened many areas of research in Wireless Sensor Networks. In the last few years these networks have been successfully used in many applications such as localization, tracking, surveillance, battlefield monitoring, structural health monitoring, routing etc. Most of these applications need localization i.e. estimating location information either relative or absolute. In this paper we propose a computationally efficient algorithm namely, Light weight Multidimensional Scaling. This approach takes advantage of Nyström approximation for estimating location of unknown sensor nodes, using the information of available distances between neighbours and anchors. Various node densities, noise factors and radio ranges are considered for simulation. The performance of the algorithm is obtained with Monte Carlo Simulation.

KEYWORDS

Wireless sensor network, localization, Multidimensional scaling, Nyström approximation

1. INTRODUCTION

Wireless Sensor Networks (WSNs) is a fast growing field which incorporates sensing, computation, and communication into a single tiny device. With the development of MEMS technology, advancement in digital electronics, and wireless communications, it has been possible to design small size, low cost energy efficient sensor nodes. These could be deployed in different environments and serve many applications such as military [1], environmental [2], structure [3], safety and security [4] etc. WSNs are specifically important in the remote or hazardous environment. Mostly, nodes are spread across the field in hundreds or in its multiples. The localization issue becomes important in case of uncertainty about some location in critical applications. If the sensor network is used for monitoring an event in a building, the exact location of each node can be known. However, if the sensor network is used for monitoring an event in a remote area like forest, nodes could be deployed from an airplane. In such case, the precise location of most of the sensors is unknown. With the help of all the available information from the nodes, an effective localization algorithm can compute all the positions. From last few years location based services have gained considerable attention from the researchers, and a lot of contribution have been made. Currently reported services and applications consists of coverage analysis [5], location-aware applications [6], environmental monitoring [7], target tracking [8], intrusion detection [9], location based routing [10] etc. A detailed survey of localization algorithms is provided in [11]. Localization algorithms enable nodes to automatically determine their relative positions after deployment. The localization algorithm should possess following characteristics:

- Energy efficiency (less computation, and communication)
- Self-organization (independent of global infrastructure)
- Robust (being tolerant to range errors, and node failures),

We propose an algorithm based on Nyström approximation of the eigenvectors of the large matrix. It is a variant of Classical Multidimensional Scaling (CMDS) which not only preserves all of its attractive properties but also is computationally efficient. It is a two phase approach. In the first phase, initial estimates are obtained with Nyström approximation. These are further refined using least square optimization in the second phase. Later, transformation is performed to convert the local to global map. The paper is organized as follows. In Section II, we review the literature of localization in WSNs. In Section III, we give the details of our proposed algorithms. Section IV presents the simulation results and we conclude the paper in Section V.

2. RELATED WORK

The localization methods are broadly divided into range based methods that compute an estimation of the distances between two motes, or range-free methods, that depend on range measurements. Range-based algorithms are based on hardware support by applying methods such as time of arrival (TOA) [12], time difference of arrival (TDOA) [12], angle of arrival (AOA) [12], or radio signal strength indicator (RSSI) [13] technologies. On the contrary, range free algorithms are based on information of hop count or connectivity [14-15].

   To obtain the location information, the Global Positioning System (GPS) [16] could be used. However, there are limitations on use of GPS. It is not a cost effective solution for large-scale ad-hoc networks in terms of price and energy. To assist in estimation of absolute positions, a limited number of nodes whose position is known a-priory called anchors can be used. Anchors are supported by GPS or with manual configuration their location is obtained. Most of the localization research has been carried out using anchor based approach [15-16, 22-24].

   The localization algorithm initially collects the distance information, and the number of anchors, localization algorithm can be applied to estimate the location of the remaining nodes of the network. There are several approaches developed by researchers in the literature of localization. For example Doherty *et al.* [18] provides a centralized method of the convex optimization for position estimation and proposes a set of convex constraint models. Nodes which are in the range of each other lie in a proximity constraint. To estimate globally optimum solution, this convex constraint problem has been solved by semidefinite programming (SDP). The localization problem has been formulated as linear programming model for directional communication which is further solved by interior point method. This method needs to place anchors at outer boundary preferably at the corners. If all anchors are placed at interior of the network, the position estimation of outer nodes moves towards center showing large estimation error. This technique has been extended by Biswas [19] with the non-convex inequality constraints. Basically, this technique converts the non-convex quadratic distance constraints into linear constraints with introduction of relaxation to remove the quadratic term of the equation. The distance measurements among nodes have been modeled as convex constraints, and to estimate the location of nodes semidefinite programming methods were adopted. Biswas's method was further improved by Tzu-Chen Liang [20], using a gradient search technique. An approach with triangulation is presented in [21] called (APS) 'Adhoc Positioning System'. Three methods have been proposed by authors namely, DV-Hop, DV-Distance and Euclidean distance. DV - Hop method uses only connectivity information, whereas DV-Distance uses the distance measurements between neighbouring nodes, and Euclidean uses the local geometry of the nodes. Initially anchors flood their location to all the nodes in the network and every unknown receiver node performs triangulation to three other anchor nodes to estimate the position. With anisotropic or irregular network topology, these methods do not perform well. A two phase multilateration based algorithm called Hop-TERRAIN is proposed in [22]. The first phase is similar to the DV-Hop [21], which does initial estimate of positions. In second phase, with the measured ranges and the location estimates of connected neighbours, each sensor refines its estimation iteratively by triangulation. However, it only works for well connected nodes. In [23], authors have presented a triangulation based approach, where a technique of iterative multiplication is used. It provides good results with more anchors. Nodes connected to 3 or more anchors compute their position by triangulation and upgrade their location. This position information is used by other unknown nodes for their position estimation in the next iteration. The algorithm of APS is refined using trilateration by adjustment called as LATN in [24]. In this method with DV-Hop or DV-Distance the positions are estimated and trilateration is performed on unknown nodes for reducing the localization error. The least squares estimation with Kalman filter approach is used by Savvides [25] to locate the positions of sensor nodes to reduce error accumulation in the same algorithm. This method also needs more number of anchors to work well than other methods. Shang *et al* [14] presented a centralized algorithm based on MDS, namely, MDS-MAP(C). Initially, using the connectivity or distance information, a rough estimate of relative node distances is obtained. Then, MDS is used to obtain a relative map of the node positions and finally an absolute map is obtained with the help of anchor nodes. The initial location estimation is refined using least square technique in MDS-MAP (CR) [26]. Both techniques work well with few anchors and reasonably high connectivity.

Shang *et al.* [27] proposed an improved version of MDS-MAP(C,R), called MDS-MAP (P), which eliminated the requirement of global connectivity information and centralized computation. Here, an anisotropic sensor network is divided into a number of small regions, where each one is considered to be locally isotropic. Relative local maps are formed, and merged into a global map. With this step, the characteristics of anisotropic were retained in the global map, and accuracy is improved. However, due to the error propagation during the merging process, its performance is affected.

In this paper, we present an energy efficient algorithm for localization of nodes using Nyström approximation namely Light Weight MDS (LwMDS).

3. PROPOSED ALGORITHM

Multidimensional scaling methods are said to be computationally intensive since these use singular value decomposition (SVD) of full distance matrix. With the proposed method, we have shown that all unknown node locations can be estimated even with partial distance matrix. This can be achieved with Nyström approximation. The anchor-anchor node distances and anchor to non-anchor node distances are required to estimate locations approximately. The aforementioned approach also uses classical multidimensional scaling. However, there is a significant difference between Nyström approach and CMDS based approach. In CMDS based methods, SVD is required to be applied on full distance matrix. This makes MDS computationally intensive for real time use with large number of nodes, especially when these are in thousands. With Nyström approach, the SVD is applied on anchor to anchor distance matrix only, thus reducing its computational complexity. Due to this property, the algorithm is named as Light Weight MDS (LwMDS). In this section a brief review of CMDS and its relation with Nyström approximation is presented. The proposed algorithm is described later.

3.1 Review of Classical MDS

MDS has its origin in psychometrics and psychophysics [30]. In the literature of localization, a number of techniques have been reported which use Multidimensional scaling. It is a set of data analysis techniques which display distance like data as geometric structure. This method is used for visualizing dissimilarity data. It is often used as a part of data exploratory technique or information visualization technique. MDS based algorithms are energy efficient as communication among different nodes is required only initially for obtaining the inter-node distances of the network.

Multidimensional scaling algorithms map a distance matrix D between N items to a d dimensional coordinate vector for each item. Entries in dissimilarity matrix are Euclidean distances.

Classical MDS consist of four steps:

1. Obtain the distance matrix D of Euclidean distances.
2. Apply double centering to this matrix. Double centering the distance matrix, converts it to a new matrix B as shown in equation (1)-

$$B_{ij} = -1/2(D^2_{ij} - e_i \sum_i c_i D^2_{ij} - e_i \sum_j c_j D^2_{ij} + \sum_i c_i c_{ij} D^2_{ij})$$ (1.1)

Where e_i is the vector of ones.

The term $\sum_i c_i = 1$, and its parameters determine the origin of the coordinate vectors. The matrix B, also known as Gram or kernel matrix is a matrix of dot products between coordinate vectors in that space [30].

3. Extract coordinate vectors from kernel matrix B using eigenvector decomposition.

$$B = Q \Lambda Q^T$$ (1.2)

Where Q is a matrix whose columns are orthonormal eigenvectors and Λ is a diagonal matrix of eigenvalues. The d dimensional coordinate vectors that form kernel matrix B can be seen as scaled rows of Q. From the matrix Λ, highest d values are retained in order to minimize the difference between original and embedded distances for d dimension. For d dimensional extraction of coordinate vectors, we get two equations.

$$x_{ij} = \sqrt{\lambda_j} Q_i \qquad (i=1, 2.....N; j=1,2,...p)$$ (1.3)

Where λ_1 and λ_2 are first and second eigen values of the matrix Λ respectively.

3.2 Sensor Position Estimation by Nyström Approximation and MDS

We use an approach from physics called Nyström approximation [31]. This approximation assumes that Gram matrix is positive semidefinite.

Consider a network of N sensors with n beacons of known positions and (N-n) of unknown positions, present in a d dimension (2D or 3D). Let X and Y be the coordinate matrices of beacons and sensors of size $n \times d$ and (N-n) $\times d$ respectively, and $[X^T Y^T]^T$ be the overall coordinate matrix. The inner product between their coordinates is given by KM = [X Y] [$X^T Y^T$], which can be decomposed into four block sub-matrices [32], A = XX^T, Z = YX^T, C= YY^T as below.

KM=	A	Z		DM=	E	F		(1.4)
	Z^T	C			F^T	G		

KM=	XX^T	X^TY	(1.5)
	Y^TX	YY^T	

Where A and E have dimension $n \times n$; Z and F have dimensions $n \times (N-n)$ and C and G have dimension $(N-n) \times (N-n)$. As mentioned before matrix KM is a product of dot matrix, thus can be expressed in terms of dot products of columns of matrices X and Y as in (1.5).

As the matrix $A=X^TX$ is a standard form of CMDS, the solution for X can be obtained by eigen decomposing A:

$$A = UVU^T \tag{1.6}$$

Where U is a matrix whose columns are orthonormal eigenvectors and V is a diagonal matrix of eigenvalues. The coordinates can be obtained using (1.7)

$$X = \sqrt{V_d}U_d^{\ T} \tag{1.7}$$

Where the subscript d is corresponding to dimension, first d largest positive eigen values. The coordinates of Z can be derived by solving the linear system as shown in (1.8)

$$\left.\begin{array}{ll} x_{ij} = \sqrt{v_j}U_{ij} & \text{if } i \leq n \\[2ex] \quad = \sum_d Z_{di}U_{dj} / \sqrt{v_j} & \text{Otherwise} \end{array}\right\} \tag{1.8}$$

Where U_{ij} is the ith component of jth eigenvector of A and v_j is j^{th} eigenvalue of A. Subscript J varies from 1 to d, according to d dimensions.

$$B= \begin{array}{|c|c|} \hline A & Z \\ \hline Z^T & Z^T A^{-1} Z \\ \hline \end{array} \tag{1.9}$$

To find relation between Nyström approximation, and MDS method, one has to consider sub-matrices E and F in place of A and Z. The centering coefficients in equation (1.1) can now be changed as in (1.5).

$$\left.\begin{array}{lll} c_i & = & 1/n \\ & = & 0 \end{array}\right\} \begin{array}{l} \text{if } i \leq n \\ \text{Otherwise} \end{array} \tag{2.0}$$

Thus the centering formulas for A and Z in terms of E and F are :

$$A_{ij} = -\frac{1}{2}(E^2_{\ ij} - e_i\frac{1}{n}\sum_d E^2_{\ dj} - e_j\frac{1}{n}\sum_q E^2_{\ iq} + \frac{1}{n^2}\sum_{d.q} E^2_{\ ij}) \tag{2.1}$$

$$Z_{ij} = -\frac{1}{2}(F^2_{\ ij} - e_i\frac{1}{n}\sum_d F^2_{\ qj} - e_j\frac{1}{n}\sum_q E^2_{\ id}) \tag{2.2}$$

Constant centring term is dropped from (2.2) as it causes irrelevant shift of origin. The dimension, i varies from 1 to N, d varies from (N-n) to N. Thus, as can be seen from equation (1.8), advantage of Nyström method is, coordinates of sensors can be computed using only the information in A and Z matrices.

Since these coordinates are determined in the space by the eigenvectors, it is necessary to perform a final step of mapping them, with an affine transformation, into the initial space of anchors.

3.3 Light Weight MDS Algorithm (LwMDS)

Assume that nodes are randomly scattered in the region of consideration. Let p_{ij} refer to the proximity measure between objects i and j. The simplest form of proximity can be represented using Euclidean distance between two objects. This distance in m dimensional space is given by (2.3)

$$d_{ij} = \sqrt{\sum_{k=1}^{m} (X_{ak} - X_{bk})^2} \qquad (2.3)$$

Where, X_a $(x_{a1} ; x_{a2} ;.... x_{am})$ and X_b $(x_{b1} ; x_{b2} ;.... x_{bm})$ are distance vectors.

The steps of LwMDS are illustrated as below:

Phase 1:

Step 1:

For a given region compute the shortest paths between every pair of nodes. These shortest path distances are used to construct the distance matrix to be used for LwMDS. In this step we assign distances to the edges of graph, if available. Remaining entries in distance matrix are marked as infinity. The matrix completion is accomplished with Floyd's algorithm.

$$D^2 = \begin{bmatrix} 0 & d^2_{12} & .. & .. & d^2_{1n} \\ d^2_{21} & 0 & d^2_{23} & .. & d^2_{2n} \\ .. & .. & 0 & .. & .. \\ .. & .. & .. & 0 & .. \\ d^2_{n1} & d^2_{n2} & d^2_{n3} & .. & 0 \end{bmatrix} \qquad (2.4)$$

The time complexity of the aforementioned algorithm is $O(N^3)$, where N is number of nodes.

Step 2:

Apply MDS to the anchor-anchor distance matrix. Retain first two or three largest eigenvalues and eigenvectors according to the requirement of dimensions (2D or 3D). Let A be such distance vector. Applying SVD to A yields,

$$A = UVU^T \qquad (2.5)$$

Obtain the local coordinates with $X = \sqrt{V_d}U_d^T$ with first d largest positive eigen values .

This step reduces complexity with large factor. As complexity of CMDS is dominated by eigenvalue problem, for a fully symmetric matrix $[NxN]$, it costs $O(N^3)$ for N number of nodes, whereas for LwMDS it is $O(n^3)$ (n is number of anchors).

Step 3:

Using Nyström approximation to anchor node's distance matrix, obtain the embedding of $(N-n)$ sensor nodes, and relative map of N nodes with equations expressed in (1.8).

Phase 2:

Step 4:

Apply least square minimization to refine estimated positions.

This step includes refinement of estimated positions to improve the absolute map. The least square minimization algorithm is used to minimize the error between estimated and true positions. The complexity of this method is $O(n^3)$. Least square minimization problem has lots of local minima and with random starting points a wrong solution may be obtained. However, positions provided by CMDS have proved to be good starting points for minimization problems [26].

Step 5:

With the information of absolute positions of anchor nodes transform relative map to absolute map. This step converts relative map to absolute map. This transformation includes scaling, rotation, reflection etc. This step is necessary to minimize the sum of squares of errors between actual positions of anchors and their relative positions in the

MDS map. For n number of anchor nodes, this transformation computation takes $O(n^3)$ time. Applying this transformation to the whole relative map takes $O(N)$ time.

4. SIMULATION RESULTS

The density of nodes is varied from 50-200, and are considered to be placed randomly with a uniform distribution in a square area with some ranging errors. In a square area of l unit length n^2 nodes are placed in nl by nl square. The radio range is R. The actual distance is blurred with Gaussian noise. Uniform radio propagation is considered. The noise is added to radio range from 5% to 20%. Number of anchors is varied from 4 to 10. The Monte Carlo simulation have been performed, and the number of simulations for each experiment is set to 30. We have compared our results with MDS-MAP(C,R). Results of LwMDS for 200 nodes with isotropic topology are presented here.

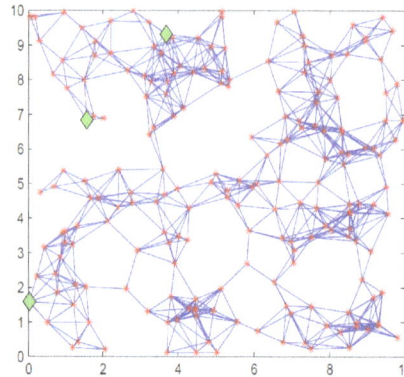

Figure 1: Isotropic network with average connectivity of 12

Fig.(1) shows a similar scenario for 200 nodes spread randomly in a square area of 10 units X 10 units. Stars (*) depict nodes and lines are the radio link between nodes forming the connectivity of 12. Diamonds are anchors selected randomly. With various radio ranges the connectivity level has been estimated with 200 nodes, and it is observed that the network becomes fully connected from the connectivity of greater than 12.

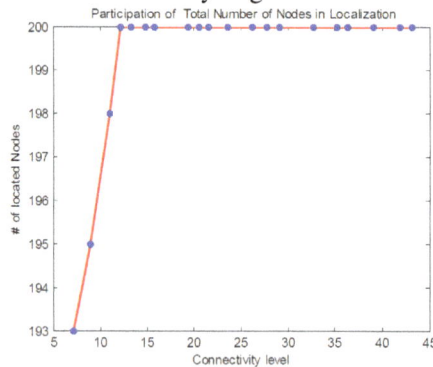

Fig.2: Participation of nodes in the network connectivity

The fraction of nodes participating in the network connectivity with increasing radio range is shown in Fig (2). As can be seen, with the average connectivity of 10, for radio range of 1.3, about 195 nodes take active part in forming the network. As the radio range is increased to 1.5 all the nodes come into the network. The full connectivity of network for all possible scenarios of simulation is obtained from connectivity of 20 onwards. We have estimated the embedding of nodes and the error in estimation for these connectivities. The performance measure used for comparison of errors is Root Mean Square Error (RMSE).

$$RMSE = \sqrt{\sum[(X_i - X_j)^2 + (Y_i - Y_j)^2]} / \sqrt{N}$$

Where (X_i, Y_i) are estimated locations, (X_j, Y_j) are corresponding true locations, and N is number of nodes.

The LwMDS in its first phase computes initial locations with Nyström approximations. Instead of working on full [N x N] distance matrix, as discussed earlier, only [$nx(N-n)$] part of the matrix is used to find embedding, where N is number of nodes and n is number of anchors. As can be seen from Fig.(3), the RMSE is about 20% for connectivity of 21.6. This graph shows the RMSE of embeddings after converting it to absolute map with the help of available anchor nodes. The error goes on decreasing as the connectivity increases. After refinement with least square optimization the error reduces to 0.81R%.   The second and third graph shows the performance of LwMDS and MDS-MAP (C, R). The error difference between the last two graphs is very less, as can be seen from Fig.(3) and Fig.(4).

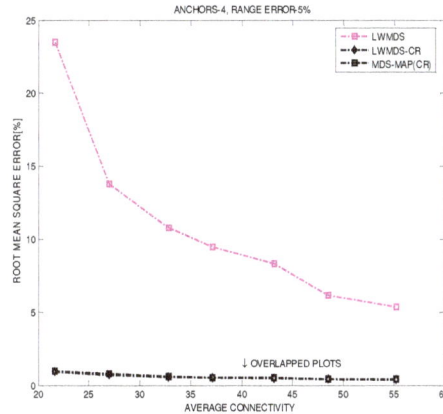

Figure 3(a): Average connectivity Vs RMSE with 5% Range error

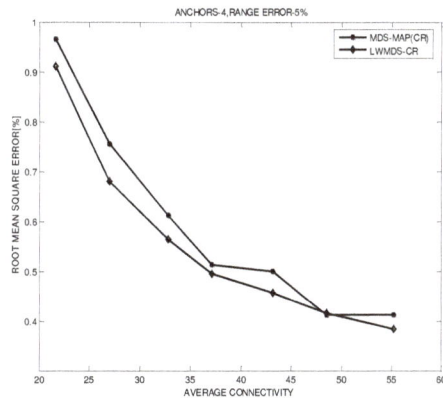

Figure 3(b): Average connectivity Vs RMSE with 5% Range error

From these two graphs, it is apparent that whenever there is full connectivity, using only the partial distance matrix [n X (N-n)] also, almost same results are obtained as that of MDS-MAP(C,R). The main bottleneck of computation in CMDS is of computation of eigen values. The complexity reduces at this step drastically.

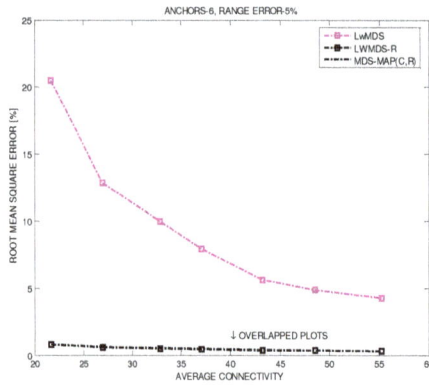

Figure 4(a): Average connectivity Vs RMSE with 5% Range error, and 6 anchors

The performance of LwMDS and MDS-MAP is observed with increasing number of anchors to 6, keeping the range error to 5% as shown in Fig.(5). The difference in the RMSE of both the algorithms is very narrow as shown in Fig. (6). It proves here that even though the complete distance matrix is not used for computation of embedding, the performance of the algorithm is not hampered.

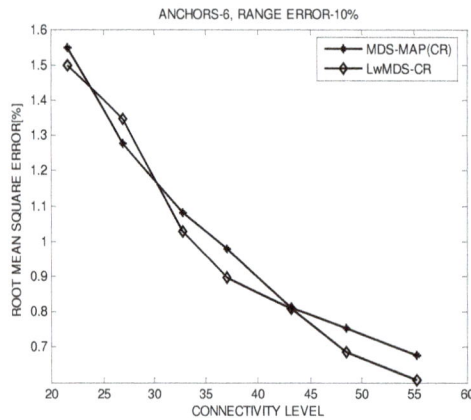

Figure 4(b): Average connectivity Vs RMSE with 5% Range error

With increase in anchors the RMSE reduces. With LwMDS it is reduced to 0.81R% from 0.91R% for the average connectivity of 21.6 as shown in Fig.(5).

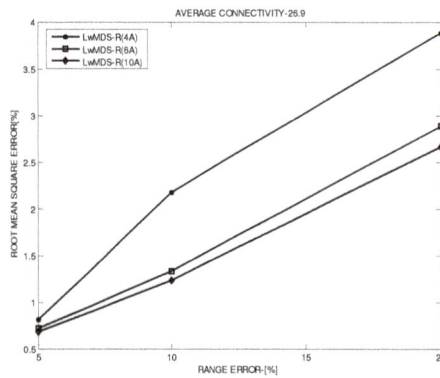

Figure 5: Range Error Vs RMSE with average connectivity of 26.9

The effect of varying range error on RMSE can be observed from Fig. (6). As the range error is increased the RMS error also goes on increasing. This figure shows for the average connectivity of 26.9, the effect of increasing range error with increasing anchors.

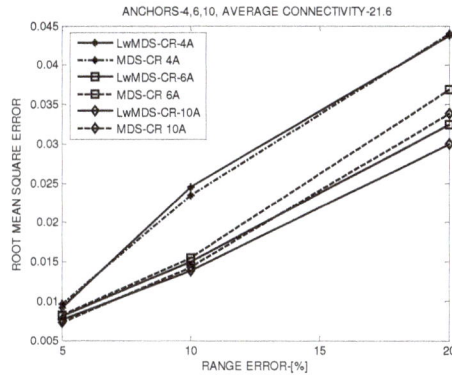

Figure 6: Average connectivity Vs RMSE with 5% Range error

With four number of anchors, and 5% range error, the average estimated location error obtained for LwMDS is 0.8R% , and for MDS-MAP it is 0.85R% . As the range error is increased to 20% this error increases to 3.8R%, and 3.9R% respectively. As the number of anchors is increased to six and ten, the RMS error with 5% range error reduces to 0.72%R and 0.68R% respectively. Observing all these figures, it can be noted that there is no much difference in the average estimation errors of the LwMDS and MDS-MAP with increasing range error, or number of anchors. Showing clearly, the Light Weight MDS can also be used for estimation with same effectiveness as that of MDS-MAP. This reduces the complexity as discussed before.

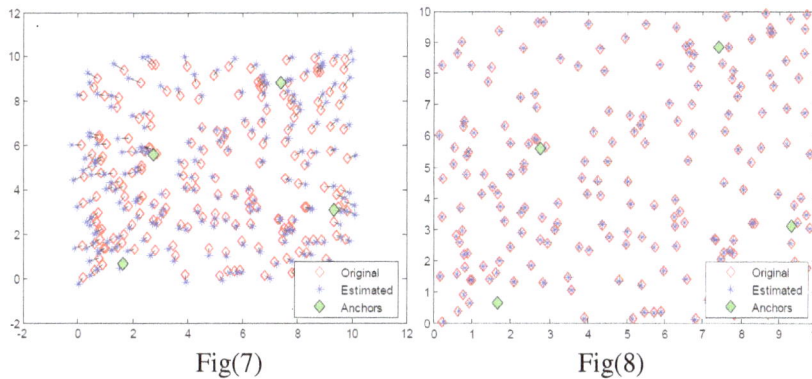

Fig(7)   Fig(8)

Figure (7,8): Scatter plot of localized nodes before and after refinement

The scatter plot of original nodes and its estimated locations is shown in figure (7) before refinement. Original location of nodes is shown by diamond (◊), and star (*) shows estimated locations. Solid diamonds are anchors. The line joining these two is error in estimation. Longer the line larger is the error. The RMSE is 0.266 R% for this scenario. This estimation has been refined with least square optimization. The scatter plot of the final localized nodes after refinement is shown in the Fig. (8). The RMSE is 0.0138R % for this arrangement of nodes.

CONCLUSIONS

Multidimensional scaling based localization methods are one of the robust methods in the literature. However, these methods are computationally intensive. In this paper, we have proposed a comparatively less complex approximation namely Light weight MDS using Nyström approximation for finding location estimation of unknown sensor nodes. Through extensive simulations, we have shown that, even if the distance matrix of anchors is available, with the help of Nyström approach, the full embedding can be obtained, which reduces the complexity drastically without much affecting the location estimation.

REFERENCES

[1]   N. Alsharabi, L.R. Fa, F. Zing, and M. Ghurab, "Wireless sensor networks of battlefields hotspot Challenges and solutions", *Proc. Sixth International conf on Modeling and optimization in Mobile, Adhoc and Wireless Net-works and workshops,*2008, pp.192-196

[2]   I.Hakala, M. Tikkakoski, and I. Kivela, "Wireless Sensor Network in Environmental Monitoring - Case Fox," *Proc. of Second International Conference on house Sensor Technologies and Applications SENSORCOMM '08,*25-31 Aug. 2008,pp-202-208

[3]   M. Li and L. Yunghao, "Underground Structure Monitoring with Wireless Sensor Networks", *Proc. of International symposium on Information Processing in Sensor Networks* , 25-27 April 2007,pp.69-78.

[4]   A.S.K Pathan.; L.yung-Woo, and H. C. Seon, "Wireless sensor networks - a security perspective", *Proc. Of 8th International Conference on Advanced Communication Technology (ICACT),* Sept. 2006.Vol.2 pp. 1048 -1054

[5]   W. Wang, V. Srinivasan, B. Wang, and K. C. Chua, "Coverage for target localization in wireless sensor networks," in *Proc. IPSN,* Apr. 2006, pp.118–125

[6]   U. Varshney, "Location management for wireless networks: issues and directions", *International Journal of Mobile Communications,* 2003 Vol. 1, Nos.2, pp.91–118.

[7]   H. Liu ; Z. Meng and S.Cui," A Wireless Sensor Network Prototype for Environmental Monitoring in Greenhouses", *Proc. of International Conference on Wireless Communications, Networking and Mobile Computing,* 2007. WiCom 2007., 21-25 Sept. 2007 pp- 2344

[8]   Z. Guo, Mengchu Zhou, "Optimal Tracking Interval for Predictive Tracking in Wireless Sensor Network", *IEEE Communication Letters,*Vol.9,No9,Sept.2005, pp.805-807

[9]   M. Estiri , and A. Khademzadeh, "A game-theoretical model for intrusion detection in wireless sensor networks," *In Proc. of 23rd Canadian Conference on Electrical And Computer Engineering (CCECE),* 2010, 2-5May 2010,pp.1-5

[10]   X. Hong, K Xu, and M. Gerla M, "Scalable routing protocols for mobile *ad hoc* networks", *IEEE Network Magazine,* Vol. 16, No. 4, (2002),pp.11-21

[11]   G.Mao, B.Fidan, and Anderson, "Wireless sensor network localization techniques", *The Int. Journal of Computer and Telecommunications Networking Computer net-works,* Vol.51, No.10, July 2007, pp.2529-2553

[12]   P. Xing, H. Yu and Y. Zhang, "An assisting localization method for wireless sensor networks," *In Proc. Of second International Conference on Mobile Technology, Applications and Systems,* 15-17 Nov. 2005, pp.1-6

[13]   X. Li, H. Shi and Yi Shang, "A Sorted RSSI Quantization Based Algorithm for Sensor Network Localization", *In Proc. Of 11th International Conference on Parallel and Distributed Systems, 2005.* pp.557 - 563.

[14]   Y. Shang, W. Ruml, Y.Zhang, M. Fromherz, "Localization from Mere Connectivity" *In 4th ACM international symposium on Mobile and Ad-Hoc Networking & Computing symposium on Mobile and Ad-Hoc Networking & Computing,* (2003), pp. 201–212.

[15]   Premaratne, K., Zhang, J. and Doguel, M. , "Location information-aided task-oriented self-organization of *ad hoc* sensor systems", *IEEE Sensors Journal,* February 2004, Vol. 4, No. 1,pp.85-95

[16]   S. Kumar, and J. Stokkeland, "Evolution of GPS technology and its subsequent use in commercial markets", *International Journal of Mobile Communications* (2003), Vol. 1, Nos. 1–2, pp.180–193.

[17]   X. Ji and H. Zha, "Multidimensional scaling based sensor positioning algorithms in wireless sensor networks" , *Proceedings of the 1st Annual ACM Conference on Embedded Networked Sensor Systems*, November 2003, pp.328–329.

[18]   L.Doherty, K.pister, and L. El Ghaoui,   "Convex position estimation in wireless sensor networks," in *IEEE INFOCOM 2001*, vol. 3, pp.1655-1663

[19]   P.Biswas and Y.Ye, "Semidefinite programming for ad hoc wireless sensor network localization", *Third international symposium on Information processing in sensor network*, April 2004, pp.46-54

[20]   T. C. Liang, T. C. Wang, and Y. Ye, "A gradient search method to round the semidefinite programming relaxation solution for adhoc wireless sensor network localization," Stanford University, formal report 5, 2004. Availale:http: //www.stanford. edu/-yyye/ formal-report5. Pdf

[21]   D.Niculescu and B. Nath, "Adhoc positioning system", *In Proc. of the Global Telecommunications Conference*, San Antonio, CA, USA, (2001). pp. 2926– 2931.

[22]   C.Savarese, J.Rabay and K. Langendoen, "Robust positioning algorithms for distributed ad-hoc wireless sensor networks", *In USENIX Technical Annual conference*, June 2002, pp.1-10.

[23]   A. Savvides, C. Han, and M. B. Srivastava, "Dynamic fine-grained localization in ad-hoc networks of sensors". *In Mobile Computing and Networking,* 2001, pp. 166-179

[24]   F.Tian, W.Guo, C. Wang and Q. Gao, "Robust Localization Based on adjustment of Trilateration Network for Wireless Sensor Networks", *In proc. of 4th Inter-national Conference on Wireless Communications, networking and Mobile Computing*, 2008. WiCOM '08,pp.1-4.

[25]   A. Savvides, H. Park, and M. B. Srivastava, "The bits and flops of the n-hop multilateration primitive for node localization problems," *In International Workshop on Sensor Networks Application*, 2002, pp. 112-121.

[26]   Y. Shang, W Ruml, Y. Zhang, and M. Fromherz, "Localization from connectivity in sensor networks", *IEEE Transactions on Parallel and Distributed Systems* ,2004,Vol.15 No.11, pp. 961– 974.

[27]   Y. Shang and W. Ruml, "Improved MDS-based localization," *in IEEE INFOCOM* 2004,pp.2640-2651

[28]   J.A.Costa, N.Patwari, andA.O.Hero,III, "Distributed weighted multidimensional scaling for node localization in sensor networks," *Transactions on Sensor Networks (TOSN)* ,February 2006.Vol2(1),pp.1-26

[29]   H. Lim and J. C. Hou, "Localization for anisotropic sensor networks", in *IEEE INFOCOM* 2005,PP.138-149

[30]   I. Borg and P. Groenen *Modern Multidimensional Scaling, Theory and Applications*. Springer-Verlag,New York, 1997

[31]   B. Scholkopf, "The kernel trick for distances", *In Proc. of NIPS*,pp-301-307,2000

[32]   C. Williams, and M. Seeger, " Using the Nyström method to speed up kernel machines", *In Proc. of Advances in Neural Information Processing Systems,* Vol.13,pp.682-688,2001

16

MULTIPLE SINK BASED COMPRESSIVE DATA AGGREGATION TECHNIQUE FOR WIRELESS SENSOR NETWORK

Mohamed Yacoab M.Y.

Research Scholar, Karpagam University, Coimbatore, India
MEASI Institute of Information Technology, Chennai, India
yacoab@yahoo.com

Dr.V.Sundaram

Director, Karpagam College of Engg, Coimbatore, India
dr_vsundaram@yahoo.com

ABSTRACT

In a wireless sensor network, the single sink based data aggregation technique leads to inefficiency due to imbalance in energy consumption. Moreover it induces scalability problems and overload at the sink, since it is a "many to one" pattern. Hence, in this paper, we develop a multiple sink based data aggregation technique, assuming the sinks are static. In this technique, initially a sink oriented tree is determined for each sink. If the amount of data in the network becomes large, the data is transmitted in the slots allocated for the specific part of the data such that interference is avoided in the data transmission. As data gets aggregated at the nodes which are nearer to the sink it will be compressed and then forwarded to the next level. This way data is efficiently transmitted to the sink without any loss and interference. By simulation results, we show that our proposed technique achieves good packet delivery ratio with reduced energy consumption and delay.

KEYWORDS : Data Aggregation, Compressive, Multiple Sinks

1. INTRODUCTION

1.1 Wireless Sensor Network

A wireless sensor network is an upcoming technology which is being given major consideration by the research community. Several small, low cost devices constitute the sensor network which is actually a self-organizing ad-hoc system. Its main function is monitoring the physical environment and consequently collects and dispatch information to one or more sink nodes [1]. In wireless sensor network, the main operations performed are related to monitoring the physical environment, sensed information processing and result delivery to the particular sink nodes. To perform these tasks, the sensor nodes are powered by the batteries which are resources of limited energy. Hence, designing an energy efficient protocol for increasing the network lifetime is the major dispute in this energy constrained system [2].

1.2 Data Aggregation

A common function of the sensor network is data gathering. Here the sampled information at each of the sensor nodes has to be transferred, for the sake of further processing and analysis to the central base station [3]. In wireless sensor networks, data aggregation is considered as the most primary distributed data processing procedures which saves energy and reduces medium access layer contention [4]. For wireless routing in sensor networks, data aggregation is considered as a vital paradigm. It involves merging the data from various sources along the route avoiding the redundancy, reducing the transmission numbers and hence saving the energy. Thus the focus is shifted from the customary address-centric approach for networking to a more data-centric approach [5].

1.3 Issues related to the Single Sink Based Data Aggregation

1] In sensor data gathering, the "many-to-one" traffic pattern may lead to imbalance in energy expenditure to a higher extent in the complete network, causing early termination of the network lifetime. Hence the open challenge is to extend the network lifetime by balancing the energy consumption and at the same time maintaining energy efficiency [8].

2] Scalability problem is seen in the single sink network architecture. In single sink architecture, the data aggregated at the sink may become more than its communication capacity due to the large number of sensors in the network. In addition, the radio channel capacity close to the sink may become overloaded if the average number of hops between the source and the sink becomes excess [10].

3] In single sink network architectures, in case the sink is heavily loaded, then the data will not reach the destination, resulting in transmission failure.

1.4 Multi Sink based Data Aggregation

The scalability of sensor networks increases with the introduction of the multiple sinks [11]. By a pseudo link, all sink nodes are connected to pseudo destination. In multiple sink network which in turn changes it to a single reverse multicast tree [12]. Efficient data gathering trees are formed by the nodes and then the best sink is chosen for transferring the data in multiple sink networks. The mean distance between the nodes and the sink will be decreased in the multi sink network leading to energy savings and increased lifetime. The sink acts as a gateway in multi sink network, forwarding the sensed data towards the storage system in the network as in the case of internet. Only the data produced by a particular set of devices is collected by the sink and then the entire phenomenon which is monitored will be reconstructed at the data storage system [10].

1.5 Previous Works

In our first work [6], we have proposed an adaptive traffic aware aggregation technique for wireless sensor networks. A traffic monitoring agent is used to monitor the load status of the event traffic. If the total traffic load of the system is less than a threshold value, then the structured lossless aggregation is applied. When the traffic load crosses the threshold, then the aggregation technique is adaptively changed to structure-free lossy aggregation.

In our second work [7] as an extension of the first work, we provide a cost effective compressive data gathering technique to enhance the traffic load, by using structured data aggregation scheme. The use of compressive data gathering process provides a compressed sensor reading to reduce global data traffic and distributes energy consumption evenly to prolong the network lifetime.

In this work as an extension to our previous two works, we propose to develop an effective data aggregation technique which includes multiple sinks in WSN.

2. RELATED WORK

Khalid N Chaaran et al [13] proposed a scheme in which the all nodes except sinks, act also as message forwarding nodes. These messages are received from any one of the neighbor nodes and are needed to be forwarded to one or many neighbor nodes. The forwarding decision is based on node's own knowledge, sender's guidance and neighborhood knowledge. The ultimate goal, while making these forwarding decisions at nodes, is to find shortest possible route with maximum path aggregation but not at the cost of delay. A scalable multi-path routing approach called Neighbor Sink Nexus (NSN) routing algorithm is presented in this paper to address propagation energy constraints. Security features are not incorporated in this technique and robustness need to be enhanced.

Luca Mottola et al [14] in this paper presented MUSTER (MUlti-Source MUlti-Sink Trees for Energy-efficient Routing), a routing protocol expressly designed for many-to-many communication. First, we devise an analytical model to compute, in a centralized manner, the optimal solution to the problem of simultaneously routing from multiple sources to multiple sinks. Next, we illustrate heuristics approximating the optimal solution in a distributed setting, and their implementation in MUSTER. To increase network lifetime, MUSTER minimizes the number of nodes involved in many-to -many routing and balances their forwarding load. This technique results in uneven energy consumption when the nodes along merged paths experience an increased routing load.

Sixia Chen et al [15] in this paper considered Multiple-Sink Data Collection Problem in wireless sensor networks, where a large amount of data from sensor nodes needs to be transmitted to one of multiple sinks. An approximation algorithm is designed to minimize the latency of data collection schedule and show that it gives a constant-factor performance guarantee. A heuristic algorithm is also presented based on breadth first search for this problem.

Waleed Alsalih et al [16] in this paper proposed a mobile base station placement scheme for extending the lifetime of the network. In our scheme the life of the network is divided into rounds and base stations are moved to new locations at the beginning of each round. In this paper, a more general problem in which a base station can be placed anywhere in the sensing field is defined and solved. The problem is formulated as an Integer Linear Program (ILP) and an ILP solver (with a constant time limit) is used to find a near-optimal placement of the base stations and to find routing patterns to deliver collected data to base stations.

Zoltan Vincze et al [17] in this paper give a mathematical model that determines the locations of the sinks minimizing the sensors average distance from the nearest sink. First an iterative algorithm called global that is able to find the sink locations given by the mathematical model is presented. However, it uses global information about the network that is impractical in wide area sensor networks, thus they have proposed a novel iterative algorithm called 1hop that carries out the sink deployment based only on the location information of the neighboring nodes while the location of the distant nodes is being approximated. The two algorithms are compared and show that 1hop approaches the performance of global very closely. Another important issue is that the neighboring nodes of the sinks have a high traffic load, thus the lifetime of the network can be elongated by relocating the sinks from time to time. Based on the 1hop algorithm they have proposed the 1hop relocation algorithm for the coordinated relocation of multiple sinks.

3. PROPOSED WORK

3.1 Compressive Data Gathering

In our previous paper [7], we have proposed a compressed data gathering technique which is adopted in this paper for the transmission of the data towards multiple sinks. The perception behind CDG in our previous work is that joint transmission of the correlated sensor readings instead of transmission of the readings separately will increase the efficiency to a higher level. On the sub tree basis data gathering and reconstruction of the CDG are performed.

We assume that the multiple sinks in the network are static. Each node should know its local routing structure so as to combine the sensor readings when it is being transmitted. That is, if the given node in the routing tree is a leaf node or not or if the node is an inner node then how many children does it have. To the standard routing protocol, a small modification is done so as to facilitate proficient aggregation: when a parent node is chosen by the node, it transmits a "subscribe notification" to that node and an "unsubscribe notification" is sent to the old parent, when the node changes the parent.

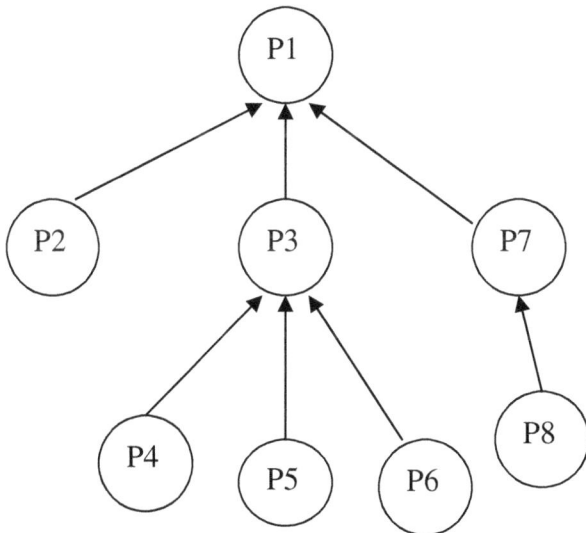

Fig.1 data gathering process of CDG

The example shown in fig. 1 illustrates the data gathering process of CDG. The leaf nodes will initiate the transmission only after all nodes receive their readings.

In this example, a random number β_{i2} is generated by P2 and it computes $\beta_{i2}u_2$ and then the value is sent to P1. The ith weighted sum is denoted by the index i which ranges from 1 to

M. Likewise $\beta_{i4}u_4$, $\beta_{i5}u_5$ and $\beta_{i6}u_6$ is transmitted to P3 by P4, P5 and P6. After the three values are received by P3 it will compute the value $\beta_{i3}u_3$ and then it adds to the sum of the relayed value. It then transmits to P1 the value. Next $\beta_{i1}u_1$ is computed by the node P1 and is transmitted. Lastly, to the sink, the message which contains the weighted sum of all readings in a sub tree is forwarded.

In a specific tree, if it is assumed to have N nodes and M measurements are intended to be collected by the sink. Then regardless of the hop distance of the node to the sink, all nodes will send the same number of $O(M)$. $O(NM)$ will be the overall message complexity. If M << N, then less messages are transmitted by CDG when compared with the baseline data collection when $O(N^2)$ is the worst case message complexity. More importantly, for the extension of the lifetime of the bottleneck sensors as well as the entire network, the transmission load is spread uniformly.

The ith weighted sum can be represented by:

$$U_i = \sum_{j=1}^{N} \beta_{ij} u_j \qquad (1)$$

The sink obtains M weighted sums $\{U_i\}$, i = 1, 2 ...M.
Mathematically, we have:

$$
\begin{pmatrix} U_1 \\ U_2 \\ . \\ . \\ . \\ U_M \end{pmatrix} =
\begin{pmatrix} \beta_{11} \beta_{12} ...... \beta_{1N} \\ \beta_{21} \beta_{22} ...... \beta_{2N} \\ . \\ . \\ \beta_{M1} \beta_{M2} ... \beta_{MN} \end{pmatrix}
\begin{pmatrix} u_1 \\ u_2 \\ . \\ . \\ u_N \end{pmatrix} \qquad (2)
$$

In the above equation, series of random numbers are placed in each column of $\{\beta_{ij}\}$ which is produced at the corresponding node. A simple strategy is used for preventing the transmission of the random matrix from sensors to the sink: a random seed is broadcasted to the entire network before transmission. Using this global seed and its unique identification, each sensor will generate its own seed. Each sensor generates a corresponding series of coefficients from a pre-installed pseudo random number generator. Given that the sink knows the identifications of all sensors, the coefficients can be reproduced at the sink.

In (2), u_i (i = 1, 2 ...N) is a scalar value. Each node is possibly attached with a few sensors of different type, e.g., temperature sensor and a humidity sensor in a practical sensor network. Then from each node, the sensor readings become a multi dimensional vector. In this case, in each dimension we may separate the readings and process them. Alternatively, since for the sensor readings, the random coefficients β_{ij} are irrelevant, u_i is treated as a vector. Ai which is a weighted sum becomes vectors of the same dimension too.

3.2 Tree Construction for Multiple Sinks

As an extension to our previous work, we propose to develop an effective data aggregation technique which includes multiple sinks in WSN.

For determining the routing structure towards the sinks, we use the Link Reversal Algorithm. The main objective of this proposed algorithm is to construct and maintain links to multiple sinks to seamlessly aggregate data.

A $REQ_{\sin k}$ message is flooded periodically by the sink node to the sensor nodes. The request message, $REQ_{\sin k}$ includes information like Sink ID, Timestamps, Period, Max_Height, Hops and Root ID. Based on the density of deployment of the sensor nodes, the field period is fixed. During this period, the $REQ_{\sin k}$ message is flooded. The nodes that are at one hop distance from the sink are called as root nodes. The connection or the disconnection status of the sink with its descendent nodes is dependent on the root node. Until the hop of the $REQ_{\sin k}$ becomes equal to the Max_Height value, the nodes continue to forward the request. Then the nodes start sending its response, $RPY_{\sin k}$ message to its ascendant node in the tree. Thus, through an established reverse path, a sink oriented tree by height is established.

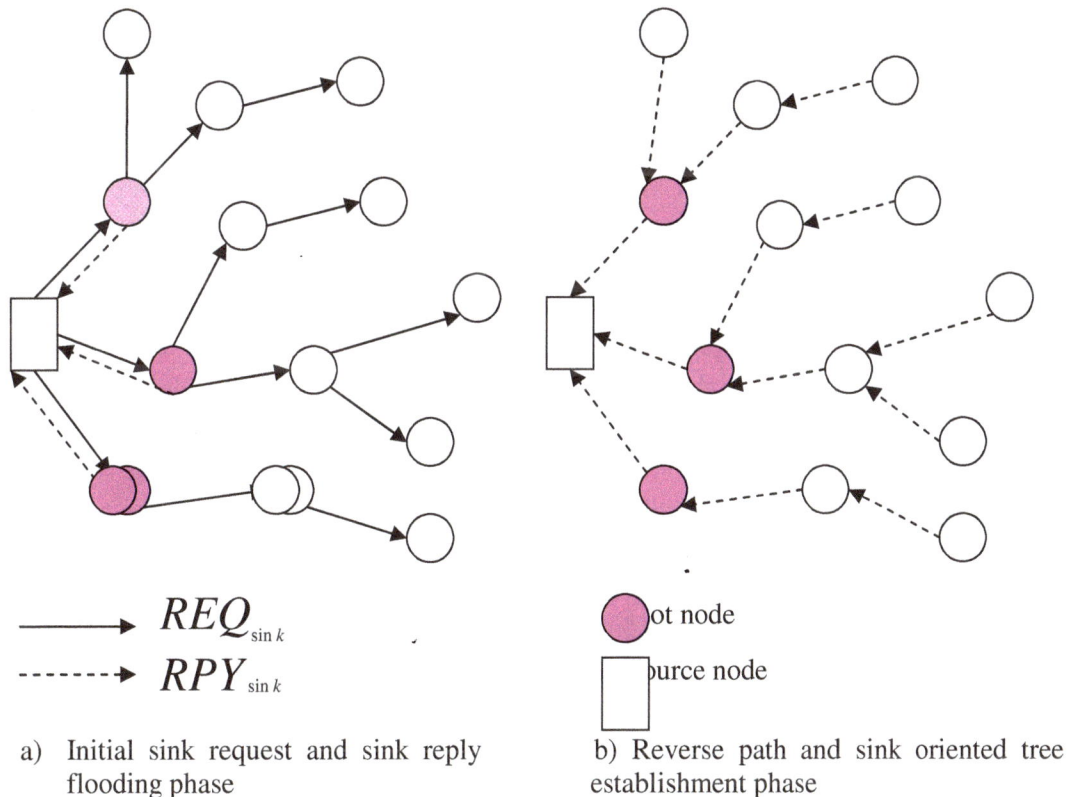

$\longrightarrow$ $REQ_{\sin k}$

$----\rightarrow$ $RPY_{\sin k}$

● ot node

□ ource node

a) Initial sink request and sink reply flooding phase

b) Reverse path and sink oriented tree establishment phase

Another important function of our protocol is the information exchange between sinks. Due to the wide coverage, generally difficulty is faced in developing tree based routing protocols. But due to the merge of independent sink oriented trees our protocol has wider coverage when compared with single sink case. In case a sink sends a request $REQ_{\sin k}$ to a node which is already a member of some other sink, then the sink id is stored by the node in its memory for passing the data collected to own sink to the new sink which wishes to join it. Hence, to the own sink a query of the sink is sent and in this way the merging of the sink oriented tree of various sinks takes place.

3.3 Data Collection Scheduling

For scheduling the data to be aggregated at the sink we develop a scheduling technique for each time interval. We assume that the amount of data to be collected is sufficiently large, and since there are multiple sinks, we split the data into pieces as small as necessary.

In this technique, the entire network is divided into subparts, and then the data is allocated slots for efficient transmission without getting overlapped with the other data. For an interval, [t, t+1], we develop an technique such that basically, for each node $v_i \in V$, a duration is scheduled s_i^t as early as possible without causing any interference.

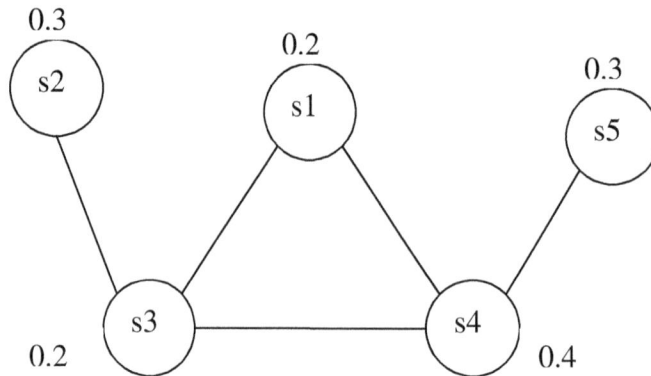

(a) $G_t = (V_t, E_t)$

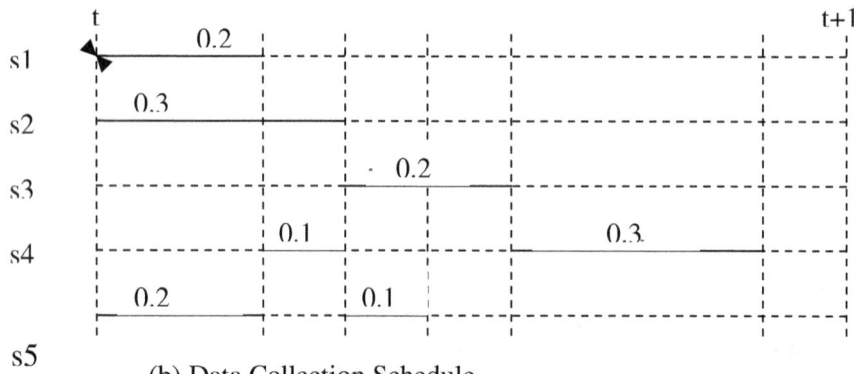

(b) Data Collection Schedule

Fig.2

Fig 2(a) indicates a link interference graph, G_t in which the amount of flow to be transferred over the link is indicated by the number beside each node. For the graph in fig 2(a), a corresponding data collection schedule is shown in fig 2(b). The scheduling starts from time t for the flow at nodes s1 and s2. The scheduling of data transmission for node s3 starts only after t+0.3 in order to prevent the overlap of its data with that of s2 leading to interference. For s4, the scheduling is such that 0.1 is transmitted before the transmission of s3 starts and 0.3 is transmitted after t + 0.5. In a similar process the scheduling of s5 can be performed.

We have developed a heuristic algorithm for a case in which the network has links with same capacity. In this algorithm, initially based on the distance to the set of sinks, the nodes are divided into layers. The nodes at far distance from the sink transfer their data to the nodes which are comparatively at a smaller distance from the sink, which in turn transfers it to the sink.

In the graph G, let the distance between p and q be denoted by $d(p,q)$, which indicates the shortest length in hop counts of the path which connects p and q in the graph G. Let the distance between the node p and set of nodes P be denoted by $d(p,P)$ and is defined by $\min_{p \in P} d(p,q)$. Based on the distance of the nodes to the sink set SS, we divide the nodes into different levels. The set of nodes whose distance i to SS is denoted as n_i. The index of the level that is farthest away from SS is denoted by n_{max}. Using the breadth first search, n_i is found where, i=1,....., n_{max}. The amount of data at node p at time t is denoted by $data(p,t)$. We consider $Func(t) = \sum_{p \in Q} d(p,SS) data(p,t)$, a heuristic function which will be equal to zero at the end of the schedule i.e., when all the data have reached the sink. In order to reduce $Func(t)$, the greedy approach maximizes the nodes used for transferring the data at time t, by taking into consideration that (i) interference is not created due to transmissions and (ii) transmission to nodes in n_i is done by nodes in n_i+1. This process is repeated until the function $Func(t)$ is reduced to zero. The algorithm given below shows the pseudo code of the algorithm. Let LS_t denote the set of transmitting links at time t.

Greedy BFS based Algorithm

1: Classify nodes into different levels using breadth first search
2: Let $t = 0$
3: while $Func(t) > 0$
   do
4: Let $LS_{t=0}$
5:   for $i = n_{max}$ $to1$ do
6:     sort $data(p,t)$ in decreasing order, $p \in n_i$.
7:     for each $p \in n_i$ such that $data(p,t)>0$ do
8:       if p has neighbor $q \in n_{i-1}$ such that the link (p,q) does not interfere with
         any link in LS_t then

9:              $LSt \leftarrow LSt \cup \{(p,q)\}$
10:        end if
11:      end for
12:    end for
13:    Transmit data on every link in LS_t
14:    Update data(p,t) for all $p \in Q$
15:    $t \leftarrow t+1$
16: end while

Thus by using this algorithm, the data from numerous nodes can be transmitted to multiple sinks efficiently. Due to the slotting process, the data is split into parts and sent over the channels so as to ensure the complete data transmission without any loss due to overlapping of the data leading to interference in the network.

So the data is transmitted effectively towards the sink through the sink oriented tree established in section 3.2, and as the nodes start aggregating at the higher nodes as it moves nearer to the sink, the data gets compressed as explained in the section 3.1. So at the sink, the data is aggregated such that the compressed data includes slots of data satisfying the sink requirements.

3.4 Multiple Sink Data Aggregation Algorithm

1. Sink broadcasts a request message, $REQ_{\sin k}$ to the sensor nodes periodically based on the interval up to the maximum height.

    Sink $\xrightarrow{REQ_{\sin k}}$ sensor nodes

2. The sensor nodes send the reply message, $RPY_{\sin k}$ to the sink after the request message reaches the maximum height.

    Sink $\xleftarrow{RPY_{\sin k}}$ sensor nodes

3. The sink oriented tree is established based on the path followed for the transmission of the reply message.

4. The last node in the sink tree starts transmitting the data after computing $\beta_{ij} u_j$ using a random number β_{ij}, to the next possible node in the tree lying towards the sink.

5. The data is transmitted in the slot allocated, so that the overlap between the data being transmitted is prevented.

6. The data gets compressed as it starts getting accumulated from several nodes at the higher nodes.

7. The compressed data is collectively transferred to the sink in interference free slots using the Greedy BFS based algorithm (given in last section).

4. SIMULATION

4. 1. Simulation Setup

Multiple Sink Based Compressive Data Aggregation Technique is evaluated through NS2 [18] simulation. A random network deployed in an area of 500 X 500 m is considered. We vary the number of nodes as 20, 40….100. Initially the nodes are placed randomly in the specified area. The base station is assumed to be situated 100 meters away from the above specified area. The initial energy of all the nodes is assumed as 3.1 joules. In the simulation, the channel capacity of mobile hosts is set to the same value: 2 Mbps. The distributed coordination function (DCF) of IEEE 802.11 is used for wireless LANs as the MAC layer protocol. The simulated traffic is CBR with UDP source and sink. The number of sources is varied from 1 to 5.

Table 1 summarizes the simulation parameters used

Table 1: Simulation Parameters

No. of Nodes	20,40,….100
Area Size	500 X 500
Mac	802.11
Simulation Time	50 sec
Traffic Source	CBR
Packet Size	512
Transmit Power	0.660 w
Receiving Power	0.395 w
Idle Power	0.335 w
Initial Energy	3.1 J
Transmission Range	75m
No. of Sinks	2

4. 2. Performance Metrics

The performance of Multiple Sink based Compressive Data Aggregation (MSCDA) technique is compared with our previous Cost Effective Compressive Data Aggregation CECDA [] protocol, which is based on single sink. The performance is evaluated mainly, according to the following metrics.

Average end-to-end Delay: The end-to-end-delay is averaged over all surviving data packets from the sources to the destinations.
Average Packet Delivery Ratio: It is the ratio of the number of packets received successfully and the total number of packets transmitted.
Energy Consumption: It is the average energy consumed by all the nodes in sending, receiving and forwarding operations
The simulation results are presented in the next section.

4. 3. Simulation Results

In our experiment, we vary the number of nodes as 20, 40, 60, 80 and 10 in which the sources are sparsely deployed.

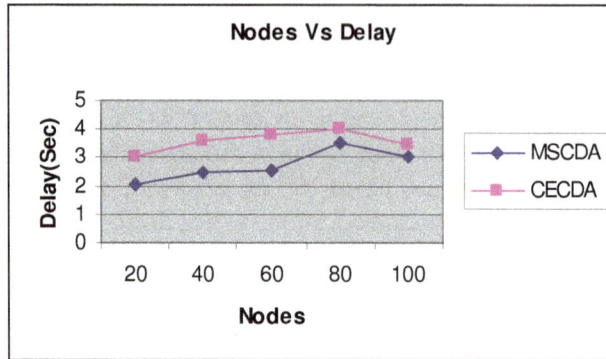

Fig 3: Nodes Vs Delay

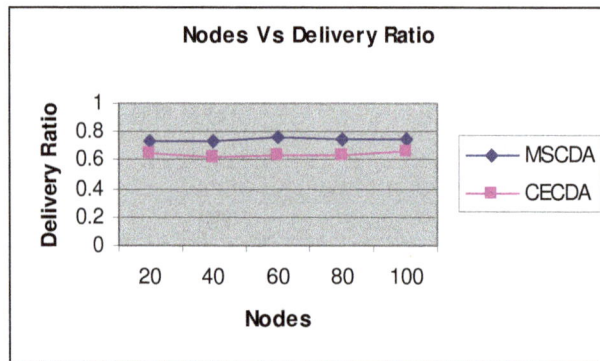

Fig 4: Nodes Vs delivery Ratio

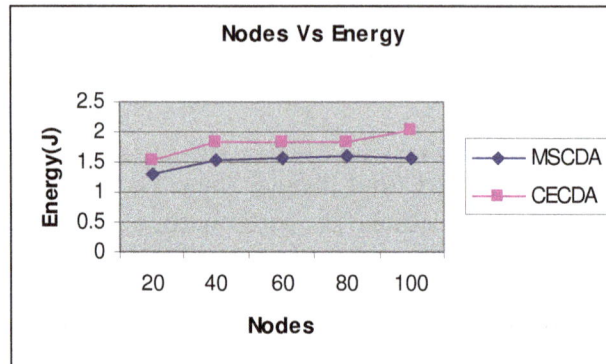

Fig 5: Nodes Vs Energy

Since the aggregation involves compressed data, the delay incurred in sending the data from sensors to the sink, will be significantly reduced.

Fig 3 gives the average end-to-end delay when the number of nodes is increased. From the figure, it can be seen that the average end-to-end delay of the proposed MSCDA technique is less when compared with CECDA.

Fig 4 presents the packet delivery ratio when the number of nodes is increased. MSCDA achieves good delivery ratio, compared to CECDA. The compressed data aggregation eliminates the packet drops at the intermediate nodes and hence increases the packet delivery ratio.

Fig 5 shows the results of energy consumption when the number of nodes is increased. Compressing the data during data aggregation reduces the number of data packets to be aggregated at the aggregator nodes. Hence the total energy consumption involved in the aggregation process will also be reduced. From the results, we can see that MSCDA technique has less energy consumption when compared with CECDA, since it has the energy efficient tree

5. CONCLUSION

In this paper, we have developed an efficient data aggregation technique for multiple sinks. The data is aggregated based on the assumption that all the sinks are static. We initially develop a sink oriented tree for each sink depending on the requirement of the sink like Timestamps, Interval, Max_Height, Hops, etc. Each sink broadcasts its request to the network and based on the response to the request, the tree is set up. Once the tree is established then the nodes starts transmitting its data to the sinks. The transmission of data is initiated by the nodes which are at a larger distance from the sink, to the following nodes in the tree. The data is transmitted by splitting it into smaller parts. The data is then allocated slots such that the overlapping between the consecutive data is avoided. Then the data is aggregated at nodes nearer to the sink in the compressed form. It is then transmitted similarly to the next level till the sink is reached. This technique is proficient since it allows complete transmission of data even in the presence of multiple sinks and large amount of data. By simulation results, we have shown that our proposed technique achieves good packet delivery ratio with reduced energy consumption and delay.

References

1] Dorottya Vass and Attila Vidacs, "Distributed Data Aggregation with Geographical Routing in Wireless Sensor Networks", Pervasive Services, IEEE International Conference, 08 August 2007.
2] Cunqing Hua and Tak-Shing Peter Yum, "Optimal Routing and Data Aggregation for Maximizing Lifetime of Wireless Sensor Networks", IEEE/ACM TRANSACTIONS ON NETWORKING, VOL. 16, NO. 4, AUGUST 2008.
3] Tao Cui, Lijun Chen, Tracey Ho, Steven H. Low, and Lachlan L. H. Andrew, "Opportunistic Source Coding for Data Gathering in Wireless Sensor Networks", TAPIA '07 Proceedings of the 2007 conference on Diversity in computing, 2007.
4] Zhenzhen Ye, Alhussein A. Abouzeid and Jing Ai, "Optimal Policies for Distributed Data Aggregation in Wireless Sensor Networks", INFOCOM 2007. 26th IEEE International Conference on Computer Communications. IEEE, 29 May 2007.
5] Bhaskar Krishnamachari, Deborah Estrin and Stephen Wicker, "The Impact of Data Aggregation in Wireless Sensor Networks", ICDCSW '02 Proceedings of the 22nd International Conference on Distributed Computing Systems, IEEE Computer Society Washington, DC, USA ©2002.
6] M.Y. Mohamed Yacoab and V. Sundaram, "An Adaptive Traffic Aware Data Aggregation Technique for Wireless Sensor Networks", American Journal of Scientific Research ISSN 1450-223X Issue 10 (2010), pp. 64-77, © EuroJournals Publishing, Inc. 2010.

7] M.Y. Mohamed Yacoab1, Dr.V.Sundaram2 and Dr.A.Thajudeen, "A Cost Effective Compressive Data Aggregation Technique for Wireless Sensor Networks", International Journal of Ad hoc, Sensor & Ubiquitous Computing (IJASUC) Vol.1, No.4, December 2010.

8] Feng Wang and Jiangchuan Liu, "Networked Wireless Sensor Data Collection: Issues, Challenges, and Approaches", IEEE Communications Surveys and Tutorials, 2010.

9] Frank Yeong-Sung Lin, Hong-Hsu Yen and Shu-Ping Lin, "A Novel Energy-Efficient MAC Aware Data Aggregation Routing in Wireless Sensor Networks", ISSN 1424-8220, Sensors 2009.

10] Emanuele Cipollone, Francesca Cuomo, Sara Della Luna, Ugo Monaco, Francesco Vacirca, "Event detection capabilities of IEEE 802.15.4 Multi-Sink Wireless Sensor Networks", GTTI 2007, Rome (Italy), June 2007.

11] Olga Saukh, Robert Sauter and Pedro Jose Marron, "Convex Groups for Self-organizing Multi-sink Wireless Sensor Networks", Industrial Electronics, 2009. IECON '09. 35th Annual Conference of IEEE, ISSN: 1553-572X,2009.

12] Frank Yeong-Sung Lin and Yean-Fu Wen, "Multi-sink Data Aggregation Routing and Scheduling with Dynamic Radii in WSNs", IEEE COMMUNICATIONS LETTERS, VOL. 10, NO. 10, OCTOBER 2006.

13] Khalid N Chaaran, Munzza Younus and Muhammad Younus Javed, "NSN based Multi-Sink Minimum Delay Energy Efficient Routing in Wireless Sensor Networks", European Journal of Scientific Research, ISSN 1450-216X Vol.41 No.3 (2010), pp.399-411, © EuroJournals Publishing, Inc. 2010.

14] Luca Mottola and Gian Pietro Picco, "MUSTER: Adaptive Energy-Aware Multi-Sink Routing in Wireless Sensor Networks", IEEE Transactions on Mobile Computing, 2010.

15] Sixia Chen, Matthew Coolbeth, Hieu Dinh, Yoo-Ah Kim, and Bing Wang, "Data Collection with Multiple Sinks in Wireless Sensor Networks", WASA '09 Proceedings of the 4th International Conference on Wireless Algorithms, Systems, and Applications, Springer-Verlag Berlin, Heidelberg ©2009.

16] Waleed Alsalih, Selim Akl, and Hossam Hassanein, "Placement of multiple mobile base stations in wireless sensor networks", Signal Processing and Information Technology, 2007 IEEE International Symposium, ISBN: 978-1-4244-1835-0.

17] Zoltan Vincze, Rolland Vida, Attila Vidacs, "Deploying Multiple Sinks in Multi-hop Wireless Sensor Networks", Pervasive Services, IEEE International Conference, ISBN: 1-4244-1325-7, 2007.

18] Network Simulator, http://www.isi.edu/nsnam/ns

17

ADAPTIVE CONGESTION CONTROL PROTOCOL (ACCP) FOR WIRELESS SENSOR NETWORKS

James Dzisi Gadze[1], Delali Kwasi Dake[2], and Kwasi Diawuo[3]

[1]Department of Electrical & Electronic Engineering, Kwame Nkrumah University of Science and Technology, Kumasi, Ghana
[2]Department of Computer Engineering, Kwame Nkrumah University of Science and Technology, Kumasi, Ghana
[3]Department of Computer Engineering, Kwame Nkrumah University of Science and Technology, Kumasi, Ghana

ABSTRACT

In Wireless Sensor Networks (WSN) when an event is detected there is an increase in data traffic that might lead to packets being transmitted through the network close to the packet handling capacity of the WSN. The WSN experiences a decrease in network performance due to packet loss, long delays, and reduction in throughput. In this paper we developed an adaptive congestion control algorithm that monitors network utilization and adjust traffic levels and/or increases network resources to improve throughput and conserve energy. The traffic congestion control protocol DelStatic is developed by introducing backpressure mechanism into NOAH. We analyzed various routing protocols and established that DSR has a higher resource congestion control capability. The proposed protocol, ACCP uses a sink switching algorithm to trigger DelStatic or DSR feedback to a congested node based on its Node Rank. From the simulation results, ACCP protocol does not only improve throughput but also conserves energy which is critical to sensor application survivability on the field. Our Adaptive Congestion control achieved reliability, high throughput and energy efficiency.

KEYWORDS

Energy Efficient Congestion Control, Event Detection, Traffic Control, Resource Control, Wireless Sensor Network.

1. INTRODUCTION

The emerging field of wireless sensor network (WSN) has potential benefits for real-time system monitoring. A wireless sensor network consists of remotely deployed wireless sensor nodes in a physical phenomenon. The embedded sensors continuously monitor the physical process and transmit information in a multi-hop fashion to a special node called the sink. WSN therefore has three basic characteristics: centralized data collection, multi-hop data transmission, and many-to-one traffic patterns. It means that the nodes closer to the base stations need to send more data packets and their traffic burden will be more severe. This leads to severe packet collisions, network congestion, and packet loss. In most severe cases it even results in congestion collapse. Within the framework of (WSNs), there are many application areas where sensor networks are deployed: for environmental monitoring, battlefield surveillance, health and industrial monitoring control. We use the recent oil find in Ghana and its associated environmental impact as example of one application and describe congestion problem in WSNs in this context. Residents of oil and gas field communities often report incidents of: asthma, respiratory and cardiovascular illnesses, autoimmune diseases, liver failure, cancer and other ailments such as headaches, nausea, and sleeplessness. These health effects could be the result of air contaminant such as Volatile Organic Compounds (VOCs). We envisage the use of wireless sensor network in Ghana to monitor the quality of air in oil and gas field communities. One of the main problems to the success of the monitoring scheme might be congestion in the wireless sensor network.

A typical WSN deployment scenario is illustrated in figure 1. The VOC measurement in the oil and gas field is done by sensors deployed in the field. Information gathered from the sensor nodes is transmitted in a multi-hop fashion to the sink node and then to the gateway computer as shown in Figure1.

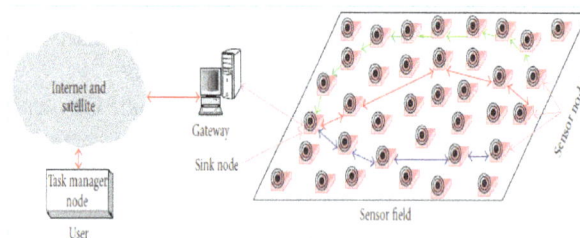

Figure 1: WSN deployment scenario

The traffic in the network is low under light load conditions. When an event is detected, the load increases and transmission of traffic through network approaches the packet handling capacity of the network. Congestion is a state in a network when the total sum of demands on network resources is more than its available capacity. Mathematically:

$$\sum \text{Demand} > \text{Available Resources} \quad (1)$$

In WSNs, congestion happens due to contention caused by concurrent transmission, buffer overflows and dynamic time varying wireless channel condition [1][2][3].

At low levels of VOC pollution when the load is light, throughput and hence the wireless sensor network utilization increases as the load increases thus before point A in figure 2a.

As the load increases, a point A in Figure 2a is reached beyond which the wireless network utilization (throughput) increases at a rate lower than the rate the load is increased and the wireless sensor network enters into moderate congestion state. As the load continues to increase and the queue lengths of the various nodes continue to grow, a point B is reached beyond which the throughput drops with increased load.

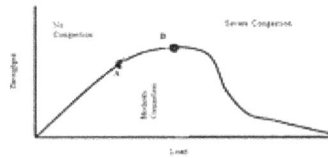

Figure 2a: Levels of Congestion

Figure 2b: Effect of Congestion

As seen in figure 2a, the sharp decline of throughput from point B is a state where wireless sensor network is rendered inefficient since critical network resource such as energy and bandwidth are wasted. This is a state where bandwidth and buffer overflow have reached a threshold value that any further packet arrival leads to severe network performance degradation. In terms of VOCs monitoring from the oil field, the gateway computer will receive and report data that is *misleading* due to the severe congestion. It is therefore imperative to control congestion in the WSN.

Existing congestion control protocols solve the severe congestion problem using unilateral congestion control approach [4][5][6]. That is the use of either traffic control or resource control strategy. There exists, for instance, an inverse relationship between energy and throughput in these approaches.

$$E = \frac{1}{TP}, E = energy, TP = Throughput \qquad (2)$$

A good energy and throughput performance cannot be achieved through this unilateral approach. In this paper, we proposed the development of ACCP a protocol that uses a switch algorithm to trigger DelStatic (traffic) or DSR (resource) protocol that results in both good energy and throughput performance in WSNs.

The remainder of the paper is organized as follows: Section II provides review on literature, Section III on ACCP Design, Section IV on ACCP NR and Feedback Calculation and the final section on Simulation and Results.

2. LITERATURE REVIEW

A number of previous works have addressed the issue of congestion control in Wireless Sensor Networks [7]. In these works congestion alleviation is either by traffic control or resource control approach.

2.1. Traffic Control

Congestion at a node happens when the incoming traffic volume exceeds the amount of resources available to the node. To alleviate congestion, the incoming data is throttled (referred to as traffic control) mostly using hop-by-hop backpressure. Congestion Detection and Avoidance (CODA) uses Buffer size and Channel condition to detect congestion and then reduces the rate of incoming traffic into the network[8]. The algorithm used to adjust traffic rate works in a way like additive increase multiplicative decrease (AIMD).

The reduction in traffic affects the accuracy level when congestion is transient. In such transient congestion condition, increasing resources such as hop-by-hop bandwidth or additional data path may be a better option to cope with congestion. In addition CODA used Closed Loop end-to-end upstream approach to control congestion. This approach generally has a longer response time when congestion occurs, and in-turn results in dropping of lots of segments [9]. The drop of segments results in the waste of limited energy resources.

Event-to-Sink Reliable Transport (ESRT) is traffic congestion scheme that provides reliability from sensors to sink in addition to congestion control [10]. It jointly uses average local packet service time and average local packet inter-arrival time in order to estimate current local congestion degree in each intermediate node. The use of hop-by-hop feedback control removes congestion quickly.

Priority based congestion control protocol (PCCP) [11] defines a new variable, congestion degree as a ratio of average packet service time over average packet interval arrival time at each sensor node. PCCP uses a hop-by-hop rate adjustment technique called priority-based rate adjustment (PRA) to adjust scheduling rate and the source rate of each sensor node in a single-path routing WSN. The idea is to allow data flow generated by a source node to pass through the nodes and links along with single routing path. With PCCP control, it is easy for sensor nodes to learn the number of upstream data sources in the sub tree roots and measure the maximum downstream forwarding rate. This helps the congested nodes to calculate the per-source rate based on priority index of each source node.

Dynamic Predictive Congestion Control (DPCC) [12] predicts congestion in a node and broadcasts traffic on the entire network fairly and dynamically. To achieve a high throughput value, DPCC uses three rate adjustment techniques: backward and forward node selection (BFS), predict congestion detection (PCD) and dynamic priority-based rate adjustment (DPRA), which are introduced with responsibility for precise congestion discovery and weighted fair congestion control.

In a hop by hop mitigation congestion control protocol [13], congestion detection is based on the following parameters: buffer size, hop count and MAC overhead. When congestion is detected at a node, each downstream node is set with a congestion bit using Node Rank threshold policy. With hop by hop traffic control mitigation strategy, each source node will adjust their transmission rate dynamically based on the RAF feedback from a congested node.

2.2. Resource Control

In resource congestion control schemes when data traffic at a node increases beyond the resources available at that node, the resources to the node are increased to enable the node cope with the excess traffic. In [14] a resource congestion control scheme is developed where large numbers of sensor nodes are turned off during normal traffic. When congestion is detected some nodes are woken up to form one or more additional routing paths called multiplexing paths. The congestion is taken care of by distributing the incoming traffic over the original path and the multiplexing paths. The challenge is that precise network resource adjustment is needed to avoid over- or under- provision of resources. The paper established that when congestion is transient, increasing resources by creating multiple paths around the hotspot effectively increases the number of delivered packets (accuracy level), and saves a lot of energy by avoiding collisions and retransmissions.

3. ACCP DESIGN

Sensitive wireless sensor applications such as medical and hazardous environmental monitoring are loss-intolerant. In such applications it is required to have accurate data (high fidelity) from the source nodes to the sink node for reliable analysis. The use of traffic congestion control in particular during transient congestion may result in the reduction in the accuracy level of data reaching the sink. ACCP uses a sink switching algorithm to switch between traffic congestion control logic (DelStatic) and resource congestion control logic (DSR). DelStatic is developed by introducing backpressure mechanism into NO Ad-Hoc Routing Agent (NOAH). We analyzed various routing protocols and established that DSR has a higher resource congestion control capability.

As shown in Figure 3, an intermediate node receives data traffic from multiple source nodes. The many-to-one modelposes threat to the resources available to the intermediate nodes. If bandwidth and buffer occupancy usage of the node exceeds a threshold, congestion sets in. It is important to control congestion at the intermediate node to guarantee reliable transmission of data generated by the source nodes to the sink node (base station)

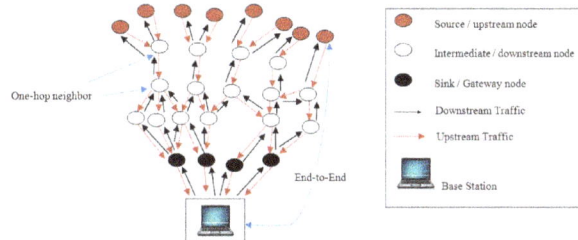

Figure 3: Wireless Sensor Network and data transmission

ACCP is thus based on two mechanisms:

1) Detect congestion at a node using Buffer Occupancy & Channel Utilization Strategy
2) Control the detected congestion by initializing ACCP protocol

3.1. Congestion Detection

To effectively detect congestion, we implement in ACCP a double congestion detection mechanism: channel utilization strategy and buffer occupancy.

3.3.1. Buffer Occupancy

In ACCP the instantaneous buffer occupancy of intermediate nodes is compared to a threshold value. If the threshold value is reached, then congestion may be about to set in. This kind of detection is also seen in [15][16]. In order to avoid late buffer threshold detection, the buffer growth rate is also monitored. An exponentially weighted moving average of the instantaneous queue length is used as a measure of congestion [17]:

$$AVG_q = (1 - w_q) * AVG_q + w_q * inst_q \qquad (3)$$

Where: 1) $AVG_q = average queue length of node$

2) w_q = weighted queue length of node

3) $in\ st_q = queuelenghtofnodeatthecurrentinstant$

The average queue length is updated whenever a packet is inserted into the queue. Thus, if AVG_q exceeds a certain upper threshold U, the node is said to be congested. The node remains in a congestion state until AVG_q falls below a lower threshold L. In practice, a single threshold is too coarse-grained to effectively react to congestion. Buffer occupancy alone is not a reliable congestion indicator because packets can be lost in the channel due to collision or hidden terminal situations and have no chance to reach a buffer [18]. ACCP implements both Channel Utilization Strategy and buffer occupancy for effective congestion detection.

3.3.2. Channel Utilisation and Detection Strategy

Channel Utilization is the fraction of time the channel is busy due to transmission of frames. High channel utilization is used as an indication of congestion. When a sensor node has a packet to be sent, it samples the state of the channel at regular interval. Based on the number of times the channel is found to be busy, the node calculates a utilization factor which when above a certain level indicates congestion.

Channel Utilization technique uses Carrier Sensing Multiple Access (CSMA) algorithm to listen to the communication medium before data is transmitted. This is accomplished with the help of the MAC Protocol. In ACCP we used IEEE 802.11 MAC with collision avoidance. The channel occupation due to MAC contention is computed as:

$$C_{occ} = t_{RTS} + t_{CTS} + 3t_{SIFS} \tag{4}$$

where
t_{RTS} and t_{CTS} are the time spent on RTS and CTS, exchanges and t_{SIFS} is the SIFS period.

Then the *MAC* overhead is represented as

$$OH_{MAC} = C_{occ} + t_{acc} \tag{5}$$

where t_{acc} is the time taken due to access contention. That is,
OH_{MAC} is strongly related to the congestion around a given node.

The channel busy ratio is another factor used in ACCP

$$CHBR = \frac{Blnt}{Ttot} \tag{6}$$

Where *Blnt* represents the time interval that the channel is busy due to successful transmission or collision, and *Ttot* represents the total time.

To maximize channel utilization detection strategy, the channel detection delay of wireless medium is of essence [19].

$$S_{max} \approx \frac{1}{(1 + 2\sqrt{\beta})} \left(for \beta = \frac{\tau C}{L} \ll 1 \right)$$

1) S_{max} = CSMA maximum theoretical throughput approximation

2) β = the measure of radio propagation delay and channel detection delay

3) τ = the sum of both radio propagation delay and channel idle detection delay in seconds

4)C = the raw channel bit rate

5)L = the expected number of bits in a data packet

In achieving a high S_{max} value for good performance, we lowered the β value, which determines how quickly a node can detect idle periods. CSMA efficiency is highly dependent on the β value.

3.2. ACCP Algorithm

To effectively detect congestion, we implement in ACCP a double congestion detection mechanism: channel utilization strategy and buffer occupancy.

The NS2 modular implementation of ACCP is based on DSR and DelStatic protocols. The Dynamic Source Routing (DSR) is an on-demand routing protocol that is based on the concept of source routing where the sender of a packet determines the complete sequence of nodes through which, the packets are forwarded.

Our DSR logic has the capability to reduce the sleeping interval of backup nodes near the congested node so that they become active. One's they become active, they can communicate with each other using DSR Route Discovery and Route Maintenance algorithm to trace the sink node. The goal of DRS is:

- To increase resource provisioning as soon congestion occurs;
- To reduce the resource budget as soon as congestion subsides

The DelStatic logic has a backpressure mechanism that allowsa sensor node that has its channel utilization level and buffer occupancy above a threshold broadcasts a suppression message to upstream nodes. It uses an open-loop hop-by-hop data flow strategy until the message reaches the source nodes. Each node reacts to the suppression message by throttling its sending rate or drop packets using packet drop and Additive Increase Multiple Decrease (AIMD) policy. The DelStatic traffic reduction scenario is as shown in Figure 4.

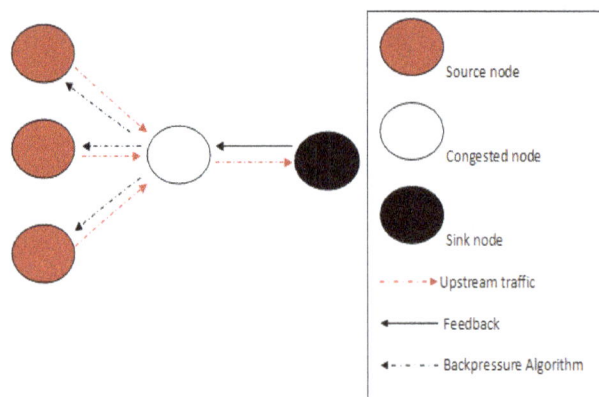

Figure 4: DelStatic Traffic Reduction Scenario

The goal of this work is to allow DelStatic and DSR congestion control protocols to work in one framework by implementing a switching mechanism that is controlled by the sink node. From the framework diagram in Figure 5, congestion detection is done at the node level with a feed sent to the sink to report congestion in the network.

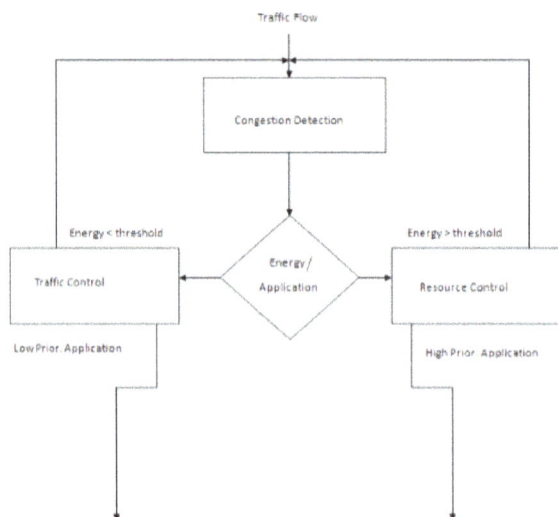

Figure 5: Framework for Adaptive Protocol

The sink uses two important parameters, Energy Remaining (*ER*) and Sensor Application Priority (*SP*) to determine the control to the sending node. Either Traffic Control (DelStatic) or Resource Control (DSR) is triggered by the sink depending on the algorithm below:

Switch (congestion control){
Case 0:
If (Energy > threshold){resource control}
Case1:
If (Energy > threshold){traffic control}
Case2:
If (High- fidelity/priority application){resource control}
Case3:
If (Low- fidelity/priority application){traffic control}
default: congestion control not triggered}

3.3. ACCP NR and Feedback Calculation

With adaptive control, each node calculates its node rank (NR) based on its Buffer level and Channel Utilization Ratio. Here the channel busy ratio represents the interference level and is defined as the ratio of time intervals when the channel is busy due to successful transmission or collision to the total time. After estimating its rank, a node forwards this to its downstream node. When NR crosses a threshold T, then the node will set the congestion bit (CB) in every packet it forwards. On receiving this node rank, the downstream node will first check for the congestion bit. If it is not set, it will simply compute its node rank and adds it to the rank obtained from its previous node and passes on to the next node. On the other hand, if the congestion bit is set, the node sends a feed to the sink node to confirm whether to trigger DelStatic or DSR control. Depending on the energy remaining to the network and the sensor application priority, the sink will send a feedback control to the sending node. This node will in-turn based on the information received from the sink set Rate Adjustment Feedback (RAF) either triggers an alternate path (DSR Control) or transmit it towards the source as a feedback. On receiving the feedback, the source nodes will adjust their transmission rate (DelStatic).

Each node calculates its node rank NR based on the following parameters *BSize, HC, CBHR* and OH_{MAC}, where *BSize* is the buffer size of the node and *HC* is the hop count value.

$$NR = \alpha1.BSize_{n1} + \alpha2.H + \alpha3.CHBR + \alpha4\,OH_{MAC} \qquad (8)$$

Here $\alpha1, \alpha2, \alpha3\ and\ \alpha4$ are constant weight factors whose values between 0 and 1

Each data packet has a congestion bit (CB) in its header. Every sensor node maintains a threshold T. When NR crosses the value of T the node will set its CB in every packet it forwards.

In Figure 6, for example nodes n1, n2, n3 estimates their rank NR1, NR2, NR3 and forwards this to its downstream node n4 along with data packets. On receiving NR1, NR2 and NR3, n4 will first check their CB value. If it is not set, it will simply compute its node rank NR4 and adds it to the rank obtained from nodes n1, n2, and n3 and passes it to the next node or sink. On the other hand, if the CB is set at n1, n2 and n3, the node n4 will send a feed to the sink about the congestionscenario. The sink node uses energy remaining (*ER*) to the network and priority of sensor application (*SP*) to determine which control to use for feedback

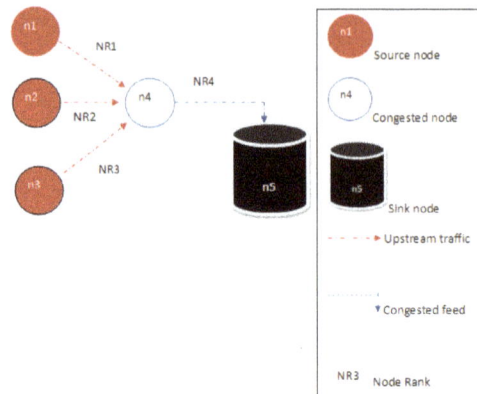

Figure 6: Node Rank (NR) Propagation

The sending node receives either *N1* (DSR Control) or *N0* (DelStatic Control) from the sink as a feedback signal. The sink operates the feedback with request from directed diffusion protocol to disseminate *N1* and *N0* information to the congested nodes.

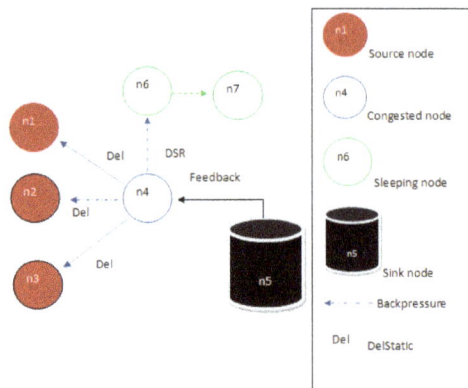

Figure 7: Feedback Propagation

The sending node based on the *N1* or *N0* feedback information from sink calculates its Rate Adjustment Feedback (*RAF*) based on the rank as:

$$RAF = \left(\frac{Arate}{HC}\right) - \Sigma\,{OH}/{MAC_i} - \Sigma CBHR_i \qquad (9)$$

Where *Arate* is the arrival rate of packets at node *n* which is given as:

$$Arate = \frac{NP}{t} \qquad (10)$$

Here NP – is the number of packets received and t is the time of interval for the packet transmission.

When feedback (N1) is received, the RAF is propagated to a neighboring node by waking up a sleeping node (DSR Control) as shown in figure 7. On receiving the feedback packet, a neighboring node n6 is signaled to continue data transmission.

When feedback (N0) is received, backpressure algorithm is initiated to the upstream nodes n1, n2, n3 to adjust their transmission rate by dropping packet or triggering a delay in milliseconds. To adjust the rate dynamically, DelStatic protocol uses formulae on node

$$Nrate = Nrate - RAF \qquad (11)$$

Thus the traffic rate is adaptively adjusted according to the MAC contention and buffer size. The procedures are repeated for all the hops towards the sinks which are congested.

4. SIMULATION RESULTS

We use ns2 to simulate ACCP protocol. For the purpose of achieving our objective, we modified the routing agent class of NOAH to 'start-DelStatic'. The ACCP agent interface linkage to MAC component is also modified to enable fix routing in DelStatic Control.

In testing ACCP protocol, sensor nodes of sizes 25, 64 and 150 nodes were deployed using grid and random topology for 10 seconds of simulation time. We set all nodes to have transmission range of 250 meters, interference range of 550 meters and maximum packet in queue of 50. The simulated traffic source is Constant Bit Rate. The performance of DelStatic compared with NOAH protocol using the metrics of throughput and energy consumption.

The result of the simulations is an output trace file that can be used to do data processing and to visualize the simulation with a program called Network Animator (NAM). For this project, NS version 2.34 is used.

4.1. Simulation Parameters

Table 1: Simulation Parameters

No. of nodes	Transmission Range	Interference Range	Mac Protocol	Maximum packet in Queue	Source Agent	Sink Agent
25	250	550	Mac/802_11	50	UDP	null

Simulation Time (seconds)	Source Traffic
10	CBR

Ad-hoc Protocol	Interface Queue Type
AODV	Queue/DropTail/PriQueue
DSDV	Queue/DropTail/PriQueue
DSR	CMUPriQueue
TORA	Queue/DropTail/PriQueue

4.2. Energy Parameters

Set opt(initialenergy) 900   ;# Initial energy in Joules
-rxPower 0.3 \   ;# power consumption in state (watt)
-txPower 0.6 \   ;# power consumption in sleep state (watt)

4.3. Part 1: TADD Protocols

Simulation is conducted on Ad hoc On-demand Vector Protocol (AODV), Destination-Sequenced Distance Vector (DSDV), Dynamic Source Routing (DSR) and Temporarily Ordered Routing Algorithm (TORA) which forms the TADD protocols. The aim is to compare these routing protocols and their performance in terms of traffic bytes generated, end to end delay and throughput. The protocol with a higher metric value, DSR is used for the resource controlling part of ACCP.

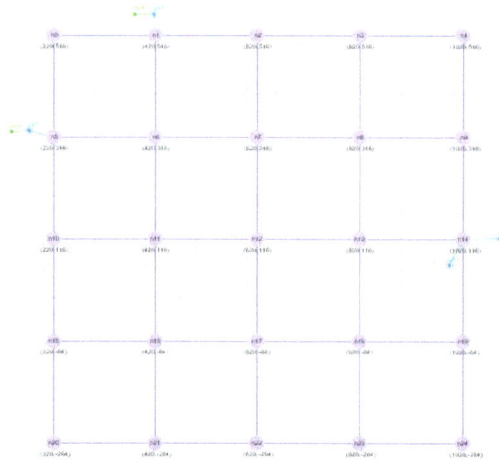

Figure 8: Grid Topology 25 nodes [2 Source / 1 Sink]

In Figure 9, the average end to end delay has a maximum value at '5.5' and a minimum at '0.0'. This is a positive result from DSR protocol. Though 5.5 seconds is on the high side, it is an indication that CBR source data is transmitted over a significant period. Congestion is effectively dealt with during the transmission process.

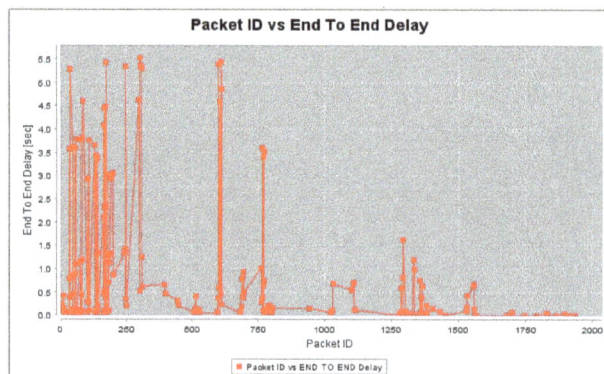

Figure 9: End to End Delay of DSR

4.4. Part 2: NoAH and DelStatic

Our proposed traffic control (DelStatic) protocol is compared with NOAH protocol using metrics of end to end delay and throughput. The generated graph shows a significant throughput increase in DelStatic protocol over a NOAH protocol.

It is observed in Figure 10 that the delay performance of DelStatic is better than that of NOAH. The better delay performance is due to the backpressure algorithm we introduced in DelStatic. The backpressure algorithm is significant in controlling congestion during data packet transmission from source nodes to the sink node.

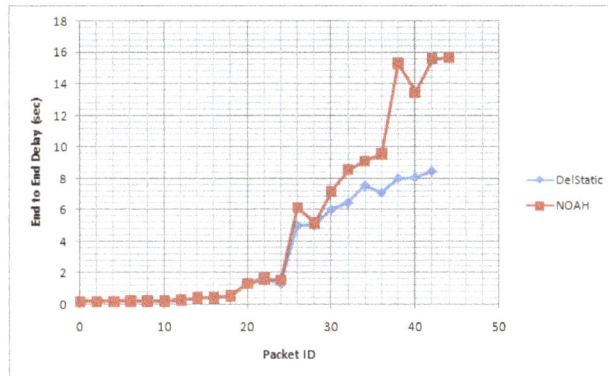

Figure 10: End to End Delay of DelStatic and NOAH

Throughput is the average rate of successful packet delivered to the sink over a sensor network. As seen in Figure 11, the throughput performance of DelStatic is better than that of NOAH during congestion. An average throughput value of 8 is recorded during the initial stages of simulation as compared to 7 of NOAH protocol.

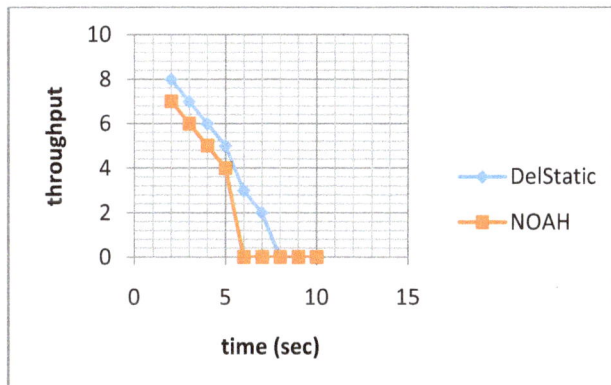

Figure 11: Throughput, DelStatic and NoAH

4.5. Part 3: ACCP

The proposed adaptive congestion control seeks the gains of DSR and DelStatic protocol when used in sensor networks. The metrics of essence are the energy remaining and throughput value which are essential for a successful data delivery and prolonged sensor life deployment.

We first examine the energy performance of DSR and DelStatic. The energy threshold was set at 500 Joules and the simulation runned for 10 seconds. It was noted as seen in figure 12 that, DelStatic after controlling congestion had 600J of energy remaining. This is above the set threshold value. DSR after congestion control had only 100J of energy left.

The sink node of an application using ACCP will use the switching algorithm to trigger DelStatic anytime energy consumption is below a constant threshold. ACCP protocol is energy efficient.

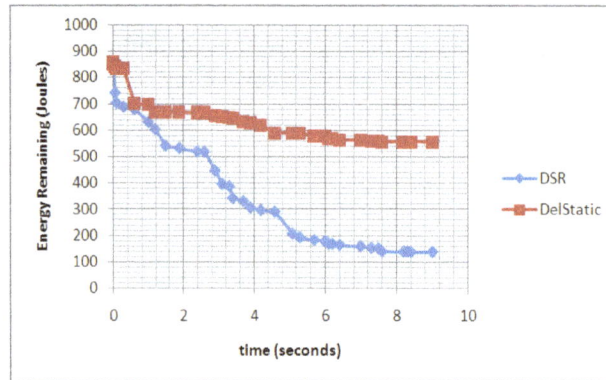

Figure 12: Energy Remaining, ACCP

Figure 13 shows the throughput performance of ACCP. As expected, the figure shows a higher throughput value for DSR as compared to DelStatic.

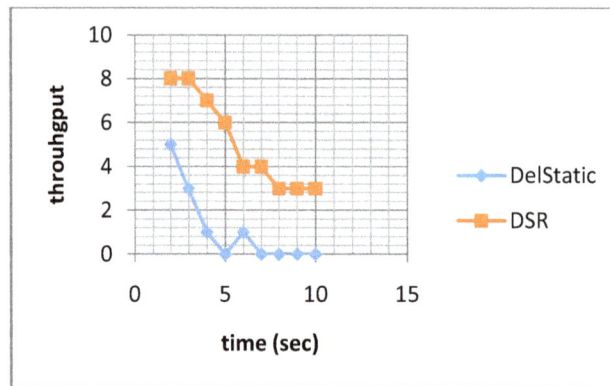

Figure 13: Throughput, ACCP

ACCP protocol is recommended for a wireless sensor network application that requires high data accuracy level and energy efficiency.

5. CONCLUSION

In this paper, we proposed Adaptive Congestion Control protocol which switches between DSR and DelStatic to control congestion in WSNs based on the nodes remaining energy and priority of the application. We developed DelStatic by introducing backpressure mechanisms into NOAH. We chose DSR after our analysis of various on-demand routing protocols showed that DSR has the highest resource congestion control capability. The simulated results show that ACCP is both energy and throughput more efficient than existing traffic only or resource only protocols.

REFERENCES

[1]   Wan, C.Y., Eisenman,S.B. and A. T. Campbel, "CODA: Congestion Detection and Avoidance in Sensor Networks," in the proceedings of ACM SenSys, Los Angeles, pp. 266279. ACM Press, New York (2003).

[2]   Ee, C.T., Bajcsy, R., "Congestion Control and Fairness for Many-to-One Routing in Sensor Networks," in the proceedings of ACMSenSys, Baltimore, pp. 148161. ACM Press, New York (2004).

[3]   Wang, C., Sohraby, K., Li, B., Daneshmand, M., Hu, Y.: "A survery of transport protocols for wireless sensor networks", IEEE Network Magazine 20(3), 3440 (2006).

[4]   D. B. Johnson and D. A. Maltz, "Dynamic Source Routing in Ad hoc Networks," *Mobile Computing*, ed. T. Imielinski and H. Korth, Kluwer Academic Publishers, 1996, pp. 153-181.

[5]   W. Ye, J. Heidemann, and D. Estrin, "An Energy-Efficient Mac Protocol for Wireless Sensor Networks," in *Proceedings of IEEE INFOCOM'02*, June 2002.

[6]   Wan, C.Y., Eisenman, S.B. and A. T. Campbel, "CODA: Congestion Detection and Avoidance in Sensor Networks," in the proceedings of ACM SenSys, Los Angeles, pp. 266279. ACM Press, New York (2003).

[7]   Wamg. C., Li, B., Sohraby, K., Daneshmand, M., Hu, Y., "A Survey of transport protocols for wireless sensor networks", IEEE Network, vol. 20, no. 3, May-June 2006, 34-40.

[8]   S. Chen and N. Yan, "Congestion avoidance based on lightweight buffer management in sensor networks," in *Proceedings of* ICPADS, pp. 934 – 946,Sep. 2006.

[9]   Wan, C.Y., Eisenman, S.B. and A. T. Campbel "CODA: Congestion Detection and Avoidance in Sensor Networks," in the proceedings of ACM SenSys, Los Angeles, pp. 266279. ACM Press, New York (2003).

[10]   Y. Sankarasubramaniam, O. B. Akan, and I.F. Akyidiz, "ESRT: Proceedings of ACM Mobihoc'03, June 1-3, 2003, Annapolis, USA.

[11]   Chonggang Wang, KazemSohraby, Victor Lawrence, Bo Li, Yueming Hu, "Priority-based Congestion Control in Wireless Sensor Networks", IEEE International Conference on Sensor Networks, Ubiquitous, and Trustworthy Computing, Vol 1 (SUTC'06), 2006,pp. 22-31.

[12]   Saeed RasouliHeikalabad, Ali Ghaffari, Mir AbolgasemHadian, and HosseinRasouli, "DPCC: Dynamic Predictive Congestion Control in Wireless," IJCSI International Journal of Computer Science Issues, Vol. 8, Issue 1, January 2011, ISSN (Online):1694-0814.R.

[13]   Kamal , K.S., Harbhajan, S and, R.B Patel 2010, "A Hop by Hop Congestion Control Protocol to Mitigate Traffic Contention in Wireless Sensor Networks", International Journal of Computer Theory and Engineering, vol. 3, No. 6,pp. 1793-8201.

[14]   R. Zheng, J.C. Hou, and L.Sha. Asynchronous Wakeup for Ad Hoc Networks: Theory and ProtocolDesign.

[15]   S. Chen, N. Yang Congestion Avoidance based on Light-Weight Buffer Management in Sensor Networks, University of Florida, Gainesville, FL 32611,USA.

[16]   Hull, B., Jamieson, K., and Balakrishnan, H, Bandwidth management in wireless sensor networks, Tech. Rep. 909, MIT Laboratory for Computer Science, July2003.

[17]   Rangwala, R. Gummadi, R. Govindan, and K. Psounis, "Interference-Aware Fair Rate Control in Wireless Sensor Networks, SIGCOMM2006."

[18]   Wan, C.Y., Eisenman, S.B., Campbel, A.T.: CODA: Congestion Detection and Avoidance in Sensor Networks. In: the proceedings of ACM SenSys, Los Angeles, pp. 266279. ACM Press, New York (2003).

[19]   Bertsekas, D. and Gallagher, R. 1991. Data Networks, 2nd ed. Prentice Hall, Englewood Cliffs, NJ, USA,1991.

[20]   Kamal , K.S., Harbhajan, S and, R.B Patel 2010, "A Hop by Hop Congestion Control Protocol to Mitigate Traffic Contention in Wireless Sensor Networks", International Journal of Computer Theory and Engineering, vol. 3, No. 6, pp.1793-8201.

Permissions

List of Contributors

S.P.V.Subba Rao
Sreenidhi Institute of Science and Technology, Dept of Electronics and Communication Engineering, Hyderabad, Andhra Pradesh, India

Dr.S. Venkata Chalam
CVR Engineering College, Dept of Electronics and Communication Engineering, Hyderabad, Andhra Pradesh, India

Dr.D.Sreenivasa Rao
JNTU CE Dept of Electronics and Communication Engineering, Hyderabad, Andhra Pradesh, India

Rajeev Mathur
Department of ECE, Suresh Gyan Vihar University, Jaipur, Rajasthan, India

Sunil Joshi
College of Engineering & Technology, MPUAT, Udaipur, India

Krishna C Roy
Pecific Institute of Technology, Udaipur, India

Arathi.R.Shankar
Department of Electronics &Communication BMS College of Engineering, B'lore

Adarsh Pattar
BMS College of Engineering, B'lore, India

V.Sambasiva Rao
Department of Electroncics, PESIT, B'lore, India

PrasenjitChanak
Department of Information Technology Bengal Engineering and Science University, Shibpur, Howrah-711103, India

TuhinaSamanta
Department of Information Technology Bengal Engineering and Science University, Shibpur, Howrah-711103, India

Indrajit Banerjee
Department of Information Technology Bengal Engineering and Science University, Shibpur, Howrah-711103, India

Ayan Kumar Das
Department of Information Technology, Calcutta Institute of Engineering and Management, Kolkata, India

Dr. Rituparna Chaki
Department of Computer Science & Engineering, West Bengal University of Technology, Kolkata, India

Dr.G.Padmavathi
Professor and Head, Department of Computer Science, Avinashiligam University for Women, Coimbatore – 641 043

Dr.P.Subashini
Associate Professor, Department of Computer Science, Avinashilingam University for Women, Coimbatore – 641 043

Ms.D.Devi Aruna
Project fellow, Department of Computer Science, Avinashiligam University for Women, Coimbatore – 641 043

Yasir Malik
Department of Computer Science, University of Sherbrooke, Quebec, Canada

Kishwer Abdul Khaliq
Center of Research in Networks and Telecom (CoReNeT), Mohammad Ali Jinnah University, Islamabad, Pakistan

Bessam Abdulrazak
Department of Computer Science, University of Sherbrooke, Quebec, Canada

Usman Tariq
Department of Information Systems, College of Computer and Information Sciences, Al-Imam Mohammed Ibn Saud Islamic University, Riyadh, Saudi Arabia

Abbas Asosheh
Faculty of Technical Engineering, Tarbiat Modares University, Tehran, Iran

Nafise Karimi
Faculty of Technical Engineering, Tarbiat Modares University, Tehran, Iran

Hourieh Khodkari
Faculty of Technical Engineering, Tarbiat Modares University, Tehran, Iran

Ahmed E. El-Din
Computer Engineering Department, Cairo University Cairo, Egypt

Rabie A. Ramadan
Computer Engineering Department, Cairo University
Cairo, Egypt

Niranjan Kumar Ray
Department of Computer Science and Engineering,
National Institute of Technology Rourkela, India

Ashok Kumar Turuk
Department of Computer Science and Engineering,
National Institute of Technology Rourkela, India

Bhuvan Modi
Center of Excellence for Communication Systems
Technology Research Department of Electrical and
Computer Engineering, Prairie View A & M University,
TX 77446 United States of America

A. Annamalai
Center of Excellence for Communication Systems
Technology Research Department of Electrical and
Computer Engineering, Prairie View A & M University,
TX 77446 United States of America

O. Olabiyi
Center of Excellence for Communication Systems
Technology Research Department of Electrical and
Computer Engineering, Prairie View A & M University,
TX 77446 United States of America

R. Chembil Palat
Nokia Research Center, Berkeley, CA 94304 United States
of America

Femi-Jemilohun Oladunni .Juliet
School Of Computer Science and Electronic Engineering
University of Essex Colchester, Essex, United Kingdom

Walker Stuart
School Of Computer Science and Electronic Engineering
University of Essex Colchester, Essex, United Kingdom

Joanne Mun-Yee Lim
Department of Engineering, UCTI (APIIT), Malaysia

Chee-Onn Chow
Department of Electrical Engineering, University Malaya,
Malaysia

Kaushik Ghosh
Faculty of Engineering & Technology, Mody Institute
of Technology & Science (Deemed University),
Lakshmangarh, Dist. Sikar, Rajasthan – 332311, India

Partha Pratim Bhattacharya
Faculty of Engineering & Technology, Mody Institute
of Technology & Science (Deemed University),
Lakshmangarh, Dist. Sikar, Rajasthan – 332311, India

Pradip K Das
Faculty of Engineering & Technology, Mody Institute
of Technology & Science (Deemed University),
Lakshmangarh, Dist. Sikar, Rajasthan – 332311, India

Shailaja Patil
Department of Computer Engineering, Sardar Vallabhbhai
National Institute of Technology, Surat, India

Mukesh Zaveri
Department of Computer Engineering, Sardar Vallabhbhai
National Institute of Technology, Surat, India

Mohamed Yacoab M.Y.
Research Scholar, Karpagam University, Coimbatore,
India
MEASI Institute of Information Technology, Chennai,
India

Dr.V.Sundaram
Director, Karpagam College of Engg, Coimbatore, India

James DzisiGadze
Department of Electrical & Electronic Engineering,
Kwame Nkrumah University of Science and Technology,
Kumasi, Ghana

DelaliKwasiDake
Department of Computer Engineering, Kwame Nkrumah
University of Science and Technology, Kumasi, Ghana

Kwasi Diawuo
Department of Computer Engineering, Kwame Nkrumah
University of Science and Technology, Kumasi, Ghana

www.ingramcontent.com/pod-product-compliance
Lightning Source LLC
Chambersburg PA
CBHW080529200326
41458CB00012B/4379